# Agricultural Zoology Reviews

## Volume 4

# Agricultural Zoology Reviews

## Volume 4

Editor:

K. Evans

*Department of Entomology and Nematology,*
*AFRC Institute of Arable Crops Research,*
*Rothamsted Experimental Station, Harpenden, Herts, UK*

Intercept

Andover

**British Library Cataloguing in Publication Data**
Agricultural zoology reviews.—Vol. 4–
A CIP catalogue record for this book is available from the British Library.
ISBN 0–946707–31–6
ISSN 0269–0543



Published in February 1991 by Intercept Limited,
P.O. Box 716, Andover, Hants SP10 1YG, England.

Filmset in 'Linotron' Times by
Ann Buchan (Typesetters), Shepperton, Middlesex

Printed in Great Britain by
Athenaeum Press Ltd, Newcastle upon Tyne.

# Preface

The first three volumes of Agricultural Zoology Reviews, edited by Gordon Russell, were enthusiastically received, providing as they did a very varied mixture of information. Gordon's objective in founding the series was to present specialists in any one field with the opportunity of keeping abreast of developments in other aspects of the very broad subject matter covered by the series, and he decided that the journal should encompass information about all agriculturally important animals. I intend to continue with the same objectives and thereby draw together review articles, the basic information for which is scattered through many primary research journals.

Within volume 4, readers will find articles on insects, nematodes and earthworms, on subject matter as varied as insect and nematode pest management strategies to the physiology of anhydrobiosis in nematodes, plant resistance mechanisms to earthworm biology. Future volumes will be even more diverse and will include items on vertebrates, both as pests and as useful species. Some reviews each year will address items of topical interest, such as the likely impact of climatic changes on the importance and severity of pest attack, the problems of new instances of resistance to pesticides, and the ever-changing faces of pest-management stragegies. Of particular interest here will be the management of pest complexes associated with crops newly introduced to an area, either as a result of climate change or in response to market forces.

Whatever the content of each volume, I hope that everyone will find something of interest. Sometimes reviews will appeal more to specialists in particular fields but many will also provide teachers and students of applied biology and environmental science with authoritative coverage of important topics. I have already, in my role as editor, benefitted greatly from the opportunities to read on subjects outside my own immediate field and I hope that you will do so too.

K. Evans
December 1990

# Contributors

JOHN BARRETT, *Department of Biological Sciences, University College of Wales, Aberystwyth, Dyfed SY24 3DA, UK*

ROGER COOK, *AFRC Institute of Grassland and Environmental Research, WPBS, Plas Gogerddan, Aberystwyth, Dyfed SY23 3EB, UK*

K. E. FLETCHER, *'Ecocare', 15 Grange Gardens, Wellesbourne, Warwickshire CV35 9EF, UK*

JOSEPH E. FUNDERBURK, *University of Florida, Quincy, FL 32351, USA*

R. T. GAHUKAR, *Pest Control Consultant, Plot 220, Reshimbag, Nagpur 440009, India*

H. S. GAUR, *Division of Nematology, Indian Agricultural Research Institute, New Delhi 110012, India*

SCOTT H. HUTCHINS, *DowElanco, Indianapolis, IN 46268, USA*

ATAUR RAHMAN KHAN, *Department of Zoology, Rajshahi University, Rajshahi–6205, Bangladesh*

PETER LANGLEY, *ODA/University of Bristol, Tsetse Research Laboratory, Langford, Bristol BS18 7DU, UK*

ABDUL MANNAN, *Department of Zoology, Rajshahi University, Rajshahi–6205, Bangladesh*

R. N. PERRY, *Entomology and Nematology Department, AFRC IACR, Rothamsted Experimental Station, Harpenden, Hertfordshire AL5 2JQ, UK*

RICHARD WALL, *ODA/University of Bristol, Tsetse Research Laboratory, Langford, Bristol BS18 7DU, UK*

# Contents

**Preface** v

**List of Contributors** vii

1 **INJURY GUILDS: A PRACTICAL APPROACH FOR MANAGING PEST LOSSES TO SOYBEAN** 1
Scott H. Hutchins, *DowElanco, Indianapolis, IN 46268, USA* and Joseph E. Funderburk, *University of Florida, Quincy, USA*
Introduction 1
Pest-management decision algorithms 3
Injury equivalency and guilds 9
Management of soybean injury guilds 13
Future opportunities 17
References 18

2 **RECENT DEVELOPMENTS IN SORGHUM ENTOMOLOGY RESEARCH** 23
R. T. Gahukar, *Pest Control Consultant, Plot 220, Reshimbag, India*
Introduction 23
Soil pests 24
Seedling pests 25
Foliage pests 28
Stem borers 37
Earhead pests 41
Stored-grain pests 47
Insecticide residues 48
Perspectives 48
References 49

3 **STORED-PRODUCTS ENTOMOLOGY IN THE TROPICS** 67
Ataur Rahman Khan and Abdul Mannan, *Department of Zoology, Rajshahi University, Bangladesh*
Introduction 67
Storage pests 68
Taxonomy of storage pests 73
Damage by storage pests 74

Control of storage pests 76
Guidelines for future research 84
Conclusions 85
Acknowledgements 85
References 85

**4 CURRENT APPROACHES TO CASSAVA PEST MANAGEMENT IN SUBTROPICAL AFRICA** 97
R. T. Gahukar, *Pest Control Consultant, Plot 220, Reshimbag, India*
Introduction 97
Cassava mealybug 98
Other insect pests 116
Cassava green mite 119
Red mites 124
Limitations to and perspectives of cassava pest management in traditional agriculture 125
Conclusions 126
Acknowledgements 127
References 127

**5 FROM BEHAVIOUR TO CONTROL: THE DEVELOPMENT OF TRAP AND TARGET TECHNIQUES FOR TSETSE FLY POPULATION MANAGEMENT** 137
Richard Wall and Peter Langley, *ODA/University of Bristol, Tsetse Research Laboratory, Langford, Bristol, UK*
Introduction 137
The tsetse life cycle 138
Early approaches to tsetse control 139
The lesson of Principe Island 139
Insecticidal methods of tsetse control 140
The sterile male technique 141
Modern trap development 141
Exploiting tsetse behaviour 147
The development of autosterilizing techniques 151
Tsetse control with traps and targets 152
Conclusions 154
Acknowledgement 154
References 154

**6 ANHYDROBIOTIC NEMATODES** 161
John Barrett, *Department of Biological Sciences, University College of Wales, Aberystwyth, Dyfed, UK*
Introduction 161
Characteristics of anhydrobiotic nematodes 162
Structural adaptations 164

Metabolic changes during desiccation 165
Metabolic changes during rehydration 169
Membrane function in anhydrobiotes 169
Commercial aspects 170
Conclusions 171
References 172

7 **THE BIOLOGY AND CONTROL OF THE PLANT PARASITIC NEMATODE *ROTYLENCHULUS RENIFORMIS*** 177
H. S. Gaur, *Division of Nematology, Indian Agricultural Research Institute, New Delhi, India* and R. N. Perry, *Entomology and Nematology Department, AFRC IACR, Rothamsted Experimental Station, Harpenden, Hertfordshire, UK*
Introduction 177
Gross morphology and systematics 178
Biological races 180
Life history 181
Infectivity and histopathology 183
Duration of the life cycle 184
Pathogenicity and crop losses 184
Ecology 186
Interaction with other soil micro-organisms 189
Control 190
References 200

8 **RESISTANCE IN PLANTS TO CYST AND ROOT-KNOT NEMATODES** 213
Roger Cook, *AFRC Institute of Grassland and Environmental Research, WPBS, Plas Gogerddan, Aberystwyth, Dyfed, UK*
Introduction 213
Occurrence and use of resistance 213
Definition and specificity of resistance 215
Resistance to cereal cyst nematode 216
Resistance phenomena 217
Non-host resistance 226
Multiple resistance 226
Plant–nematode recognition 227
Novel resistance 230
Concluding remarks 232
References 233

**9 THE UTILIZATION OF ORGANIC WASTES BY EARTHWORMS** 241

K. E. Fletcher, *Ex Entomology Department, Horticultural Research International, Wellesbourne, UK*

Introduction 241
The problems of organic waste disposal 242
The origins of vermiculture 243
Soil improvement and reclamation 244
Vermicomposting and the stabilization of organic wastes 244
Earthworms as animal feed protein 245
Types of earthworms used in vermiculture 246
Methods of production in vermiculture 246
Conclusions 249
References 250

**Index** 255

1

# Injury Guilds: A Practical Approach for Managing Pest Losses to Soybean

SCOTT H. HUTCHINS* AND JOSEPH E. FUNDERBURK[†]

* *DowElanco, Indianapolis, IN 46268, USA and* [†]*University of Florida, Quincy, FL 32351, USA*

Introduction
Pest-management decision algorithms
Economic-injury level models—Gain threshold—Biological variables of the EIL
Injury equivalency and guilds
Size-class differences—Species differences
Management of soybean injury guilds
Leaf-mass consumers—Fruit feeders
Future opportunities
References

## Introduction

Soybean, *Glycine max* (L.) Merr., meal and oil products are the most produced and exchanged commodities among the eight major oilseeds. This position as leader within the oilseed industry is solid, with cottonseed, the nearest competitor, produced at less than half the annual volume of soybeans (Foreign Agricultural Service-USDA, 1983). Although production trends probably will not increase at past levels, the American Soybean Association (1983) predicts annual production increases of 4% over the next decade, with marketable product in 2002 estimated to be about 191 Mt. The geographical distribution of soybean production, however, has changed markedly among the major production centres over the past 10–15 years (Smith and Huyser, 1987). Specifically, production in North America has stabilized or declined, while production in Brazil, Argentina and the European Economic Community has steadily increased (Foreign Agricultural Service-USDA, 1983).

Irrespective of the geographical production centre, soybeans are inhabited by a diverse community of arthropods. Indeed, although pest pressure differs in form and/or intensity among the various geographical regions, managing injury induced by arthropods represents a considerable endeavour. Kogan and Turnipseed (1987) compiled a list of the principal arthropod pests across

*Agricultural Zoology Reviews* — Vol. 4, February 1991
0269–0543 $20.00 + $ 0.00 

major agroecological regions and determined their relative impact to global soybean production. Their conclusions clearly indicate that many arthropod pests have evolved to utilize specifically soybean plant parts (*see also* Kogan, 1981) as a dedicated source for nourishment or shelter. From a global standpoint, pest species that are more ubiquitous in distribution, or more general in feeding behaviour, produce the highest levels of total injury.

Based on an impact index developed by Kogan and Turnipseed (1987), the most significant losses to soybean result from injury either to leaf tissue or directly to the fruit. Key species in the leaf-mass consuming category include the velvetbean caterpillar, *Anticarsia gemmatalis* Hubner, and the soybean looper, *Pseudoplusia includens* (Walker). Both of these pests inhabit key soybean production centres over much of North, Central and South America. Predominant fruit-feeding pests include podworms, *Heliothis* spp., and stink bugs, especially *Nezara viridula* (L.). Pests injuring stems, germinating seeds and seedlings are less important to global soybean production. Significant pests of this type are of concern only in a few agroecological regions, with some production centres having no representative species. Similarly, relatively few arthropod species attack roots, nodules and blossoms, with the impact of injury to these plant structures eliciting a minimal loss in yield.

The amount of published information on the ecology and control of most soybean pests, even minor pests, is considerable; still, there are important informational shortfalls. Adequate information to predict pest outbreaks, for example, is generally not available. Perhaps most important, the necessary agronomic information on soybean response to insect feeding is frequently restricted to final yield without regard to the impact on a continuum of physiological processes (Poston, Pedigo and Welch, 1983). When such information does exist, typically it is based on the influence of a singular pest species at specific stages of crop growth.

Soybean leaf-mass consuming and fruit-feeding pest species receive a major emphasis in pest-management programmes. Ideally, soybeans are periodically inspected during periods of risk to monitor pest development and population density relative to crop stage and susceptibility (Newsom *et al.*, 1980; Pedigo *et al.*, 1981; Kogan and Turnipseed, 1987). The value of using a specific control tactic to suppress a pest population is determined on the likelihood of its density exceeding the economic injury level (Kogan, 1976). Current soybean pest-management programmes continue to utilize the economic injury level concept, albeit primarily for the management of *individual* pest species.

In this review we address the development of decision-making algorithms for insect pests, with particular emphasis on soybean leaf- and fruit-feeding pests. Our focus, however, will be on managing insect-induced injury, and will include recent advances in managing the collective impact of insect feeding from *different* species or size classes. We will argue that minimizing plant damage can best be achieved through management of *injury guilds*, which are logical groupings of arthropod species based on how they affect host (soybean) physiology. The use of injury guilds to manage pest complexes offers a practical approach for refining pest-management decisions in order to reduce the possibility of error. In reviewing past developments, we hope to identify and propose new directions for pest-management research, with soybeans serving as a model agroecosystem and commodity orientation.

## Pest-management decision algorithms

The objectives of insect pest management are manifold. Indeed, a concise definition of pest management remains elusive because of the multiple objectives involved and how they should be ranked. Perhaps a great deal of confusion would be alleviated if the focus were placed on the objectives of those persons making pest-management decisions; namely, farmers or their agents. With this approach, there can be little doubt that the microeconomic theories of production efficiency are applicable, and the refined objective becomes one of consistently minimizing economic losses to pests. This focus is not intended to dismiss the *potential* advantages that frequently result from pest management, such as fewer inputs and enhanced environmental quality. It does, however, underscore the fact that agricultural enterprises are small businesses which have their own set of objectives which may or may not coincide with those of the larger community. Therefore, the current operating procedure involves microeconomic decision-making by producers, constrained by the wishes of the larger community via government regulation. For example, pesticides are available for use by growers at their discretion, so long as they are used according to labelled requirements (which are regulated by public agencies). The segregation of private v. public prerogatives is important as it directs the attention of pest management toward utilizing available resources solely on the basis of economic optimization, with public concerns and policy intrinsic to the decision-making process.

With the above characterization of pest-management decision-making, it becomes clear that the primary limitation for all decisions is knowledge. Indeed, in instances where management tactics are overused (or underused), the reason is because of inadequate knowledge about the biology of the host–pest relationship. For example, when pesticides are overused, it is probably because the damage function to the crop or the presence of the pest cannot be predicted with accuracy. Therefore, pesticides are used as a risk averse strategy to hedge against economic uncertainty and eliminate the possibility of a large financial loss. It is equally conceivable, however, that pesticides may be underused because of the same reasons. Clearly, additional knowledge about pests and their relationships to the host will provide for more efficient decisions. The economic-injury level (EIL) incorporates much of the necessary knowledge and has, by far, been the most extensively used concept in pest management for minimizing losses to pests.

### ECONOMIC-INJURY LEVEL MODELS

Stern *et al.* (1959) defined the EIL as 'the lowest population density that will cause economic damage'. Further, they advocated the need to monitor pest density through time in order to determine the necessity and timing of a control tactic. This approach permits natural control factors to impact pest density and, possibly, reduce the need for artificial controls. Indeed, Stern *et al.* developed the concept as a means for more rational use of pesticides. Therefore, the original concept was designed to be suitable for situations in which pest control was utilized therapeutically to reduce insect density.

The EIL concept was widely adopted by the scientific community as a reasonable means for managing pesticide inputs while reducing insect-induced crop-losses. The integrated control concept of Stern *et al.* was expanded to include interactions with other pest species and subsequently referred to as integrated pest management (IPM). This conceptual expansion, however, maintained a focus on the EIL as a fundamental decision-making criterion.

Although Stern *et al.* had established the theoretical framework for pest management, they failed to outline the specific components for determining EILs. The first calculated EIL did not appear in the literature until 1972, when Stone and Pedigo (1972) developed a method of calculating EILs for insects, which included both economic and biological components. They accepted the premise that an EIL was, in business terms, a break-even point, and developed the concept of a gain threshold (GT). The GT expresses economic damage as physical commodity loss, and requires data on the cost of the proposed management tactic, as well as the market value of the product:

$$\mathrm{GT} = C/V \qquad \text{Eq.(1)}$$

where $C$ is the cost of management for the anticipated control tactic; and $V$ is the anticipated market value for the commodity.

Thus, if management costs totalled US\$27.00/ha and the predicted value of soybeans is US\$0.2205/kg, then the GT = 122.45 kg/ha. In other words, insect pests would have to consume (collectively) the equivalent of 122.45 kg/ha before the cost of control per ha would be justified economically. Stone and Pedigo (1972) used the GT, and estimates of yield loss per individual insect (determined from defoliation measurements and agronomic data on yield response to defoliation), to determine EILs for the green cloverworm, *Plathypena scabra* (Fab.), at several stages of soybean growth.

Norton (1976) expanded the general EIL model of Stone and Pedigo (1972) to account for the expected level of control from a curative control tactic:

$$\theta = C/VDK \qquad \text{Eq.(2)}$$

where $\theta$ is the level of pest attack (functionally equivalent to the EIL); $D$ is the total damage per insect over its life; and $K$ is the proportionate reduction in pest attack from a selected tactic. Norton's contribution provided a means to evaluate the impact of different control tactics on the final EIL.

Pedigo, Hutchins and Higley (1986), in addition to reviewing the progress of EILs to date, introduced a refined version of the EIL model. Their contribution was the distinction between injury and damage, achieved by creating separate variables for each:

$$\text{EIL} = C/VIDK \qquad \text{Eq.(3)}$$

where $I$ is the expected injury, e.g. tissue consumption, per individual over its life; and $D$ is the expected yield loss per unit of injury.

Distinguishing between insect-induced injury and host response (damage) provided the impetus for entomologists to focus on quantifying the impact of insects on the basic processes of plant development. Similarly, it allowed plant scientists to focus their attention on defining yield losses resulting from the disruption of key physiological processes, e.g. leaf production.

We will use a modified equation of Pedigo, Hutchins and Higley (1986) to review the most recent advances in estimating values for the two biological variables ($I$ and $D$) of the EIL. While recognizing the utility of Norton's $K$ variable, to simplify the analysis we will assume that most control tactics are effective, i.e. $K$ approaching 1.0. Therefore, for the purposes of this review, the working EIL formula is simplified to concentrate on the biological variables of $D$ and $I$:

$$\text{EIL} = \text{GT}/DI \qquad \text{Eq.(4)}$$

Each of the variables within equation (4) have a significant impact on the final EIL and warrant further definition.

## GAIN THRESHOLD

The two economic variables contained within the GT represent both the cost and benefit considerations of the economic analysis. Market values should reflect the anticipated selling price of the commodity, not necessarily the current market value. In the case of feeds used for on-farm animal production, the value used in the EIL should equal the nutrient substitution cost of an alternative feed (Hutchins and Pedigo, 1990). Management costs also should be determined accurately. Only variable costs, such as the cost of the control agent, application cost and labour expense, should be included. Fixed costs, e.g. equipment amortization, should not be included in the management cost variable because they do not represent a direct expense of the control decision.

The accuracy of the final EIL will be diminished if care is not taken in estimating the GT. Although we will not expand the discussion of GT variables further, they are critical components of the final EIL and should be given serious consideration.

BIOLOGICAL VARIABLES OF THE EIL

*Injury*

Injury is defined as the effect of pest (arthropod) activities on host (soybean) physiology that is usually deleterious (Pedigo, Hutchins and Higley, 1986). Soybean has been exploited as a food source by numerous insect species (Kogan, 1981). Within a specific region, the functional and spatial characteristics of pest composition change temporally and are typically correlated to the specific stage of soybean development. Therefore, a frequent approach in describing insect pests of soybean has been to collate species based on their primary site of feeding during discrete periods of crop development (Turnipseed and Kogan, 1976; Kogan, 1981; Kogan and Turnipseed, 1987). For example, insects that feed predominately on the leaf foliage would be collectively grouped as defoliators, or leaf-mass consumers. Using this approach, Turnipseed and Kogan (1976) categorized soybean pests by their association with the plant. The primary feeding sites included pods and seeds, blossoms, stalks and upper stems, leaf blades, lower stems, roots/nodules and germinating seeds.

Injury, rather than insects *per se*, should be the focus of pest management. This orientation is important because the seasonal development of injury is quite different from the seasonal development of insects (*Figure 1*). Intra-

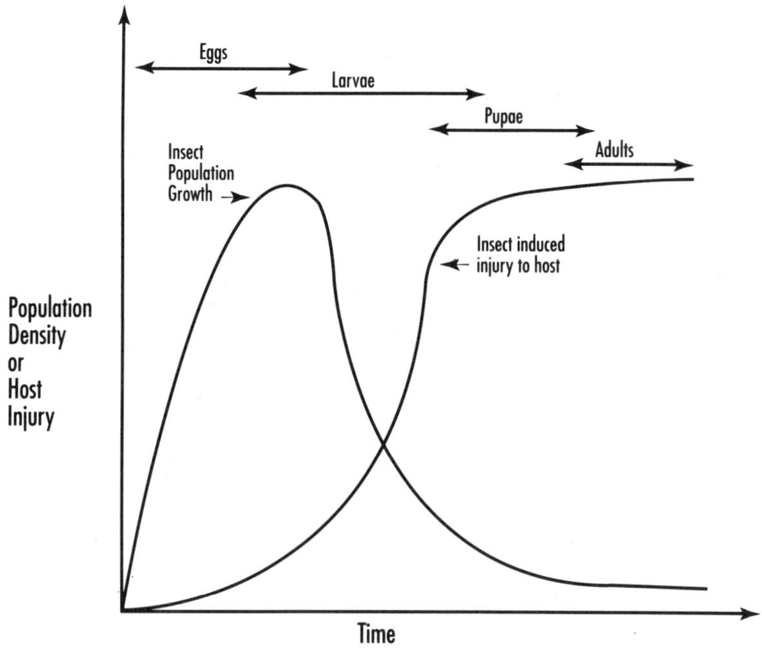


**Figure 1.** Stylized drawing of lepidopteran insect population development versus host injury development over time.

seasonal changes in the consumption potentials of insects account for the differences between population development and consumption. A key fallacy of the original EIL concept of Stern *et al.* (1959) was the focus on insect populations rather than on host injury. Insects only represent a surrogate for injury, and should be managed in proportion to their total contribution to final commodity loss.

Detailed characterizations of insect-induced injury to plants have been provided by several workers (e.g. Bardner and Fletcher, 1974). Boote (1981) classifies insects that induce plant injury into five distinct categories: leaf-mass consumers, fruit feeders, stand reducers, assimilate sappers and turgor reducers. Pedigo, Hutchins and Higley (1986) added a sixth category, called architecture modifiers, to describe insects that reduce the physiological yield of plants by disrupting normal morphological development. Each of these categories represent a primary and critical form of physiological disruption in the plant which may produce a measurable loss in yield. Pedigo, Hutchins and Higley (1986) provide examples of insect species known to produce injury for each of the injury types, and discuss the plant physiological mechanisms affected.

Categorizing insect-induced injury simplifies the research process by directing experimentation to describe, in standard terms, the relative impact that each individual or known density of insects has upon a particular physiological function of the plant. Plant response to injury, i.e. damage, is the topic of agronomic research and is represented by the $D$ variable of the EIL.

## *Damage*

Damage is distinguished from injury as being the measurable loss of host utility, most often including yield quantity, quality or aesthetics (Pedigo, Hutchins and Higley, 1986). This definition, which implies that injury must accumulate to a certain level before damage accrues, is consistent with the assertion of Stern *et al.* (1959) that crops can sustain a certain level of insect-induced injury before control is necessary. The exact nature of the host response, however, is a function of several characteristics of injury. Specifically, damage is dependent upon the intensity of injury, the type of injury, the plant part injured and the time of injury.

Some estimate of the functional form of yield or quality reduction resulting from a designated level of physiological injury (referred to as the damage curve) is necessary to calculate an EIL (Bardner and Fletcher, 1974; Fenemore, 1982; Poston, Pedigo and Welch, 1983). Host response to injury, however, typically proceeds through several phases (*Figure 2*). Pedigo, Hutchins and Higley (1986) described these phases as: tolerance, compensation, linearity, desensitization and inherent impunity.

*Tolerance* represents a situation in which there is no detectable host response to injury. Individual plants tend to exhibit less tolerance than do crop stands, the latter being able to compensate for spatial non-uniformity of injury. Some level of tolerance to injury is common among most plants, but the actual degree varies both with genetic make-up and with environmental

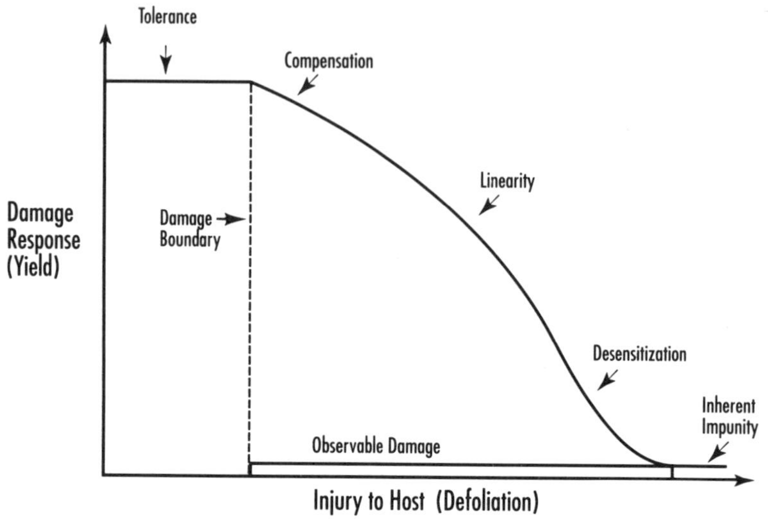


**Figure 2.** General form of damage curve demonstrating tolerance, compensation, linearity, desensitization and inherent impunity phases of host response to insect-induced injury.

circumstances. Soybean crops, for example, have been shown to withstand as much as 35% defoliation under normal growing conditions before any yield reduction can be detected (Higgins, Pedigo and Staniforth, 1984). If, however, the crop is subjected to drought stress, the tolerable level of defoliation is reduced by more than 50% (Hammond and Pedigo, 1982; Ostlie and Pedigo, 1985). Pedigo, Hutchins and Higley (1986) defined the level at which tolerance ends and damage begins as the damage boundary. Plants that increase growth and yield with insect-induced stress are considered to exhibit a special form of tolerance known as *overcompensation*. Soybeans, for example, can show an overcompensatory crop response to injury induced by seedcorn maggot, *Delia platura* (Meigen), (Funderburk and Pedigo, 1983). In these situations, the slope of the damage curve is positive, although usually only for a short range of injury.

Once the level of injury exceeds the damage boundary, plants experience increased yield loss with each additional increment of injury. The exact rate of increase is dependent upon genetically determined resistance to injury by the host, as well as environmentally mediated conditions. Because the maximal amount of damage per unit of injury is not achieved, the plants (crop) are said to be in the *compensation* phase. Additional increases in injury intensity ultimately reach a level that overwhelms the host's ability to compensate, and results in a state of sustained maximal damage (*linearity*) per additional unit of injury. A linear damage response is characteristic of direct consumers of yield [e.g. corn earworm, *Heliothis zea* (Boddie)], but also is a distinct portion of the damage curve for indirect damage such as that caused by the potato leafhopper, *Empoasca fabae* (Harris), (Ogunlana and Pedigo, 1974).

Inasmuch as physical yield for a given stand is finite, host response will inevitably decline per unit of additional injury. The *desensitization* phase of the damage curve, therefore, represents an incrementally decreasing host response to injury. Finally, *inherent impunity* is characterized as complete host insensitivity to additional increases in injury. Desensitization and inherent impunity frequently occur only after damage to the crop has reached economic status, and therefore represent a portion of the damage curve that is less applicable to managing insect-induced injury. Indeed, only the critical range of injury, i.e. beyond the damage boundary and before the EIL, is pertinent to calculating EILs.

A separate damage curve should be constructed for each type of insect-induced physiological injury (categories described previously) to the crop (Hutchins, Higley and Pedigo, 1988). The developmental stage of the crop also will have a significant impact on the final degree of damage, so separate damage curves are frequently developed for injury at discrete growth stages (Stone and Pedigo, 1972; Thomas *et al.*, 1974b). Differences in host response to injury to different plant parts also have been documented. For example, defoliation within different strata of the soybean canopy elicits unique host responses (Higgins, Pedigo and Staniforth, 1984). A more detailed discussion of the factors affecting host response are provided by Fenemore (1982) and Pedigo, Hutchins and Higley (1986).

## Injury equivalency and guilds

Future progress with pest-management decision-making will involve refinement of the primary variables associated with the EIL, especially the biological variables of injury and damage. With the current management approach, there is no accurate way to integrate the collective effect of injury, e.g. defoliation, from several insect species, or to account for consumption differences among different size classes of the same species. Indeed, segregating injury, which is a function of the insect, from damage, which is a function of the plant, allows for efficient management of several size classes and species affecting the host in a similar fashion. A logical refinement, therefore, would include the development of a system to compare the relative injuriousness of different insect sizes and species to overall crop injury.

### SIZE-CLASS DIFFERENCES

Phytophagous pests typically consume plant biomass in relation to their own size. Lepidopterous pests of soybean, for example, may consume as much as 90% of their total lifetime consumption during their final two larval instars (Reid and Greene, 1973; Boldt, Biever and Ignoffo, 1975; Greene, 1976). Therefore, if small and large larvae, feeding at the same time, are given equal consideration in relation to their impact to host injury, then the estimate of final crop damage certainly will be overvalued. This will result in an unrealistically low EIL, and possible overuse of a management tactic.

To account for differential consumption potentials of size classes, Pedigo, Hutchins and Higley (1986) proposed a system of consumption equivalents.

They defined an *injury equivalent* as the amount of injury that could be produced by one pest through its complete life cycle. In addition, they referred to the total number of injury equivalents for a population as its *injury equivalency*. In its simplest form, therefore, an equivalency system weights each individual insect by its current contribution to total injury (Hutchins, Higley and Pedigo, 1988). For example, a small larva (instar 1 or 2), which can consume only about 1% of its total lifetime potential, would be considered the equivalent of 1/100th of a large larva. Hence, leaf tissue loss from 100 small larvae is required to equal the foliage consumption of one large larva. An obvious advantage of the equivalency system for different size classes is the intrinsic ability to delineate *current* injury from *future potential* injury.

Hutchins, Higley and Pedigo (1988) developed an injury equivalency system for primary leaf-mass-consuming pests of determinate soybeans to incorporate consumption differences among three size classes. Three of the lepidopterous pests considered in their system were the soybean looper, velvetbean caterpillar and green cloverworm. Data for the consumption potential (cm$^2$) of each of these stages are presented in *Table 1*. The fraction of total consumption for each size class and the equivalence coefficient (determined as the cumulative fraction of total consumption) are also presented. The equivalence coefficient for a large larva always is 1.0 because it has consumed foliage through the previous instars and will, presumably, reach its total lifetime consumption potential.

Estimating size-class differences from a pest population within the same species, based on differential consumption potential, represents a significant refinement for measuring current, or predicting future, crop injury induced

**Table 1.** Consumption potential and injury-equivalence coefficients for soybean loopers, velvetbean caterpillars, and green cloverworms

| Larva size* | Leaf-mass consumption (cm$^2$) | Fraction of total consumption | Equivalence coefficient |
|---|---|---|---|
| Soybean looper† | | | |
| Small | 0.81 | 0.01 | 0.01 |
| Medium | 8.60 | 0.09 | 0.10 |
| Large | 88.46 | 0.90 | 1.00 |
| Total | 97.87 | 1.00 | |
| Velvetbean caterpillar‡ | | | |
| Small | 2.40 | 0.03 | 0.03 |
| Medium | 4.56 | 0.05 | 0.08 |
| Large | 85.24 | 0.92 | 1.00 |
| Total | 92.20 | 1.00 | |
| Green cloverworm§ | | | |
| Small | 1.18 | 0.02 | 0.02 |
| Medium | 4.13 | 0.08 | 0.10 |
| Large | 47.79 | 0.90 | 1.00 |
| Total | 53.10 | 1.00 | |

* Size categories: small (larval stages 1 and 2), medium (larval stages 3 and 4), large (larval stages 5+).
† Values for soybean looper consumption are based on Reid and Greene (1973) and Boldt, Biever and Ignoffo (1975).
‡ Values for velvetbean caterpillar consumption are based on Boldt, Biever and Ignoffo (1975) and Greene (1976).
§ Values for green cloverworm consumption are based on Hammond, Pedigo and Poston (1979).

by a single pest species. Soybeans, however, are commonly infested with many species of insect pests.

SPECIES DIFFERENCES

Pest species producing a common form of injury should be managed collectively. Root (1967) first proposed the creation of *guilds* as a method to logically group insect pests. He defined an insect guild as 'a group of species that exploit the same class of environmental resources in a similar way', and later used the guild approach to categorize herbivorous insects that feed on collards (*Brassica oleracea* var. *acephala*) as either pit feeders, strip feeders or sap feeders (Root, 1973). Hawkins and MacMahon (1989) reviewed the concept of animal guilds from an ecological resource adaptation standpoint, but did not address agricultural management situations directly. Pedigo, Hutchins and Higley (1986) endorsed the creation of guilds for agricultural pests and argued that they should represent primary forms of physiological injury. Hence, they proposed the formation of *injury guilds* as a means to account for and manage arthropods, based on the similarities of their physiological injury to the crop. Hutchins, Higley and Pedigo (1988) determined that injury guilds could significantly improve the estimation of EILs and proposed techniques for managing a multiple-species insect complex.

To predict the host response and final damage accurately, an insect species must meet two key requirements for inclusion in an injury guild (Hutchins, Higley, and Pedigo, 1988). First, each guild species must produce an injury that elicits a similar physiological response from the host. Specifically, all member species must:

1. produce a similar type of injury, e.g. individuals cannot be leaf-mass consumers and fruit feeders at the same point in time (but individuals can be classified in different guilds intraseasonally);
2. produce injury within the same phenological time-frame of the host, e.g. insect species consuming leaf tissue in the early vegetative stages of soybean growth cannot be compared to late-season leaf-mass consumers;
3. produce injury of a similar intensity, e.g. the slope of the damage curves for each member species should be functionally similar; and
4. affect the same plant part, e.g. leaf-mass consumption in the lower canopy stratum may elicit a different host response than consumption of terminal leaves.

If these four inclusion criteria are not met for each prospective guild member, then the combined effects of the species probably will not produce a homogeneous host response. The second key requirement for establishing injury guilds is the development of a system to account for different consumption potentials of member species.

A system of injury equivalency for different insect species, similar in principle to the size-class equivalency system, provides the basis for making adjustments in consumption potential among guild species. Data for three

**Table 2.** Matrix of equivalency coefficients for three members of a soybean leaf-mass-consuming guild composed of soybean loopers (SL), velvetbean caterpillars (VBC) and green cloverworms (GCW). The equivalency coefficients are adjusted on the basis of the choice of a standard equivalency species and represent the relative contribution to injury for each species

| Guild species | Larva size* | Injury equivalence coefficients for standard equivalency species† | | |
|---|---|---|---|---|
| | | SL | VBC | GCW |
| SL | Small | 0.01 | 0.01 | 0.01 |
| | Medium | 0.10 | 0.15 | 0.21 |
| | Large | 1.00 | 1.04 | 1.85 |
| VBC | Small | 0.03 | 0.03 | 0.04 |
| | Medium | 0.05 | 0.08 | 0.11 |
| | Large | 0.96 | 1.00 | 1.78 |
| GCW | Small | 0.01 | 0.01 | 0.02 |
| | Medium | 0.05 | 0.07 | 0.10 |
| | Large | 0.54 | 0.56 | 1.00 |

* Size categories: small (larval stages 1 and 2), medium (larval stages 3 and 4), large (larval stages 5+).
† Select a standard species, multiply the number of larvae for each size class of a guild species by its injury equivalency coefficient (in columns) to estimate the total injury from the guild.

soybean defoliating species, all of which meet the established criteria for inclusion in a leaf-mass-consumption guild, are presented in *Table 2*. The matrix of equivalency coefficients (calculated from *Table 1*) relates the injuriousness for each species (and size class) to other member leaf-mass consumers. A single *standard equivalency species* must be selected for comparison to all other species. In all likelihood the standard equivalency species would represent the dominant pest species of the geographical region, but the selection could differ among sample locations or periods. The primary consideration is the relationship of tissue consumption among species. For example, if the soybean looper is chosen as the standard equivalency species in a management system, the matrix of equivalency coefficients could be consulted for comparison to other species. Hence, based on data in *Table 2*, a medium velvetbean caterpillar larva consumes the equivalent of 5% of the leaf tissue that a large soybean looper larva consumes. Similarly, a large green cloverworm larva only consumes *c.* 54% as much as a large soybean looper larva.

The equivalency adjustment system, when based on accurate size-class consumption data (*Table 1*) and utilized for comparison to guilds with multiple pest species, provides a pragmatic method of refining EIL-based management systems. As with any predictive system, however, there are possible pitfalls of the equivalency approach. For example, the range of host response for combined injury must remain near the linear portion of the damage curve to ensure that stress is additive among individuals and size classes. In addition, the relative contribution to injury that each individual guild member makes to total injury should not be significantly altered with increases or decreases in pest density. Additional considerations for injury guild construction are presented and discussed by Pedigo, Hutchins and Higley (1986) and Hutchins, Higley and Pedigo (1988).

**Management of soybean injury guilds**

Using the injury guild approach, the EIL, as the fundamental decision criterion, can be utilized to assess the collective economic impact of several insect species. For soybean leaf-mass consumers, Hutchins, Higley and Pedigo (1988) reviewed the existing literature to quantify the relationship of soybean yield response to leaf-mass removal. They developed linear equations (*Table 3*), based on work by Todd and Morgan (1972) and Hinson, Nino and Boote (1978), that predict the *total* amount of leaf removal ($cm^2$ per row metre) at each of four soybean growth stages that results in final yield losses of 123.3 kg/ha, which was the estimated gain threshold (GT). These values, therefore, represent the critical (economic) level of damage as a function of total injury [$D(I_t)$]. The distinction between injury per insect ($I$) from equation (4) and total allowable injury for a population or guild ($I_t$) is important for determination of the number of injury equivalent units from multiple species equalling the EIL.

**Table 3.** Damage functions relating yield loss (kg/ha) to injury ($cm^2$ leaf area/m of row removed), based on data from Todd and Morgan (1972) and Hinson, Nino and Boote (1978) (regressions forced through the origin), and economic injury levels based on these functions

| Soybean stage | Regression equation | $r^2$ | $F_{1,3}$ | $P>F$ | $EIL_t$* |
|---|---|---|---|---|---|
| V8–10† | $D(I_t) = 0.09408I_t$ | 0.95 | 59.08 | 0.0046 | 1310.2 |
| R2 | $D(I_t) = 0.09030I_t$ | 0.96 | 79.82 | 0.0030 | 1365.1 |
| R4 | $D(I_t) = 0.07869I_t$ | 0.98 | 149.74 | 0.0012 | 1566.5 |
| R6 | $D(I_t) = 0.12140I_t$ | 0.99 | 467.21 | 0.0002 | 1015.4 |

* $EIL_t$ in $cm^2$ leaf area/row-m removed; calculated from $I_t = C/Vb$, where $C$ = US$27.18/ha, $V$ = US$0.2205/kg and $b$ = the regression coefficient.
† Soybean stages are based on Fehr *et al.* (1971).

A separate EIL can be calculated for each individual species ($EIL_i$) within an injury guild, based on its relative contribution to $I_t$:

$$EIL_i = I_t/I_i \qquad \text{Eq.(5)}$$

where $EIL_i$ is the number of injury equivalents (a full-grown larva equals one injury equivalent) for species $i$ that corresponds to economic damage; $I_t$ is the total level of injury ($cm^2$ per m of row) that corresponds to economic damage; and $I_i$ is the total consumption potential of an individual of species $i$ (*Table 1*).

Therefore, each of the three species from *Table 1* comprising the leaf-mass-consuming guild will have a unique EIL for soybeans of different developmental stages.

The matrix of equivalency coefficients (*Table 2*) can be used to determine the total injury ($I_t$) from a known density of individuals for each species within the injury guild. The model, therefore, calculates $I_t$ as the sum of specific injury equivalencies multiplied by the density of each species and size class:

$$I_t = \sum_{i=1}^{n_i} \sum_{j=1}^{n_j} N_{(i,j)} \cdot E_{(i,j)} \qquad \text{Eq.(6)}$$

where $N$ is the number of individuals for each species ($i$) and at each size class ($j$); and $E$ is the injury equivalency coefficient from *Table 2* for each species ($i$) and at each size class ($j$).

## LEAF-MASS CONSUMERS

Management of soybean leaf-mass consumers has been used throughout this review as an illustrative example to present the development of EIL and injury guild concepts. Indeed, because of an existing data-base relating the impact of defoliation on soybean physiology, and experimental techniques to measure leaf-mass consumption by individuals within a species, the conditions for constructing an injury guild were satisfied and the developments described previously were possible.

The significance of using the injury guild approach to manage multiple species of leaf-mass consumers was evaluated by Hutchins, Higley and Pedigo (1988). They sampled four distinct soybean arthropod pest communities and calculated a control solution for three scenarios:

1. all sampled individuals considered as a single injury equivalent, regardless of size or species;
2. individuals segregated by species, but not size class; and
3. individuals segregated by species and size differences.

The purpose of the study was to validate the model for estimating $I_t$ (equation 6) and to determine the relative impact on pest-management decision outcomes.

The results of the management comparisons were dramatic. In one typical situation, the estimated injury of a multiple-species population varied from 24.8 to 9.6 equivalency units. Because the calculated EIL for the standard equivalency species (soybean looper) was 13.9, a failure to incorporate differences in species and size-class consumption potentials resulted in an unnecessary management action. Hence, the use of injury guilds to manage soybean leaf-mass consumers may significantly enhance the value of the EIL model and the accuracy of management decisions.

## FRUIT FEEDERS

Fruit-feeding pests pose a major economic threat to soybeans in many production areas. The podworm and stink bug complexes represent the primary fruit-feeding species in the south-eastern US (Kogan and Turnipseed, 1987). The most economically important stink bug species is *N. viridula*, but *Acrosternum hilare* (Say) and *Euchistus servus* (Say) also are common, economically important pests. The dominant podworm species is *Heliothis zea* (Boddie), with *H. virescens* (Fab.) present in lesser frequency. Because soybean response to injury by fruit-feeding pests has been most intensively investigated in the south-eastern US, we will focus on findings from that region.

Before consideration can be given to constructing a fruit-feeding injury guild, the preconditions for species inclusion must be evaluated. The first condition is that all candidate species elicit a similar physiological response from the host. Although the physiological effects of injury induced by fruit-feeding pests have not been studied directly, considerable research has been conducted on the response of soybeans to surrogate pod or seed removal as a means to simulate fruit feeding by insects. With incomplete depodding, soybean leaves change from a source to a sink organ for assimilates (Lauer and Shibles, 1987). Hence, protein in the leaves is not mobilized to the seeds, and protein content per unit area of leaf is increased. Leaf photosynthetic rate and the time of initiation or duration of growth stages, however, are generally unaffected. With complete depodding, visual leaf senescence is delayed (Lindoo and Nooden, 1978; Nooden, Kahanak and Okatan, 1979; Nooden and Murray, 1982), but functional leaf senescence proceeds normally or is enhanced (Wittenbach, 1982, 1983). The photosynthetic rate drops and there is a loss of ribulose biphosphate carboxylase activity and level. Surgical seed removal results in visual leaf senescence which implies that the seed, not the pod, is directing the physiological response of the plant (Lindoo and Nooden, 1977). Actual feeding activities may differ from surrogate injury in that the former does not always result in complete consumption or destruction of individual seeds. Partial consumption may result in either partial or complete disruption of nutrient translocation to the seed, with the final amount of disruption differing among fruit-feeding pest species. Consequently, complete depodding or deseeding may not accurately simulate injury by some fruit-feeding pests.

Infestations from fruit-feeding pests vary spatially and temporally, but populations from each of the primary species sometimes occur concurrently in soybean fields (Kogan, 1981). The synchrony of fruit-feeding injury with soybean development must be understood to determine the degree and duration of injury. Although the crop develops through pod and seed stages over a relatively short period of time, many fruit-feeding species are able to progress through one or more life cycles during this interval. All stink bug species colonize soybean during early reproductive stages of crop growth, and reach peak population densities during pod fill (Panizzi, 1985). Detailed phenological information on the developmental rate for each nymphal stage, however, is lacking for the economically important stink bug species. By contrast, podworms may colonize soybeans during any crop developmental growth stage (Johnson, Stinner and Rabb, 1975; Dietz *et al.*, 1976; Hillhouse and Pitre, 1976) and often develop outbreak populations within a short period of time (Boyer and Duman, 1963; Boyer *et al.*, 1963; Boyer, 1965, 1967). Unfortunately, the actual response, i.e. damage curve, of soybean to injury induced by individual fruit-feeding species has not been measured. Without this information, the assumption that all species within the guild must produce a homogeneous physiological response can not be stated with confidence.

The second major information requirement for constructing injury guilds is the availability of accurate consumption potential data for each candidate species. Numerous studies have been conducted with the objective of evaluating the effects of feeding by economically important stink bug species on soybean seed yield and quality (Panizzi and Slansky, 1985). Densities of as little as one or less stink bug per metre of row measurably increase the number of damaged seeds (Todd and Turnipseed, 1974; Yeargan, 1977; McPherson, Newsom and Farthing, 1979). Germination of damaged seed is affected (Todd and Turnipseed, 1974; Todd, 1976; Yeargan, 1977), oil content of seeds decreases and leaf protein content increases (Thomas *et al.*, 1974a; Todd and Turnipseed, 1974; Todd, 1976). In some studies, treatments consisted of known densities of one or more life stages of stink bugs maintained on soybeans for specified periods of time (Todd and Turnipseed, 1974; Yeargan, 1977; McPherson, Newsom and Farthing, 1979). In other studies, cages were infested with different densities of adults that were allowed to reproduce, with the resultant number of nymphs quantified (Thomas *et al.*, 1974a). The amount of injury occurring at each soybean growth stage, or by each pest life stage, however, was not measured in any study. The duration and timing of stink bug infestation greatly affects the relationship between stink bug density and soybean yield. Densities of one adult *N. viridula* per 0.3 m of row or greater, established during early podding and maintained until crop harvest, are capable of significantly reducing yield (Todd and Turnipseed, 1974). Densities of two adult *A. hilare* per 0.3 m of row or greater can reduce yield (Yeargan, 1977). Lower densities maintained throughout these periods of crop growth, however, have not significantly reduced seed yield. Various stink bug densities maintained from early podfill to crop maturity have also failed to reduce yield (Thomas *et al.*, 1974a; Todd and Turnipseed, 1974; Yeargan, 1977).

Todd and Pitre (1979) review the published information involving studies to evaluate the effects of podworms on soybean seed yield and quality. The population dynamics of *H. zea* and *H. virescens* in soybean, including the developmental rate of each growth stage, are reported by Stinner, Rabb and Bradley (1974). Cages were also utilized in studies involving actual insects, to maintain desired densities of *H. zea* and to exclude endemic pests and natural enemies (Mueller and Engroff, 1980). Treatments consisted of several densities of young *H. zea* larvae confined to soybean plants and allowed to develop to pupation. Only limited attempts, however, were made to measure the amount of injury occurring at specific soybean growth stages. Methods have not been developed to measure adequately injury by most fruit-feeding species, and measurement undoubtedly will prove technically difficult. Experiments using simulated injury techniques (e.g. Kincade, Laster and Hartwig, 1971; Smith and Bass, 1972) suggest that podworm injury affects yield and quality of soybean, although at most soybean growth stages substantial levels of injury can be sustained with no apparent effect on yield or quality. Effects on yield increase as the soybean host develops to maturity (Thomas *et al.*, 1974b). Depodding at all growth stages decreases oil content and increases protein content (Thomas, Ignoffo and Smith, 1976). Unfortunately, the utility of these results is limited as no studies address the validity of depodding as an accurate simulation technique for podworm injury.

Obviously, important informational shortfalls exist concerning injury to soybean by fruit-feeding pests. Consumption rates for individual life stages or for total development have not been determined experimentally for most species. The effects of injury induced by stink bug and podworm insect species on soybean physiology are essentially unknown. Much research remains to be done, therefore, in order to satisfy the requirements for constructing an injury guild for fruit-feeding pests.

**Future opportunities**

Pest-management research must redirect its focus towards managing aggregate injury from all subtractive organisms. In doing so, pests will be recognized as the surrogate of injury and can then be logically categorized, based on their physiological impact on the plant (crop). This fundamental emphasis, once fully embraced, will promulgate the injury guild management approach.

The contrast in level of knowledge between the leaf-mass and fruit-feeding guilds is indicative of the intensive degree of supporting research that must be accomplished in order to satisfy the assumptions of guild construction. Nevertheless, research, if appropriately focused, can greatly refine the management of soybean pests. We suggest that initial efforts focus on the development and validation of simulated injury techniques. This approach would simplify the acquisition of agronomic data for a determination of host response and also serve as a benchmark by which to perform consumption studies.

Once the basic data are acquired to meet the conditions for establishing injury guilds, a plethora of advanced refinements can be envisioned. One particularly useful modification would be to incorporate existing knowledge of age-specific and species-specific survivorship into the injury equivalency coefficients. These data can be obtained from life-table studies (many of which already exist) and subsequently used as a quantitative method to account for natural mortality. Already, Ostlie and Pedigo (1987) have demonstrated the validity and potential of this approach with existing survivorship data on *P. scabra* in Iowa.

Another possible refinement could involve the modification of equivalency coefficients, based on current knowledge of control tactic efficacy, e.g. pesticide selectivity. This would essentially incorporate Norton's (1976) *K* value for each size class and/or species within the injury guild. A further extension may utilize knowledge about biological control effectiveness to modify the equivalency coefficients. This will provide a mechanism for augmenting insect control with inimical organisms in a fully integrated fashion.

Perhaps the ultimate opportunity for future development within the context of injury guilds is the incorporation of distinct pest types. For example, research may show that fungi causing leaf necrosis and insects consuming leaves elicit a similar physiological response from the plant. Similarly, weed competition and insect feeding both may eventually occupy the architecture modifier guild.

Clearly, the future development and eventual use of guilds to manage pest-induced injury to soybeans and other hosts will require extensive research and validation. We believe, however, that this approach holds much promise as a practical management tool to extend the utility of the economic injury level decision criterion.

## References


AMERICAN SOYBEAN ASSOCIATION (1983). *Soybean bluebook*. American Soybean Association, St. Louis, Missouri.

BARDNER, R. AND FLETCHER, K. E. (1974). Insect infestations and their effects on the growth and yield of field crops: a review. *Bulletin of Entomological Research* **64**, 141–160.

BOLDT, P. E., BIEVER, K. D. AND IGNOFFO, C. M. (1975). Lepidopteran pests of soybeans: consumption of soybean foliage and pods and developmental time. *Journal of Economic Entomology* **65**, 480–482.

BOOTE, K. J. (1981). *Concepts for modeling crop response to pest damage*. ASAE Pap. 81–4007. American Society of Agricultural Engineers, St. Joseph, Missouri.

BOYER, W. P. (1965). Survey of bollworm (*Heliothis zea*) infestations in soybeans in Arkansas 1960–1964. *Cooperative Economic Insect Reports* **15**, 462–463.

BOYER, W. P. (1967). The effect of type of plant growth on bollworm (*Heliothis zea*) infestations in soybeans in Arkansas. *Cooperative Economic Insect Reports* **17**, 83–86.

BOYER, W. P. AND DUMAN, W. A. (1963). Soybean insect survey as used in Arkansas. *Cooperative Economic Insect Reports* **13**, 91–96.

BOYER, W. P., WHITCOMB, W. H., DOWELL, G. C. AND BELL, R. (1963). Notes on *Heliothis* in Arkansas. *Cooperative Economic Insect Reports* **13**, 109–112.

DIETZ, L. L., VAN DUYN, J. W., BRADLEY, J. R., JR, RABB, R. L., BROOKS, W. M. AND STINNER, R. E. (1976). *A guide to the identification and biology of soybean arthropods in North Carolina*. North Carolina Agricultural Experiment Station Technical Bulletin No. 238.

FEHR, W. R., CAVINESS, C. E., BURMOOD, D. T. AND PENNINGTON, J. S. (1971). Stage of development descriptions for soybeans, *Glycine max* (L.) Merrill. *Crop Science* **11**, 929–930.

FENEMORE, P. G. (1982). *Plant Pests and Their Control*. Butterworths, Wellington, New Zealand.

FOREIGN AGRICULTURAL SERVICE, US DEPARTMENT OF AGRICULTURE (1983). Various issues. US Government Printing Office, Washington, DC.

FUNDERBURK, J. E. AND PEDIGO, L. P. (1983). Effects of actual and simulated seedcorn maggot (Diptera: Anthomyiidae) damage on soybean growth and yield. *Environmental Entomology* **12**, 323–330.

GREENE, G. L. (1976). Pest management of the velvetbean caterpillar in a soybean ecosystem. *Proceedings, World Soybean Conference Champaign, Illinois* (L. D. Hill, Ed.) pp. 602–610. Interstate Publishers, Danville, Illinois.

HAMMOND, R. B. AND PEDIGO, L. P. (1982). Determination of yield-loss relationships for two soybean defoliators by using simulated-insect defoliation techniques. *Journal of Economic Entomology* **75**, 102–107.

HAMMOND, R. B., PEDIGO, L. P. AND POSTON, F. L. (1979). Green cloverworm leaf consumption on greenhouse and field soybean leaves and development of a leaf consumption model. *Journal of Economic Entomology* **72**, 714–717.

HAWKINS, C. P. AND MACMAHON, J. A. (1989). Guilds: the multiple meanings of a concept. *Annual Review of Entomology* **34**, 423–451.

HIGGINS, R. A., PEDIGO, L. P. AND STANIFORTH, D. W. (1984). Effect of velvetleaf competition and defoliation simulating a green cloverworm (Lepidoptera: Noctuidae) outbreak in Iowa on indeterminate soybean yield, yield components, and economic decision levels. *Environmental Entomology* **13**, 917–925.

HILLHOUSE, T. L. AND PITRE, H. N. (1976). Oviposition by *Heliothis* on soybeans and cotton. *Journal of Economic Entomology* **69**, 144–146.

HINSON, K., NINO, R. H. AND BOOTE, K. J. (1978). Characteristics of removed leaflets and yield response of artifically defoliated soybeans. *Soil and Crop Science Society of Florida Proceedings* **37**, 104–109.

HUTCHINS, S. H. AND PEDIGO, L. P. (1990). Phenological disruption and economic consequence of injury to alfalfa induced by potato leafhopper (Homoptera: Cicadellidae). *Journal of Economic Entomology* **83**, 1587–1594.

HUTCHINS, S. H., HIGLEY, L. G. AND PEDIGO, L. P. (1988). Injury equivalency as a basis for developing multiple-species economic injury levels. *Journal of Economic Entomology* **81**, 1–8.

JOHNSON, M. W., STINNER, R. E. AND RABB, R. L. (1975). Oviposition response of *Heliothis zea* (Boddie) to its major hosts in North Carolina. *Environmental Entomology* **4**, 291–297.

KINCADE, R. T., LASTER, M. L. AND HARTWIG, E. E. (1971). Simulated pod injury to soybeans. *Journal of Economic Entomology* **64**, 984–985.

KOGAN, M. (1976). Evaluation of economic injury levels for soybean insect pests. *Proceedings, World Soybean Conference, Champaign, Illinois* (L. D. Hill, Ed.), pp. 515–533. Interstate Publishers, Danville, Illinois.

KOGAN, M. (1981). Dynamics of insect adaptations to soybean: impact of integrated pest management. *Environmental Entomology* **10**, 363–371.

KOGAN, M. AND TURNIPSEED, S. G. (1987). Ecology and management of soybean arthropods. *Annual Review of Entomology* **32**, 507–538.

LAUER, M. J. AND SHIBLES, R. (1987). Soybean leaf photosynthetic response to changing sink demand. *Crop Science* **27**, 1197–1201.

LINDOO, S. J. AND NOODEN, L. D. (1977). Studies on the behavior of the senescence signal in Anoka soybeans. *Plant Physiology* **59**, 1136–1140.

LINDOO, S. J. AND NOODEN, L. D. (1978). Correlations of cytokinins and abscisic acid with nomocarpic senescence in soybeans. *Plant and Cell Physiology* **19**, 997–1006.

MCPHERSON, R. J., NEWSOM, L. D. AND FARTHING, B. F. (1979). Evaluation of four stink bug species from three genera affecting soybean yield and quality in Louisiana. *Journal of Economic Entomology* **72**, 188–194.

MUELLER, A. J. AND ENGROFF, B. W. (1980). Effects of infestation levels on *Heliothis zea* on soybeans. *Journal of Economic Entomology* **73**, 271–275.

NEWSOM, L. D., KOGAN, M., MINER, F. D., RABB, R. L., TURNIPSEED, S. G. AND WHITCOMB, W. H. (1980). General accomplishments toward better pest control in soybean. In *New Technology in Pest Control* (C. Huffaker, Ed.), pp. 51–98. Wiley, New York.

NOODEN, L. D. AND MURRAY, B. J. (1982). Transmission of the monocarpic senescence signal via the xylem in soybeans. *Plant Physiology* **69**, 754–756.

NOODEN, L. D., KAHANAK, G. M. AND OKATAN, Y. (1979). Prevention of monocarpic senescence in soybeans with auxin and cytokinin: an antidote for self-destruction. *Science* **206**, 841–843.

NORTON, G. A. (1976). Analysis of decision making in crop protection. *Agro-Ecosystems* **3**, 27–44.

OGUNLANA, M. O. AND PEDIGO, L. P. (1974). Economic-injury levels of the potato leafhopper on soybeans in Iowa. *Journal of Economic Entomology* **67**, 29–32.

OSTLIE, K. R. AND PEDIGO, L. P. (1985). Soybean response to simulated green cloverworm (Lepidoptera: Noctuidae) defoliation: progress towards determining comprehensive economic injury levels. *Journal of Economic Entomology* **78**, 437–444.

OSTLIE, K. R. AND PEDIGO, L. P. (1987). Incorporating pest survivorship into economic thresholds. *Bulletin of the Entomological Society of America* **33**, 98–102.

PANIZZI, A. R. (1985). Dynamics of phytophagous pentatomids associated with soybean in Brazil. *World Soybean Research Conference Illinois: Proceedings* (R. Shibles, Ed.), pp. 674–680. Westview Press, Boulder, Colorado.

PANIZZI, A. R. AND SLANSKY, F., JR (1985). Review of phytophagous pentatomids (Hemiptera: Pentatomidae) associated with soybean in the Americas. *Florida Entomologist* **68**, 184–214.

PEDIGO, L. P., HUTCHINS, S. H. AND HIGLEY, L. G. (1986). Economic injury levels in theory and practice. *Annual Review of Entomology* **31**, 341–368.

PEDIGO, L. P., HIGGINS, R. A., HAMMOND, R. B. AND BECHINSKI, E. J. (1981). Soybean pest management. In *CRC Handbook of Pest Management in Agriculture* (D. Pimentel, Ed.), pp. 427–537. Boca Raton, Florida.

POSTON, F. L., PEDIGO, L. P. AND WELCH, S. M. (1983). Economic injury levels: reality and practicality. *Bulletin of the Entomological Society of America* **29**, 49–53.

REID, J. C. AND GREENE, G. L. (1973). The soybean looper: pupal weight, development time, and consumption of soybean foliage. *Florida Entomologist* **56**, 203–206.

ROOT, R. B. (1967). The niche exploitation pattern of the blue-gray gnatcatcher. *Ecological Monographs* **37**, 317–350.

ROOT, R. B. (1973). Organization of a plant–arthropod association in simple and diverse habitats: The fauna of collards (*Brassica oleracea*). *Ecological Monographs* **43**, 95–124.

SMITH, K. J. AND HUYSER, W. (1987). World distribution and significance of soybean. In *Soybeans: Improvement, Production and Uses* (J. R. Wilcox, Ed.), pp. 1–22. Agronomy Monograph 16, 2nd edition. ASA, Madison, Wisconsin.

SMITH, R. H. AND BASS, M. H. (1972). Relationships of artificial pod removal to soybean yields. *Journal of Economic Entomology* **65**, 606–608.

STERN, V. M., SMITH, R. F., VAN DEN BOSCH, R. AND HAGEN, K. S. (1959). The integrated control concept. *Hilgardia* **29**, 81–101.

STINNER, R. E., RABB, R. L. AND BRADLEY, J. R. (1974). Population dynamics of *Heliothis zea* (Boddie) and *H. virescens* (F.) in North Carolina: a simulation model. *Environmental Entomology* **3**, 163–168.

STONE, J. D. AND PEDIGO, L. P. (1972). Development and economic-injury level of the green cloverworm on soybean in Iowa. *Journal of Economic Entomology* **65**, 197–201.

THOMAS, G. D., IGNOFFO, C. M. AND SMITH, D. B. (1976). Influence of defoliation and depodding on quality of soybeans. *Journal of Economic Entomology* **69**, 437–440.

THOMAS, G. D., IGNOFFO, C. M., MORGAN, C. E. AND DICKERSON, W. A. (1974a). Southern green stink bug: influence on yield and quality of soybeans. *Journal of Economic Entomology* **67**, 501–503.

THOMAS, G. D., IGNOFFO, C. M., BIEVER, K. D. AND SMITH, D. B. (1974b). Influence of defoliation and depodding on yield of soybeans. *Journal of Economic Entomology* **67**, 683–685.

TODD, J. W. (1976). Effects of stink bug feeding on soybean seed quality. *Proceedings of the World Soybean Research Conference* (L. D. Hill, Ed.), pp. 611–618. Interstate Publishers, Danville, Illinois.

TODD, J. W. AND MORGAN, L. W. (1972). Effects of hand defoliation on yield and seed weight of soybeans. *Journal of Economic Entomology* **65**, 567–570.

TODD, J. W. AND PITRE, H. N. (1979). Economic thresholds of *Heliothis* species in soybeans. In *Economic Thresholds and Sampling of* Heliothis *Species on Cotton, Corn, Soybeans, and Other Host Plants*, pp. 66–70. Southern Cooperative Series Bulletin No. 231.

TODD, J. W. AND TURNIPSEED, S. G. (1974). Effects of southern green stink bug damage on yield and quality of soybeans. *Journal of Economic Entomology* **67**, 421–426.

TURNIPSEED, S. G. AND KOGAN, M. (1976). Soybean entomology. *Annual Review of Entomology* **21**, 247–282.

WITTENBACH, V. A. (1982). Effects of pod removal on leaf senescence in soybeans. *Plant Physiology* **70**, 1544–1548.

WITTENBACH, V. A. (1983). Effect of pod removal on leaf photosynthesis and soluble protein composition of field-grown soybeans. *Plant Physiology* **73**, 121–124.

YEARGAN, K. V. (1977). Effects of green stink bug damage on yield and quality of soybeans. *Journal of Economic Entomology* **70**, 619–622.

# 2
# Recent Developments in Sorghum Entomology Research

R.T. GAHUKAR

*Pest Control Consultant, Plot 220, Reshimbag, Nagpur 440009, India*

Introduction
Soil pests
    White grubs—Other pests
Seedling pests
    Sorghum shoot fly—Other pests
Foliage pests
    Armyworms—Grasshoppers—Leaf caterpillars—Aphids—Plant bugs—Mites
Stem borers
    Sorghum stem borer—Other stem borers
Earhead pests
    Sorghum midge——Earhead caterpillars—Earhead bugs—Earhead beetles
Stored grain pests
Insecticide residues
Perspectives
References

## Introduction

Sorghum, *Sorghum bicolor* (L.) Moench, is one of the major cereal crops, being grown in all continents on an area of 46.8 million ha and producing about 71.4 million metric tonnes (FAO, 1987). Grain yields are lowest in Africa (881 kg/ha) and highest in North America (3943 kg/ha) where slightly more than 50% of the world's sorghum grain is produced. Grains are used for human consumption and stalks are fed to animals. Generally, white grains are preferred in food recipes. In some African countries, red-brown grains with a bitter taste are used for making alcoholic drinks, and stalks are used for fabricating fences and roofs of houses and barnyards. Husk is fed to cattle or poultry. Forage sorghums as main and ratoon crops are common in India and North America.

Sorghum is attacked by about 100 insect pests in the field and during grain storage. Young and Teetes (1977) described the geographical distribution, pest status, plant damage, pest bionomics and control of key insect and mite pests attacking grain and forage sorghums. In 1979, the expert panel

*Agricultural Zoology Reviews* — Vol. 4, February 1991
0269–0543 $20.00 + $0.00 © Intercept Ltd, PO Box 716, Andover, Hampshire SP10 1YG, UK

on Integrated Pest Management (IPM) discussed the pest problems and prepared the guidelines for IPM in sorghum (FAO, 1979). Subsequently, several symposia and workshops were organized by research institutes or project-funding agencies in sorghum-growing countries. A specific workshop on sorghum entomology was held in the United States of America (USA) in July 1984 (ICRISAT, 1985). Since then, substantial progress in research on pest bionomics and control has been made. The purpose of this review is to complement the earlier information through brief discussion of research developments made during 1984–89 and their implications for improvement of IPM strategies in sorghum.

## Soil pests

### WHITE GRUBS (BEETLE LARVAE)

*Lachnosterna consanguinea* (Blanch.) and *Holotrichia serrata* F. are the most important of several species in India, and *Phyllophaga crinita* (Burmeister) is important in North America. *Anomala* spp. and *Schizonycha* spp. are occasionally injurious to grain sorghums in the Indian subcontinent and Africa. The grubs of these beetles feed on roots and cause the death of seedlings. Plant populations can be considerably reduced and patches seen in the field within a week or so of planting. Some plants are not killed but show stunted growth and the root pruning makes the plant vulnerable to lodging because of the poorly developed root system. The insects' eggs are laid in the soil and grubs hatch in 1–3 weeks. Whitish C-shaped robust grubs develop in 8–22 weeks and pupation takes a further 1–8 weeks. The adult beetles are 13–20 mm long with brownish-black bodies. The life cycle is completed in a year.

Delayed plantings generally escape pest attack but crops may suffer from seedling damage by overwintering grubs later in the season. Birds, such as the common crow, *Corvus splendens* Vieillot, and mynah, *Acridotheres tristis* Linn., predate upon the grubs which are exposed during ploughing (Shah and Garg, 1988). About 50% grub mortality was obtained 10 days after soil treatment with a fungus, *Verticillium lecanii* (Zimmerm) Viegas, at 100 ml suspension/4 kg soil (Gour and Dabi, 1988). However, preplanting soil applications of chlordane, heptachlor, aldrin or diazinon at 2 kg a.i./ha, are the treatments most commonly used to decrease grub populations. A combination of soil insecticides and interculturing systems was quite effective in India (Hasabe, Patil and Ajri, 1984). Therefore, adjustment of cultural practices and use of soil insecticides in endemic areas continue to be the best measures for pest management.

### OTHER PESTS

The soil-inhabiting cutworms, *Agrotis infusca* (Boisd.) and *A. ypsilon* Rott., cut off seedlings at ground level and feed on roots and leaves. Roots are also pruned by the grubs of root beetles, *Pterohelaeus alternatus* Pascoe, *P. darlingensis* Carter, *Gonocephalum macleayi* (Blackburn), *G. carpentariae* (Blackburn), *Cestrinus trivialis* Erichson and *Diabrotica* spp.; wireworms, *Agrypmus variabilis* Cand.; termites, *Microtermes* spp., *Odontotermes* spp.; and black field earwig, *Nala lividipes* (Duf.). Sown seeds are damaged or destroyed by ants, *Monomorium salomonis* (L.) and *Pheidole sulcatipes* Roger. All of these pests are minor, except that a severe attack of a termite, *Microtermes nadjensis* Harris, has been reported for the first time from the Yemen Arab Republic (Wood, Bednarzik and Aden, 1987). Soil inscctícides are applied before planting or are put near the plant base to control these pests (Radford and Allsopp, 1986). In Australia, seed dressing with chlorpyrifos (at 160 g a.i./100 kg seed) and soil compaction with press wheels (10 cm wide with 40 kg pressure/wheel) effectively killed the earwigs and wireworms and improved plant establishment (Allsopp and Radford, 1987; Radford and Allsopp, 1987). These practices may prove economical in many areas.

## Seedling pests

### SORGHUM SHOOT FLY

Sorghum shoot fly, *Atherigona soccata* Rondani, is a key pest of sorghum in South-East Asia, China, India and East and West Africa (Delobel and Lubega, 1984; Pawar, Jadhav and Kadam, 1984a,b; Sithole, 1985; Doumbia and Gahukar, 1986). *Atherigona yorki* Deeming is a widely distributed species in the Middle East where it becomes a serious pest on sorghum during certain seasons (Ba-Angood and Hubaishan, 1985). Maximum yield losses of 75.6% in grain and 68.6% in fodder have been reported from India (Pawar, Jadhav and Kadam, 1984a). A new technique for simulating plant infestation of different levels by cutting the growing points and producing artificial dead hearts has facilitated the estimation of yield losses (Delobel, 1984).

Plant damage is caused by the young maggot which tunnels towards the plant base and cuts through the growing point, resulting in withering of the central shoot, which is then called a dead heart. An excessive number of tillers is then produced and these have non-productive heads. Also, plant population densities and tiller size are reduced and replanting may be necessary following early attack. Generally, dead heart percentage is correlated negatively with plant height in forage crops (Sandhu, Dhaliwal and Sidhu, 1986), and with plant height, leaf area, transpiration rate, chlorophyll content and grain and fodder yields in grain sorghums (Mote, 1985a, 1986b; Mate, Phadnanwis and Mehetre, 1988). Damage assessment studies show that dead heart abundance is highly dependent on temperature and relative humidity (Taneja, SeshuReddy and Leuschner, 1986) and on the number of shoot fly eggs (Halalli, 1985). Thus, for each unit increase in plants with eggs, a unit increase in dead heart symptoms can be expected (Mohan, Jayaraj and Rangarajan, 1988). These observations should help in the determination of economic thresholds and injury levels of shoot fly populations.

Sorghum is the preferred host plant of *A. soccata* (Gahukar, 1985). The female deposits the rod-shaped white eggs singly on the underside of leaves when the plant has 1–7 leaves. Rainfall, and high soil moisture, air temperature and relative humidity inhibit oviposition, whereas plants treated with insecticide (carbofuran or isofenphos) and healthy seedlings favour shoot fly oviposition (Naitam and Sukhani, 1985). Small maggots hatch in 2–4 days and complete their development on decaying plant tissue in 7–12 days. The mature maggot is yellowish, 6–8 mm long with two posterior spiracular lobes. Pupation occurs in plant or soil and lasts for 6–11 days. The greyish, 5 mm-long fly is identified by the presence of three pairs of black spots on the dorsum of the pale pink abdomen. Flies live for 2–3 days without food or 12–15 days with food (Gujrati, Thakur and Singh, 1988). The life cycle is generally completed in 16–24 days. Pest survival during the off-season is on grasses such as *Sorghum halepense* (Linn.) Pers., *Panicum* spp., *Brachiaria* spp. and *Cynodon dactylon* (Linn.) Pers. Bene (1986) observed overwintering of maggots in Italy where the pest completes 5–6 generations per year. There is a positive correlation between fly fecundity and pupal weight, and temperatures less than 20°C adversely affect fly fecundity and fertility (Unnithan and Delobel, 1985; Taneja, SeshuReddy and Leuschner, 1986). Temperatures in excess of 40°C and rainfall are major pest mortality factors, although a relative humidity greater than 60% increases fly populations (Delobel and Lubega, 1984; Gahukar, 1987). Yellow sticky traps (Bene, 1986) and pan traps fabricated with metal or plastic and baited with fish meal (Taneja and Leuschner, 1986; Gahukar, 1987; Singh and Verma, 1988a) have been extensively used to study the species composition and population dynamics of shoot flies in diverse agroecosystems. In India, the number of flies trapped showed a significant positive and linear correlation with the numbers of eggs or dead hearts (Taneja, SeshuReddy and Leuschner, 1986; Singh and Verma, 1988a). However, these relationships were not consistent in Senegal (Gahukar, 1987).

Cultural practices play an important role in maintaining the pest at a low level. Late-planted crops suffer severely from fly attack (Mote, 1985a; Patel and Jotwani, 1986a; Gahukar, 1987). Pest incidence is lower in dense planting (180 000 plants/ha) than in thinly populated plots (Mote and Ramshe, 1987). An application of *Azospirillum* (750 mg fresh Azolla$^R$/kg soil or 250 g Azolla$^R$/10 kg seed) (Mohan *et al.*, 1987b), or fertilization with nitrogen or phosphorus at 60–90 kg/ha or with both elements (Balasubramanian *et al.*, 1986; Mote and Ramshe, 1987), reduced shoot fly attack. On the other hand, addition of cattle manure favours plant infestation (Balasubramanian *et al.*, 1986).

Varietal resistance has been found most effective and suitable in shoot fly management. Several physical, physiological and biochemical factors involved in sorghum resistance may act individually or together. Plant height, percentage tasselling, stem thickness, number of leaves per plant, and length, width and thickness of leaf are negatively correlated with plant susceptibility (Khurana and Verma, 1985). Therefore, resistant cultivars are characterized by their tallness and by narrow, glossy and yellowish-green leaves. These leaves possess trichomes on their abaxial surface which act as physical barriers to the penetration of young maggots (Kishore, Rana and Agarwal, 1985; Mote, Kadam and Bapat, 1986). Higher silica contents are also found in resistant plants (Bothe and Pokharkar, 1985).

Major mechanisms involved in sorghum resistance to shoot fly include antixenosis for oviposition and antibiosis, the two being equally important in some cultivars (Raina *et al.*, 1984; Unnithan and SeshuReddy, 1985; Singh and Rana, 1986). These mechanisms are fairly stable at all levels of pest infestation (Ba-Angood and Hubaishan, 1985; Mote, Kadam and Bapat, 1986). When an acetone extract of susceptible sorghum was sprayed on maize, a non-host plant, it elicited a high oviposition response (Unnithan *et al.*, 1987). Tolerance, expressed by rapid tillering and recovery, exists in certain cultivars, such as SPH-196, SPH-225 and CSH-8R (Mote, Kadam and Bapat, 1985). Several grain sorghums in the IS series, including 1054, 1456, 2146, 2291, 2312, 3962, 5490, 5604, 5613, 5642, 7094, 12611 and 18551, have been identified as resistant to shoot fly in India (Mote, 1984d; Biradar and Borikar, 1985; Kishore, Rana and Agarwal, 1985; Mote and Bapat, 1985; Nimbalkar and Bapat, 1987; Jadhav, Mote and Bapat, 1988) and in Africa (Raina *et al.*, 1984; Unnithan and SeshuReddy, 1985). These resistant sources and their progenies have good general combining ability for resistance. Similarly, resistant sources in forage sorghums, such as SG-59-3, SL-21 and SL-57 (Sandhu, Dhaliwal and Sidhu, 1986), and wild sorghums, such as *Sorghum sudanense* (Piper) Staff. and *S. halepense* (Khurana and Verma, 1988a) may help in breeding programmes because dual purpose (grain + fodder) sorghums, such as IS-1054, IS-2123 and IS-18551, are preferred nowadays. The genetics of resistance showed that antixenosis is governed by *NPO npo* or *NPo npo* genes and dead heart formation by *DH1* and *DH2* genes (Sharma and Rana, 1985). Obviously, genetic variability is expressed at various plant growth stages (Paul, Kulkarni and Murthy, 1984), and inheritance of characters is mainly due to additive genes (Singh and Verma, 1988a). Therefore, recurrent selection under high pest incidence should prove satisfactory in evolving resistant cultivars.

About 20 parasitoids attack the pest during its life cycle (Bene, 1986) but natural enemies do not play a major role in reducing pest populations. Systemic insecticides in the form of granules are widely and effectively used for pest control in India (Natarajan, 1985; Natarajan and Chelliah, 1986b). Spot application in furrows or rows of carbofuran granules (at 1–2 kg a.i./ha or 1.5–2 g/m row), followed by fensulfoton, isofenphos, quinalphos, disulfoton, phorate or benfuracarb was quite effective (Gandhale, Patil and Pokharkar, 1984; Pawar, Shirshikar and Sawale, 1985; Men, Mundiwale and Borle, 1986). However, carbofuran seed treatment is generally superior to other chemicals and treatments (Mote and Kadam, 1984; Pawar, Jadhav and Kadam, 1984b; Patel and Jotwani, 1986a; Balasubramanian *et al.*, 1987; Rao and Azam, 1987a). In some cases, seed pelleting with chlorpyrifos at 4 ml/kg seed (Men, Mundiwale and Borle, 1986) and mixing of seeds with a slurry of carbofuran granules (Mote, 1986a) were generally more effective than seed treatment. The slurry treatment is compatible with fungicides (captan, thiram, benomyl, carbendazim) which are used as seed dressings (Pawar,

Jadhav and Kadam, 1984a). Mixing of untreated seeds with carbofuran-treated seeds in the proportion of 4:1 or 3:2 reduced the insecticide dose and treatment cost (Mote, 1984a). However, atmospheric humidity and a long storage period lower the viability of carbofuran-treated seeds (Mogal, Mali and Rajput, 1984), whereas seeds treated with DDT + thiram remain viable for at least 1 year (Patil and Jarhad, 1988). Considering the cost and effectiveness of control methods, shoot fly is better managed by early planting, by seed treatment with systemic insecticides and by using resistant cultivars.

#### OTHER PESTS

Seedlings are occasionally devastated by leaf beetles. The common species are: *Chaetocnema india* Wsm., *C. tibialis* Illig., *Longitarsus* spp., *Phyllotreta chotanica* Duviv. and *Menolepta* spp. in India, and *Lema planifrons* Wsm. in West Africa. Thc larvae feed on leaf epidermis and the plant takes on a bleached appearance and may die. Infestation levels are high during droughts. Maggots of certain chloropids, *Anatrichus erinaceum* Loew., *Scolipthalmus mecantipennis* Duda, *Scolipthalmus* sp., *Elachipterus abessynicus* Becker, and a diopsid, *Diopsis* sp., feed inside stems already attacked by shoot fly and aggravate the damage (Gahukar, 1985; Doumbia and Gahukar, 1986). Outbreaks of grey weevil, *Myllocerus undecimpunctata maculosus* Desb., occur in India. The adult weevils eat the leaf margins and the larvae feed on roots. Spraying 0.1% carbaryl or thiodicarb, or 0.004% monocrotophos controlled this pest (Budharaja *et al.*, 1984). The current practice to control these seedling pests is to treat the soil with granules of systemic insecticides.

## Foliage pests

### ARMYWORMS

#### *Fall armyworm*

The fall armyworm, *Spodoptera frugiperda* (J.E. Smith), is an occasional pest of sorghum in south-eastern, northern and central states of the USA, and South and Central America. The caterpillars feed within the leaf whorl and often cause extensive leaf ragging. The leaves unfolding from whorls are perforated.

Several strains of this pest species are thought to occur. Females of a Mississippian strain exhibited peak calling at 23.00 h and a Honduran strain at 03.00 h. These calling rhythms were approximately paralleled by pheromone titres. The difference in peak calling is probably attributable to a difference in ambient temperature to which the females were exposed in their native habitats. Crosses of these strains were successful and both strains are genetically similar to the maize strain. The persistence of characteristic calling rhythms despite having been colonized for 2–5 generations under similar laboratory conditions suggests that calling may be genetically determined (Ramaswamy, Ma and Pitre, 1988). When these strains were reared, development time was shorter for larvae from Mississippi than those from Honduras, and adults from Mississippi were the most fecund (Castro and Pitre, 1988). Sampling procedures for moth oviposition have been simplified by the use of metal flags in surveys (Thomson and All, 1984) and pheromone traps to monitor the seasonal fluctuations of moth populations and to predict subsequent larval abundance (Silvain, 1986). Armyworms can be reared on laboratory diets containing seeds of pinto beans, sorghum, cowpea and maize (Wiseman *et al.*, 1984; Adeyeye and Blum, 1988). When sorghum panicles taken from a resistant and a susceptible plant at pre-flowering, flowering, milky and hard dough stages were mixed in a semi-synthetic diet, larval weight was significantly different but other biological characters remained unchanged (Wiseman *et al.*, 1986). Factors in the glume have a great effect on larval weight and appear to be responsible for a type of resistance which is being considered as a possible component of pest management strategies (Schwager, Pitre and Gourley, 1984). Sorghum lines from Ethiopia and Yemen showed seedling resistance to leaf feeding by armyworms (Wiseman and Lovell, 1988). Further screening of large collections and local landraces is necessary.

Although infestation levels were similar in no tillage and ploughed fields, a high mulch concentration at the soil surface reduced insect damage and moth oviposition (All, 1988). Maize generally suffers higher oviposition and larval infestation and more plant damage than sorghum, and maize-fed insects are more fecund. Thus, maize has potential as a trap crop for monitoring insect populations in low rainfall areas in the USA where sorghum is grown as a main crop (Castro and Pitre, 1988).

Between 23 and 32% of armyworm larvae are parasitized by up to 19 parasitoids, the major ones being: *Chelonus insularis* Cresson and *Archytos marmoratus* Towns. (Pair *et al.*, 1986). Pair and Gross (1984) have reported a new parasitoid, *Diapetimorpha introita* (Cresson) from the southern USA. The fall armyworm is also susceptible to at least 20 species of entomogenous pathogens, including bacteria, fungi, viruses, protozoa and nematodes. Among them, *Bacillus thuringiensis* Berl., *Nomuraea rileyi* (Farlow) Samson, *Vairimorpha necatrix* and nuclear polyhedrosis and granulosis viruses are potential biological control agents, particularly when combined with cultural practices (Gardner, Noblet and Schwehr, 1984; Gardner and Hornby, 1987), although host plants may affect the susceptibility of armyworms to nuclear polyhedrosis virus (Richter, Fuxa and Abdel-Fattah, 1987). Chlorpyrifos at 0.56 kg a.i./ha also gave satisfactory pest control (All, 1988). Methomyl and chlorpyrifos were effective against larvae from Jamaica and Mississippi, but a Florida colony appeared to be tolerant of carbaryl, permethrin, methomyl and chlorpyrifos. This suggests that Florida populations may not be the source of insects invading Mississippi (Pitre, 1988). The use of resistant cultivars and natural enemies may be further exploited to control the fall armyworms.

### *Rice armyworms*

The rice armyworm, *Mythimna separata* (Wlk.), is a sporadic but potentially serious pest of the crop during the off-season (November–April) in India (Mote, 1984b). It prefers pearl millet, sorghum, maize and Napier grass, *Pennisetum purpureum* Schum., for larval feeding and a lemon grass, *Cymbopogon* sp., for oviposition. Larvae are often concentrated within leaf whorls. Insect reproduction is greatly affected by temperature and relative humidity and a sugar-containing diet during the adult stage is essential for egg-laying (Singh, Mrig and Chaudhary, 1987a). A braconid wasp, *Apanteles* sp., is active in Punjab (India) in September, when it may kill up to 41% of larvae (Singh, Mrig and Chaudhary, 1987b). This situation is, however, rare. Therefore, the pest is controlled by applying baits of 0.25% fenthion, quinalphos, endosulfan or monocrotophos mixed with wheat bran or granules of 10% malathion, phosalone, quinalphos or HCH mixed with 25% gum arabic solution (Giraddi and Kulkarni, 1987). Three sprays of endosulfan at 0.4–0.8 kg a.i./ha applied at 30, 40 and 60 days after planting also controlled the armyworms (Patel, Shekhawat and Bhalani, 1987).

### *Other armyworms*

Among other armyworms, *Mythimna loreyi* (Dup.) is a minor pest in India (Singh and Chaudhary, 1987) but its sporadic attack on off-season crops is common in West Africa. Outbreaks of *Spodoptera exigua* (Hb.) and *S. exempta* (Wlk.) in Africa and *Pseudoletia convecta* (Wlk.) in Australia occur during vegetative plant growth. These pests are polyphagous and voracious feeders whose control is generally achieved by use of insecticides.

## GRASSHOPPERS

Several species of grasshoppers feed on sorghum foliage at all stages of plant development. *Kraussaria angulifera* (Krauss) (Bindra and Amatobi, 1985) and *Oedaleus senegalensis* Krauss (Amatobi, 1985) in West Africa, and *Chrotogonus* spp. and *Hieroglyphus* spp. in India, are the most abundant species. For oviposition, *Oedaleus senegalensis* prefers ridge tops and sides to furrows (Amatobi, 1985). Therefore, deep ploughing after the crop harvest and dusting of insecticides in target areas, coupled with collection and destruction of hoppers, are effective measures.

## LEAF CATERPILLARS

Defoliation by leaf caterpillars is of minor or moderate importance, viz. *Marasmia trapezalis* Gn., *Euproctis subnotata* (Wlk.), *E. virguncula* Wlk., *Spilosoma obliqua* Wlk., *Amsacta albistriga* Wlk. and *A. moorei* (Btlr.) in India and *A. moloneyi* Druce in West Africa. Attacks by these pests are sporadic and may pose potential threats during certain seasons only. Young caterpillars are easily killed by insecticide sprays or dustings.

APHIDS

### *Greenbug*

The greenbug, *Schizaphis graminum* (Rondani), is a key pest of grain sorghum in the USA. About eight greenbug biotypes are present but biotypes B, C and E are the common ones. A survey in Oklahoma in 1986 showed the seasonal variation in biotype composition: in spring, there was 83% biotype E, 11% biotype B, 4% biotype C and one colony of an unidentified biotype (which differed from biotype F, when biotypes C and E clones were tested for fecundity); the summer collection contained 94% of biotype E and 6% of biotype C; the autumn collection had 97% of biotype E (Kerns, Peters and Puterka, 1987). This and earlier surveys confirmed that biotype E has replaced biotype C in much of northern Texas, Oklahoma, Kansas and Nebraska (Kindler *et al.*, 1984). Clones of biotypes C and E collected in different localities did not show the differences in fecundity on host plants on which they had been reared. However, after rearing on barley, greater fecundity of biotype C was noted on Kentucky blue grass, *Poa pratensis* L., and Sudan grass, *Sorghum sudanense*, and of biotype E on oats and Sudan grass. Greenbugs did not survive for longer than a week on maize (Beregovoy, Starks and Janardan, 1988). All greenbugs developed slowly on wheat (cv. Amigo) and sorghum (cv. IS-809) resistant to biotype C but adults had higher fecundity on sorghum than on wheat. Similarly, progeny production and development periods of biotype E collected from grain sorghum in Texas and wheat in Arkansas were similar on wheat and sorghum but dissimilar on maize. The Arkansas population had a shorter fourth instar when compared with the Texas population and did not complete development on maize (Michels, 1987). Different enzyme patterns are found in different biotypes (Beregovoy and Starks, 1986), and morphometric variation exists between the alate biotypes (Inayatullah, Webster and Fargo, 1987) but the alate and apterous forms may be present in the same colony. Greenbug resistance to certain insecticides on different host plants is believed to be responsible for creating new biotypes in greenbug populations (Mayo *et al.*, 1987). Thus, on the basis of host plant reaction to isolates, Puterka *et al.* (1988) designated an Oklahoma isolate as a biotype G and a Texas isolate as biotype H. These biotypes lack the mid-dorsal dark-green abdominal stripe characteristic of biotypes B, C and E but have the general appearance of biotype F. This biotype evolution makes the control strategies rather difficult.

Greenbugs attack sorghum after seedling emergence and reach damaging levels after heading. They are found in colonies on the underside of leaves. The sucking of sap and injection of toxins by the insects into leaf tissues cause reddish spots to develop. Later, leaves turn brownish, curl and wither. The honeydew produced by greenbugs favours the growth and development of sooty mould and the insects also transmit the maize dwarf mosaic virus and predispose the plant to charcoal rot. Greenbugs are about 1.8 mm long with a light-green body and a darker-green dorsal abdominal stripe. The distal leg segments and cornicle tips are black in colour. A single greenbug produces about 80 offspring parthenogenetically during a period of 25 days. Greenbug numbers increase rapidly on sorghum during the crop season because wheat and small grains act as winter hosts and provide a substantial reservoir of aphids to infest sorghum crops, and because *Sorghum halepense* serves as an interim host. In Texas (USA), population peaks occur from late July to early August. Rain and natural enemies suppress pest increase but greenbug abundance is associated strongly with the number of leaves and weakly with leaf area. Therefore, early maturity cultivars which have fewer leaves suffer greater yield losses from greenbug attack than do the medium- or late-maturing cultivars (Archer and Bynum, 1986b). Greenbugs aggregate in compact groups on leaves located predominantly on the lower half of plants. This aggregation is not affected by sorghum cultivar or pest density. Thus greenbug population density could be determined on leaves in the lower half or third of plants (Archer and Bynum, 1986a). Greenbug populations on water-stressed plants were distributed towards the tops of plants, with 28.0% of the population above mid-plant, 57.3% at mid-plant and 14.7% below mid-plant. Populations on well-watered plants had 19.3% of the population above mid-plant, 33.6% at mid-plant and 47.1% below mid-plant. Greenbug populations were reduced on water-stressed plants. However, no evidence was found that greenbugs modified or maintained specific leaf water and osmotic potentials (Michels and Undersander, 1986).

Pectin is a barrier to stylet penetration for phloem feeding and this has proved helpful in creating resistant cultivars (Dreyer, Campbell and Jones, 1984). The probing of biotype C generally inhibits phloem circulation in resistant cultivars, but once it reaches phloem it sucks uninterruptedly (Dreyer and Campbell, 1984). Greenbug biotypes are able to hydrolyse pectin substances at a greater rate in susceptible cultivars (Campbell and Dreyer, 1985). Biotype E overcomes the host-plant barrier by displaying increased pectin methyl esterase activity. This presented the possibility of manipulating resistance by altering middle-lamella chemistry, either through breeding or by the use of plant growth regulators (Dreyer, Campbell and Jones, 1984). Furthermore, hybridization between biotypes C and E indicated that hybrids from E mothers are able to overcome resistance, since this trait is inherited by sexually produced fundatrices in an extranuclear manner from the mother parent (Eisenbach and Mittler, 1987).

The resistance mechanisms in sorghum towards the greenbug include tolerance and moderate levels of antixenosis for oviposition and antibiosis. In Brazil, an artificial infestation of 10 greenbugs/plant 11 days after planting killed the susceptible cv. BR-601 in 21 days, whereas even with 20 greenbugs/plant, the resistant cv. TX-2567 survived for at least 28 days (Cruz and Vendramim, 1988). The bloomless character (absence of a white waxy exudate on the leaf sheath) is a resistance factor (Weibel, 1986; Weibel and Starks, 1986). Therefore, bloomless restorer parents such as BOK-8, BOK-11, Wheatland Redia, DELKAB and DK-46, have been used successfully for producing bloomless hybrids and to keep the pest populations below the economic thresholds (Weibel, 1986; Harvey and Thompson, 1988).

The role of cultural practices on greenbug attack is being studied. Pest density is low in thick plant stands of 220 000 plants/ha (Harvey and Thompson, 1988). Greenbugs respond to background conditions of host plants associated with tillage operations and seem to be repelled by crop residues and the reduced tilled soil surface. Also, greenbugs seem to distinguish between the amount of cultivation and prefer plants growing in soil with the greatest amount of disturbance (Burton *et al.*, 1987).

Greenbug predators include the coccinellids, *Hippodamia convergens* Guerin-Mandeville, *H. sinuata* Mulsant, *Scymnus* spp. and *Coleomegilla maculata* Deg.; a syrphid, *Syrphis batteatus* G.; and a chrysop, *Chrysopa basalis* (Kring and Gilstrap, 1986; Rice and Wilde, 1988). Coccinellids are the predominant predators, comprising up to 89% of the total predator population (Rice and Wilde, 1988). They are abundant at panicle emergence as they shift to greenbug from corn leaf aphid (Kring and Gilstrap, 1986). Indigenous primary parasitoids include *Lysiphlebus testaceipes* (Cresson), *Aphelinus nigritus* (Howard) and *Diaeretiella rapae* (McIntosh), the first of which is the most active (Gilstrap, Kring and Brooks, 1984; Rice and Wilde, 1988). Some exotic parasitoids have also been tried, although the natural enemies are generally able to suppress pest populations from the three-leaf stage to the boot stage (Rice and Wilde, 1988). The density of natural enemies within the top 5 cm of soil is correlated with the amount of litter fall, and may be affected by site-burning (Kring and Gilstrap, 1984) which must, therefore, be carried out carefully.

Spraying bioregulators, such as chlormequat and *N-N*-dimethyl-piperidinium chloride, on to one-week-old sorghum plants reduced greenbug reproduction by 50% (Dreyer, Campbell and Jones, 1984). Greenbug control is also achieved by soil application of carbofuran or chlorpyrifos granules mixed with or without fertilizers at planting (Wilde and Mize, 1984; Wilde *et al.*, 1984; Wilde, Russ and Mize, 1986). Treating the soil (7.6 cm deep and in a 7.6 cm wide band in seed furrows) with terbufos granules also gave good control of greenbugs (Depew and Hooker, 1987). Bynum *et al.* (1988) applied 8–12 times less chlorpyrifos than the registered rate of 0.28 kg a.i./ha through multifunction irrigation system nozzles, moving within the crop canopy, to reduce the application cost. Generally, one application of insecticide will suffice if it is applied at the right time. Seed treatment with menazon (Barbulescu, 1985) or dimethoate at 200–300 g a.i./100 kg seed (Villata, 1986) protected the plants during the early part of the season. However, insecticide resistance in greenbugs may hamper further use of insecticides. At present, resistant cultivars and predators and parasitoids should be considered as major components in greenbug management and should be integrated with cultural practices whenever these are available.

### *Corn leaf aphid*

The corn leaf aphid, *Rhopalosiphum maidis* (Fitch), is a common pest of sorghum leaf whorls in the USA and the Indian subcontinent. Mature plants may support pest attack but attack during the booting stage may cause poor head emergence. Plant population density is considerably reduced if seedlings are attacked. The aphid is a vector of maize dwarf mosaic virus. The population abundance of this greenish-blue aphid is highly correlated with minimum temperature and relative humidity (Prabhakar, Rao and Rao, 1986). Usually its population density is checked by coccinellid beetles, which feed on corn leaf aphid until panicle emergence but later move to greenbugs (Kring and Gilstrap, 1986). Syrphid larvae also contribute to aphid mortality. Resistant cultivars with high grain yields are being bred.

### *Sugarcane aphid*

The sugarcane aphid, *Aphis (Melanaphis) sacchari* Zehntner, attacks sorghum from panicle emergence through seed development in southern Africa and China. Recently it has been introduced into Argentina and Uruguay (Delfino, 1985). The aphids suck sap from the underside of lower leaves. Honeydew is produced and this favours the development of sooty mould. These aphids also transmit the persistent millet red leaf virus. As many as 56 generations may be completed per year in China, with population density peaks at the boot stage (Chang and Fang, 1984). Population abundance and distribution depend upon rainfall and temperature (Chang and Fang, 1984). Since highly significant correlations between population development of wheat aphid and sugarcane aphid exist, wheat aphid development parameters could be used to predict the population trend of *A. sacchari* on sorghum (Niu, 1987). A few resistant cultivars have been identified in China, viz. R-128, R-131, R-133, 129, 3A, 71-4 and 80Ax33R (Chang and Fang, 1984). Crosses between resistant and susceptible cultivars suggested that resistance is conditioned by a single major gene which is dominant in $F_1$ (Tan *et al.*, 1985). A coccinellid, *Scymnus babai* Sassji, is an important predator during certain seasons in China (He, Lieu and Zheng, 1987). Resistant cultivars should be a major pest-management component.

### *Yellow sugarcane aphid*

The yellow sugarcane aphid, *Sipha flava* Forbes, is a pest of sorghum seedlings in the USA and Central America. Its attack during the pre-boot stage may kill the plant. This aphid secretes a plant toxin and transmits the sugarcane mosaic virus. Losses of 67, 30 and 7% have been estimated when seedlings were infested with 6 aphids/plant at plant heights of 3.5, 4.5 and 5.5 cm, respectively (Breen and Teetes, 1986a). If the duration of infestation is increased, plant growth and blooming period are more affected than are grain yields (Breen and Teetes, 1986b). More studies on pest biology and ecology, economic thresholds and control methods are necessary before pest-management strategies can properly be formulated.

PLANT BUGS

### *Chinch bug*

Chinch bug, *Blissus leucopterus leucopterus* (Say), is an occasional pest in the USA and Central America. Adults and nymphs suck the sap from leaves and stems which become pale and later wither. Plants produce shrivelled, chaffy grains. Insects are small blackish-white, greenish-white or reddish in colour and agglomerations or colonies are found behind the leaf sheaths. Investigations on varietal reaction revealed the resistance sources in grain sorghums (BCK-60-1155, DELKAB, DK-61, 1155, SC-303) and in forage sorghums (Early Sumac-c) (Ahmed, Kindler and Pruess, 1984; Mize and Wilde, 1986a, b; Wilde, Stuart and Hatchett, 1986). Antibiosis is the major mechanism of this resistance, particularly in seedlings and older plants (Mize and Wilde, 1986a, c). Antibiosis occurring in wheat could help to protect sorghum by preventing development of large migrating populations (Wilde, Stuart and Hatchett, 1986). Plant vigour is responsible for tolerance and antixenosis for reduced oviposition (Mize and Wilde, 1986a), so plants may recover from insect damage at later stages of growth (Ahmed, Kindler and Pruess, 1984). The host cultivar may affect the efficacy of a fungus, *Beauveria bassiana* (Bals.) Vuill., which is influential in controlling pest numbers (Ramoska and Todd, 1985) and, when carbofuran was applied every year in sorghum/soybean rotation, the efficacy of the insecticide was lowered (Wilde, Russ and Mize, 1986). Carbaryl at 1.66 kg a.i./ha and carbofuran at 0.56 kg a.i./ha have been found to be most effective in chinch bug control (Bauernfeind, 1987) but these insecticides degraded rapidly in biologically active soils, whereas aldicarb, phorate and terbufos did not. Thus, pesticide biodegradation is a factor to be considered in pest-control programmes. Considering recent research progress, varietal resistance seems to be promising for chinch bug management.

### *Sugarcane pyrilla*

Sugarcane pyrilla, *Pyrilla perpusilla* (Wlk.), attack becomes moderate to serious on off-season and ratoon crops in India (Mote, 1984c; Jagtap, Ghule and Deokar, 1985). The feeding of insects on tender shoots results in leaf chlorosis and wilting. During screening tests, resistant cultivars have recently been identified, viz. PC-1, PC-7, SL-44, S-178, PC-16, PC-19, SEL-79 and PC-101 (Dhaliwal, Mehndiratta and Arora, 1988). Maximum parasitism rates of 24–60% in eggs (by *Ooencyrtus papilionus* Ashm. and *Tetrastichus pyrillae* Crawf.) and 12–25% in nymphal and adult stages (by *Epiricania melanoleuca* F.) have been recorded during mid-September (Chaudhary *et al.*, 1987). Pest incidence is generally suppressed by insecticide use.

### *Shoot bugs*

Sorghum shoot bug, *Peregrinus maidis* Ashm., causes occasional damage to sorghum in central and southern India (Rao, Bhushan and Azam, 1986). Both macropterous and brachypterous forms are present in colonies. The sucking of sap from leaves and shoots by nymphs and adults leads to leaf chlorosis, stunted growth and ultimately to reduced plant populations or shrivelled and chaffy grains. Similar damage is observed by another shoot bug, *Poophilus costalis* (Wlk.), in West Africa. Pest numbers are often low and no control measures are applied.

## MITES

### *Banks grass mite*

The Banks grass mite, *Oligonychus pratensis* (Banks), has gained, although only occasionally, economic importance as a sorghum pest in the semi-arid great plains and south-western regions of the USA. Mite colonies generally settle along the midrib on the underside of lower leaves. Sap sucking by mites makes the leaf pale yellow and later reddish and brownish. Mites are also frequently encountered on sorghum heads because, although the mites principally attack several grasses, populations increase rapidly on sorghum after heading. The mites are 0.4 mm long and deep-green in colour, except for the palpi and first two pairs of legs. Sexual dimorphism exists and the life cycle is completed in about 11 days, but may take about 2 months under unfavourable conditions. Population studies have shown that the lower two-thirds of the plant harbour the maximum mite population and that leaf quality (Archer, Bynum and Peterson, 1986a) and number influence mite distribution on the plant. A sample of all leaves on the lower two-thirds of the plant is appropriate in population studies (Archer and Bynum, 1986a). Sorghum cultivar maturity also affects mite abundance and damage potential (Archer, Bynum and Peterson, 1986b) and mite population densities are higher in hot and dry weather.

A combination of fertilizers favours mite development, whereas application of nitrogen (90–180 kg/ha) or phosphorus (20–40 kg/ha) alone reduces mite numbers (Archer, Bynum and Onkan, 1988). Work on varietal resistance is under way. The resistance is a type of tolerance which is not related to non-senescence, although non-senescent cv. SC-599-6 had few mites and showed less plant damage than other cultivars (Archer, Bynum and Peterson, 1986b). A high total sugar content is thought to be the reason for its resistance. Once available, resistant cultivars may be included in pest management. Polyphagous mites, such as *Stethorus* spp. and *Amblyseius* spp., and midges are common predators present in the regions but do not provide adequate control. Therefore, chemicals have been used to which the mites have subsequently shown resistance. Further studies are needed to determine the most appropriate pest-management components.

*Sorghum mite*

Sorghum mite, *Oligonychus indicus* Hirst., is a secondary pest on main crops but sporadically increases in numbers on off-season crops in India. Mite webbings may cover leaves and earheads. It is naturally controlled by predators, the major one being the coccinellid, *Scymnus gracilis* Motsch., whose larvae feed on mite eggs, and whose adults feed on nymph and adult mites (Mohan and Karuppuchamy, 1987). Five spiders (*Uloborus* sp., *Argyopes* sp., *Theridula* sp., *Theridion* sp., *Cheiracanthium* sp.) have been found preying on sorghum mite (Manjunatha and Puttaswamy, 1989) but their effectiveness in mite control has not been studied. Acaricide use is becoming popular on high-yielding cultivars. Monocrotophos at 0.05% has been most effective in reducing mite numbers, followed by fenpropathion or thiometon (Gopalan, 1988), but a proper strategy for pest control has not yet been formulated.

## Stem borers

Stem borer complexes differ between regions, viz. *Chilo partellus* (Swinhoe) and *Sesamia inferens* (Wlk.) in the Indian subcontinent; *C. partellus, S. inferens* and *Ostrinia furnacallis* Gn. in eastern Asia; *C. partellus* and *S. cretica* Led. in the Middle East; *Acigona ignefusalis* Hmps., *Sesamia calamistis* (Hmps.) and *Busseola fusca* (Fuller) in West Africa; *C. partellus, S. calamistis, B. fusca* and *Eldana saccharina* Wlk. in East and South Africa; *Ostrinia nubilalis* (Hbn.), *Diatraea saccharalis* (Fb.), *D. grandissella* (Dyar), *D. lineolata* (Wlk.), *Elasmopalpus lignosellus* (Zeller), *Eoreuma loftini* (Dyar) in North America; *D. saccharalis, D. lineolata* and *E. lignosellus* in Central and South Americas. In November 1987, an international workshop on sorghum stem borers was held in India to review research on pest management, and reports have been published in proceedings (ICRISAT, 1989).

### SORGHUM STEM BORER

Sorghum stem borer or spotted stalk borer, *Chilo partellus*, is the most widely distributed stem borer. It is a key pest of sorghum but also attacks maize and pearl millet. Agroclimatic adaptation by this borer has resulted in four ecotypes in India (Verma and Jotwani, 1985). Infestation begins 5–6 weeks after plant emergence. Young larvae feed on leaf whorls so that leaves show parallel window holes after opening. Late instars tunnel inside stems and feed on the pith; vascular bundles are badly affected and dead hearts are produced. Chaffy earheads are also produced and plants are lodged due to internode injury. Yield loss is directly related to leaf damage, dead-heart production and percentage stem tunnelling, but losses differ according to cultivar and duration and amount of infestation (Alghali, 1986, 1987; Mote, 1986c). Generally, for each unit increase in percentage stem length tunnelled, the average reduction in yield per plant is 0.39–1.0%, depending upon the susceptibility of the cultivar (Mote, 1986c). A reduction in minimum temperature of 1°C decreased stem damage by 1.7% at 30 days after plant emergence (Mahadevan and Chelliah, 1986a).

The scale-like overlapping eggs are laid in masses on the underside of leaves about 3 weeks after plant emergence, with peaks 2–6 weeks later. Alghali (1985a) noted up to 96% of egg batches on leaf blades, 57% on the upper leaf surface and 9% on leaf sheaths because, as the plant grows, egg-laying shifts to the upper part of the plant. After an incubation period of 5–7 days, young caterpillars hatch and wander on the leaf surface. Greater larval migration was recorded when larvae were placed on stems rather than leaf whorls (Alghali and Saxena, 1988). A larval disorienting effect is attributed to leaf-edge spines and the chemical composition of leaf waxes. Larvae accumulate wax around their prolegs as they move over the plant surface and this impedes their progress. Also, positive phototaxis is of primary importance in upward climbing behaviour of larvae (Bernays, Woodhead and Haines, 1985). Erect leaves and curled leaf bases (Woodhead and Taneja, 1987) and hairiness of ligulae and the midribs (Kishore, 1987a) all influence larval establishment. Green paper was much more effective than foliage-green or cadmium-yellow paper when used as cylindrical sticky traps. These paper traps attracted and trapped a higher number of first- to third-instar larvae than did sorghum plants (Saxena and Onyango, 1987). The larvae, which are pale white with black dots, become fully grown in 3–4 weeks, during which time the plant reacts to borer feeding by producing a red-staining compound in the stem (Alghali, 1985b). The larvae pass through five instars, which last 6, 4, 5, 3 and 7 days, respectively, and can be distinguished by head-capsule width (Alghali, 1985a). Almost 45% of larvae are present in the lower third of the stem and 50% in the middle third (Kfir, 1988); over 85% of larvae present in stem galleries are the last instar (Kfir, 1988). During larval growth and development, changes in their contents of glycogen and lipid occur (Raja *et al.*, 1986a,b). Haemolymph proteins in larvae showed a rapid rise in concentration during pupation and the early adult stage (Raja *et al.*, 1986c). During the cold season, some larvae overwinter in the stem galleries in stubble. Overwintering is facultative diapause which is terminated once favourable environmental conditions are restored (Singh *et al.*, 1985; Chaudhary and Sharma, 1988). The overall concentration of amino acids is higher in diapausing larvae but lysine is absent and proline is present (Anand, Pant and Srivastava, 1986). Pupation takes place in larval galleries inside stems and lasts 7–10 days. Adults are nocturnal, buff-coloured moths. A generation is completed in about 1 month in summer and 2 months in winter (Alghali, 1985b; Mahadevan and Chelliah, 1986a). Four generations can appear in a year but two generations may attack a single crop. Low fertility and fecundity is observed in summer moths (Mahadevan and Chelliah, 1986a).

Complete destruction of stubble, weeds and stems after harvest is an effective method of borer control (Kfir, van Hamburg and van Vuuren, 1989). In south India, crops planted in the first fortnight of March are not damaged (Mahadevan and Chelliah, 1986a). Generally, late plantings are severely damaged so should be avoided. Intercropping of sorghum with cowpea or lablab reduced borer infestation by 50% and increased grain yields by 10–20% (Mahadevan and Chelliah, 1986b). More plants with dead heart and stem tunnelling were present in plots treated with cattle manure than untreated plots, but the reverse was observed in plots treated with synthetic fertilizers (Balasubramanian *et al.*, 1986).

Stem borer-resistant sorghums, including high yielders such as E-302, E-304, IS-2123, IS-2205, IS-4664, IS-5469, P-217, P-297, P-311 and P-500, have been identified in India and Africa (Kishore, 1984, 1987a, b; Raina *et al.*, 1984; Kundu, 1985; SeshuReddy, 1985; Singh, Rana and Rao, 1987). Wild sorghums (*Sorghum virgatum*) (Khurana and Verma, 1988b) and forage sorghums (S-136, S-178, PC-6, SPV-238, Improved Ramkel) (Dhaliwal, Mehndiratta and Arora, 1988) showed some degree of resistance to borer attack. Major resistance mechanisms are tolerance, antixenosis for oviposition and variable degrees of antibiosis (Singh and Rana, 1984; Singh and Verma, 1988c). When phytochemicals extracted from pulses were sprayed on sorghum plants, moth oviposition was prevented for up to 6 days (Mahadevan and Chelliah, 1986b). Reduced larval establishment in the leaf whorl, a longer time-interval between hatching and boring of larvae into stems and lower larval survival have been observed in resistant cultivars. The larval development period is the most important factor influencing the growth index and total life span (Singh and Verma, 1988c). Plant characters such as plant height, stem thickness, number of leaves per plant, leaf length, width, thickness and strength are negatively correlated with susceptibility in sorghum (Khurana and Verma, 1985). The major plant characteristics associated with resistance are early panicle initiation and faster internode elongation. Combining ability studies indicated the predominance of additive gene effects for stem tunnelling, and both additive and non-additive gene effects for dead heart and leaf injury (Rana *et al.*, 1984; Pathak, 1985; Singh and Verma, 1988d). Highest correlations for percentage stem tunnelling were obtained between successive generations (Kishore, Rana and Sharma, 1984). Pedigree breeding with single or multiple crosses has been useful for the transfer of resistance in high-yielding cultivars. Recurrent selection for resistance and yield has been suggested to increase the level of tolerance (Singh and Rana, 1989). These findings may be useful in the search for improved host plant resistance.

Among predators, the coccinellids *Coccinella undecimpunctata* L. and *Menochilus sexmaculatus* (F.) are abundant. In South Africa, an ant, *Pheidole megacephala* (F.), preys on larvae in winter (Kfir, 1988). Braconid wasps, *Apanteles flavipes* Cam., *A. sesamiae* Cam. and *Bracon* sp., are active larval parasitoids (Chaudhary and Sharma, 1987; Skoroszewski and Hamburg, 1987; Kfir, 1988). These wasps do not undergo diapause, along with the host, and they can be reared in the laboratory for release, whereas *Chelonus curvimaculatus* (Cam.), *Chelonus* sp. and *Pristomerus* sp. hibernate in their host and *Iphiaulax* sp. hibernates in a cocoon inside the stem (Kfir, 1988). Recently, Quicke and Wharton (1989) described a new braconid wasp, *Myosoma nyanzaensis* sp. n., which is a larval ectoparasitoid in Kenya.

A single application of a non-sporulating bacterium, *Serratia marcescens* Bizio, prevented egg hatch and caused total mortality of newly hatched caterpillars (Chakravorty *et al.*, 1985). Against older larvae, 2–3 applications of sprays or granules of endosulfan, quinalphos, lindane, carbaryl, endrin, DDT, phenthoate, carbofuran or fenvalerate are effective (Natarajan and Chelliah, 1986b; Patel, Shekhawat and Bhalani, 1987; Katti and Verma, 1988). However, Kishore (1988) recommended two dustings, 20 and 35 days after plant emergence, of fenvalerate at 25 g a.i./ha or endosulfan at 40 g a.i./ha. An integrated control programme is necessary for stem borers and should combine cultural practices such as the growing of resistant cultivars with one or two applications of granules of effective insecticides. Efforts to augment the efficacy of indigenous natural enemies should be intensified.

## OTHER STEM BORERS

### *Maize stem borer*

Maize stem borer, *Busseola fusca*, is usually a maize pest but attacks sorghum in East, West and South Africa. The major secondary hosts in Ethiopia are *Pennisetum purpureum* Schum. and *Sorghum verticilliflorum* (Gebre-Amlak, 1988). Damage caused by maize borer is similar to that of sorghum stem borer. In the laboratory, post-embryonic development is completed in *c.* 41 days. Feeding of adult moths on sucrose does not affect their fecundity (Unnithan, 1987). In South Africa, about 82% of larvae of the sixth instar and 16% of the fifth instar entered diapause. About 65% of diapaused larvae were in the lower third and 30% in the middle third of the stalk (Kfir, 1988). Diapause is facultative and can be prevented in the laboratory by rearing larvae on young sorghum plants (Unnithan, 1987). Diapause is terminated in the field by the continuous moisture conditions of the long rainy season (Okuda, 1988). Predators and parasitoids of the sorghum stem borer (*C. partellus*) generally also attack *B. fusca* larvae (Kfir, 1988) and releases of *A. flavipes* controlled the borers in South Africa (Skoroszewski and Hamburg, 1987), although the application techniques need to be improved and application costs reduced. The pest is currently controlled by three applications, at weekly intervals, of granules of trichlorfon, endosulfan or sprays of carbaryl (Adesiyun, 1986). Use of varietal resistance may be exploited in future integrated pest management schemes.

### *Sugarcane borer*

Southwestern corn borer or sugarcane borer, *Diatraea saccharalis*, attacks the sorghum crop and the economic injury level occurs when 10% of internodes are bored. This usually coincides with 5% of plants containing small caterpillars in their leaf sheaths (Fuller, Reagan and Flynn, 1989). High plant density increases fibre content and this can double moth emergence but does not affect pest survival (Fuller and Reagan, 1989). The stalk height, percentage sucrose and total sugars are negatively correlated with damage (Fuller, Reagan and Flynn, 1988). Plantings in October–November in Brazil (Pereira, Dacosta and Ferreira, 1987) and in July in Texas (Youm, Browning and Gilstrap, 1988) suffer heavy pest attack. Resistant sources, including CMSx601, AF-28, CMSx5157 and BR-505, have been identified in Brazil (Lara and Perussi, 1984; Pereira, Dacosta and Ferreira, 1987). Predators such as carabid larvae, coccinellids, chrysops, anthocorids and mites are responsible for a reduction in pest populations of almost 50%, and an ant, *Solenopsis invicta* Buren, is a major predator in Texas. General predation resulted in an increase in yield of about 22% (Fuller and Reagan, 1988). Parasitization is rather low, the major parasitoid being *Apanteles flavipes* (Youm, Browning and Gilstrap, 1988). Cultural practices combined with natural enemies are probably best suited to the management of this borer.

### *European corn borer*

European corn borer, *Ostrinia nubilalis*, is a key pest of maize in North America but also feeds on leaf tissues in moist areas in the whorl and inside stem of sorghum. Work on varietal resistance has been initiated. Sorghum resistance to first-generation larvae is an antibiosis type, acting on first- and second-instar larvae feeding on leaves (Dharmalingam *et al.*, 1984), whereas second-generation borer resistance is related to sheath-collar feeding (Guthrie *et al.*, 1984a, b, 1985) for which the genetic basis of resistance is not known. Both high and low HCN-p (hydrocyanic acid potential) genotypes are resistant, and resistance to leaf feeding by first-generation larvae is not considered to be due to a chemical factor, DIMBOA [(2, 4-dihydroxy-7-methoxy-2H-1,4-benzoxazin-(4H)-one)] (Guthrie, Haskins and Gorz, 1988). Kindler, Guthrie and Jarvis (1987) compared artificial infestation techniques and found significantly greater damage on plants infested with egg masses than on plants infested with young larvae; they also noted borer resistance in hybrids 35R, GS-20, GR-108 and NP-11-BR. The promise of varietal resistance warrants further research.

No work has been undertaken on pink borers, *Sesamia calamistis, S. cretica* and *S. inferens*, except that Elamin (1988) noted the preference of *S. cretica* for sorghum over sugarcane in Sudan. Activity of the moths was studied by light-trap catches, revealing a negative correlation of activity with minimum temperature. The damage could be avoided by early planting and early harvesting in Sudan. Research is needed on all borers, including those of minor importance, because their pest status may change following modifications of cultural practices and extensive use of insecticides.

## Earhead pests

### SORGHUM MIDGE

Sorghum midge, *Contarinia sorghicola* (Coq.), is a key pest in most sorghum growing areas. This wide distribution is thought to be associated with transport of poorly cleaned seeds which harbour diapausing maggots. Introduction of this pest into Puerto Rico is recent (Segara-Carmona and Barbosa, 1988). The maggots feed on ovaries, and grain formation is affected directly. Thus, there is a positive correlation between the proportion of sterile spikelets and kernel weight, and the proportion of spikelets damaged and yield loss (Alghali, 1984; Hallman, Teetes and Johnson, 1984b). Yield loss estimates by three methods (natural infestation, sequential sampling and cage experiments) showed that 0.5–1.5 g of grain per ovipositing female is lost in susceptible hybrids and 0.14–0.3 g of grain per female in resistant hybrids in the USA (Hallman, Teetes and Johnson, 1984a). In China, the economic injury level is 0.6 g per panicle (Hong, 1987). Bionomics of the sorghum midge has been studied in Kenya (Alghali, 1984), Brazil (Busoli *et al.*, 1984b), and Colombia (Ceremenol, Galvin and Labaton, 1984).

A sorghum midge deposits eggs in the flowering earhead, preferably at mid-anthesis (Modini, Page and Franzmann, 1987). Maggots hatch in 2–3 days and develop in 8–11 days. They are reddish when fully grown. Pupation occurs beneath the glume and lasts 3–5 days. Midge peak emergence is during the morning hours (Alghali, 1984; Modini, Page and Franzmann, 1987). The adult is an orange-coloured, small (2 mm) insect which lives for 1–2 days. The complete life cycle takes 17–34 days and there are usually 9–12 generations during a crop season. A portion of the larval population undergoes facultative diapause in winter, and this is favoured by low temperature. The period from diapause initiation to adult emergence has been estimated in Australia to be 6 months at 27°C and 7.5 months at 23°C (Forrester, 1987). The diapause is terminated when warm and humid weather prevails (25–30°C and about 60% relative humidity). A temperature-dependent simulation model to predict a spring emergence of overwintered midges has been demonstrated by Baxendale, Teetes and Sharpe (1984). This model gave results that were comparable with observed midge emergence from Johnson grass, *Sorghum halepense*, so may prove useful in pest forecasting (Baxendale *et al.*, 1984).

The period of peak incidence differs according to the climatic conditions in the region (Patel and Jotwani, 1986b; Nwanze, 1988). A high temperature during flowering reduces the pest population density (Patel and Jotwani, 1986b) and the time to attain 50% flowering and size of the earhead have been known to cause pest population fluctuations (Alghali, 1984). Steward, Kring and Yearian (1989) compared six methods of adult sampling and recommended direct estimates, since visual examination, as is generally used in surveys, is time-consuming and its accuracy decreases as the panicle compactness increases. According to Peterson *et al.* (1989), visual damage rating (% blasted grains/panicle) is a more effective method for evaluation when multiple locations and years are studied. This rating is correlated with both % yield loss and midge numbers. Mohan and Tajuddin (1988) recommended battery-operated suction traps for adult sampling because these traps caught 97% more midges than other sampling methods.

Among cultural practices, the most practical control method is to avoid late plantings as the pest multiplies on early flowering cultivars or wild sorghums and then moves to those flowering late in the season. Midge resistance in cultivars AF-28, DJ-6514, TAM-2566, IS-12664c, ICSV-197, TIFT-MR-88, SGIRL-MR-1, SGIRL-MR-3 and SGIRL-MR-4 has been confirmed in different environments (Teetes, Becerra and Peterson, 1986; Agrawal, Sharma and Leuschner, 1987; Agrawal, Abraham and House, 1988; Wiseman, Duncan and Widstrom, 1988; Hanna *et al.*, 1989). Recently, new sources of resistance have been identified in the IS series (Peterson *et al.*, 1989). In order to increase pest population density in screening tests, Sharma, Vidyasagar and Leuschner (1988a) recommended several techniques:

1. adjustment of planting period to synchronize flowering with population peaks of adults;
2. planting infested rows of a susceptible cultivar 20 days before test cultivars;
3. spreading damaged panicles containing diapaused maggots in rows to be infested;
4. sprinkler irrigation during flowering in the post-rainy season; and
5. planting two sets of test material at an interval of 14 days to reduce the chances of pest escape.

The use of cages to create a no-choice condition for adult midges also proved to be a suitable technique for screening tests (Sharma, Vidyasagar and Leuschner, 1988b).

Sorghum resistance to sorghum midge is mainly antixenosis for oviposition and variable degrees of antibiosis (Rossetto, Nagai and Overman, 1984; Widstrom, Wiseman and McMillan, 1984; Natarajan and Chelliah, 1986a; Singh, 1987). Non-preference for adult visitation and tolerance of sorghum also occur but are not so important (Waquil, Teetes and Peterson, 1986a, b). Ovipositional efficiency is fourfold less on resistant cultivars (Waquil, Teetes and Peterson, 1986b). Likewise, greater numbers of large larvae developed and pupated in spikelets of susceptible hybrids (Waquil, Teetes and Peterson, 1986c). Resistant hybrids produce 50–60% fewer progeny per female. This effect is cumulative in reducing pest activity through time (Melton and Teetes, 1984).

Both morphological (short and compact glumes) and biochemical (tannin content in ripening grain) factors contribute to resistance. Inheritance of resistance is quantitative in nature and is controlled by recessive genes at two or more loci (Boozaya-Angoon *et al.*, 1984; Singh, 1987). Also, the mean performance of parents and their general combining abilities are strongly correlated (Agrawal, Abraham and House, 1988), but it appears difficult to transfer genes for midge resistance into good agronomic B-lines by simple hybridization. Therefore, pedigree and modified pedigree breeding systems have been suggested (Singh, 1987).

Ovipositing midges are predated by anthocorid bugs, mites, lacewings and spiders. An earwig, *Diaperasticus erythrocephala* Oliv., in Kenya (Alghali, 1984) and a spider, *Metazygia* or *Gregalis* sp., in Colombia (Ceremenol, Galvan and Labaton, 1984), were found feeding on midges for the first time. Thontadarya and Rao (1987) studied the biology of an anthocorid bug, *Orius maxidentex* Ghauri, which is a primary predator of ovipositing midges in

India. Each bug consumes 16–24 midges during a life span of 22–42 days. Hymenopterous parasitoids, *Tetrastichus diplosidis* Crawf., *Tetrastichus* spp. and *Eupelmus popa* Gir., lower the pest populations in certain seasons, and up to 33% parasitism in Brazil (Busoli *et al.*, 1984b) and 28.3% parasitism in Argentina (Diaz, 1988) have been noted. The parasitoid : midge ratio may be as high as 22 : 1, depending upon the cultivar and crop season, but midge infestation intensity is not affected by parasitoids (Busoli *et al.*, 1984b; Gahukar, 1984a; Diaz, 1988). This makes the use of these parasitoids as control agents unlikely. Nearly 58% of the population of parasitoids was accounted for by *E. popa* on Johnson grass in Texas, where it has gained importance as the major parasitoid (Brooks and Gilstrap, 1985), whereas in Brazil (Busoli *et al.*, 1984b), Argentina (Diaz, 1988) and Senegal (Gahukar, 1984a), *T. diplosidis* is an abundant species on sorghum, particularly at the end of the crop season.

Insecticides have been tested against sorghum midge, viz. phosalone, endosulfan, quinalphos, cypermethrin, chlorpyrifos, HCH, malathion, diazinon (Mote, 1985b; Natarajan and Chelliah, 1986c; Katti and Verma, 1988). The efficacy of chlorpyrifos (280 g a.i./ha) is enhanced by mixing it with 10 l/ha of soybean oil or 7 l/ha of cottonseed oil (Treacy *et al.*, 1986). Generally, two applications of sprays or dusts are given: the first at 50% flowering and the second dose a week later (Natarajan and Chelliah, 1986c). Resistant cultivars respond better than susceptible cultivars to insecticide treatments in terms of grain yields and net return (Teetes, Becerra and Peterson, 1986). Manipulation of the planting period and the use of resistant cultivars combined with one or two insecticide applications are the best tactics for midge management (Teetes, 1985; Natarajan and Chelliah, 1986c; Nwanze, 1988).

## EARHEAD CATERPILLARS

Attack by earhead caterpillars is often severe on compact panicles. Caterpillars feed on the starchy contents of developing grains and webs with frass may be present. If damaged grains are exposed to late rains, sooty mould develops and grain quality deteriorates. The most cosmopolitan and polyphagous caterpillar is the American bollworm, *Heliothis armigera* (Hbn.), which prefers about 41 species of host plants for feeding (Singh and Balan, 1986). Besides the grain damage, early instar larvae feed on anthers and late instars destroy the ovaries of sorghum flowers (Teakle and Byrne, 1989). In the USA, corn earworm, *H. zea* (Boddie), is common on cotton, maize and legumes but its importance has recently increased on sorghum (Hayes, 1988) where almost 50% losses in grain yield have been reported (Karlen, Sadler and Roach, 1989). The moth lays 100–300 eggs singly on flowering spikelets or seeds. Larval development is completed in 2 weeks and a generation takes about a month. There may be six generations annually and seasonal incidence differs in each agroecosystem (Slosser *et al.*, 1987; Natarajan and Sundara Babu, 1988b). *Heliothis zea* exhibits a clumped within-field spatial distribution for total and parasitized eggs but no consistent preference for any portion of panicle (upper, middle or lower third) or stage of flowering (one-third, two-thirds or full bloom) was observed (Kring, Yearian and Steward, 1989). However, non-uniform plant growth creates favourable conditions for high pest incidence (Karlen, Sadler and Roach, 1989). Pheromone traps are widely used to study fluctuations in moth populations. A dry, white, funnel trap placed just above the crop canopy caught more moths than at other heights (Pawar *et al.*, 1988). Moth numbers were positively correlated with cropping patterns and rainfall in the region (Slosser *et al.*, 1987).

Production of resistant cultivars has been possible due to the significant combining ability shown by several crosses, and some resistance sources, such as IS-12537c, IS-2177c and Chencholam, are available for further improvement (Patel *et al.*, 1986). Active parasitoids generally keep the pest population at a low level. For example, *Apanteles* sp., *Bracon* sp., *Microbracon hebetor* Say, *Campoletis chlorideae* Uchida and *Euplectrus euplexiae* Roh. are potential parasitoids of *H. armigera* in India (Singh and Balan, 1986). Insecticides in India (Mote and Kadam, 1985) and nuclear polyhedrosis virus in Australia (Teakle, Jensen and Giles, 1985) are occasionally used for pest control. The best time for insecticide application seems to be at 100% anthesis (Teakle and Byrne, 1989). Further research is necessary on pest-management tactics, particularly agronomic practices and natural enemies.

Other common earhead caterpillars include: *Stenochroia elongella* Hmps., *Spodoptera frugiperda* and *Celama sorghiella* (Riley) in the USA; *Celama analis* Wil. & West., *Eublemma silicula* Swinh., *Cryptoblabes gnidiella* Mill. and *Euproctis subnotata* Wlk. in India; *Eublemma gayneri* Roths. and *Pyroderces simplex* Wsm. in Africa. A geometrid, *Pleuroprucha insulsaria* (Guenée) was found for the first time feeding on earheads of grain sorghum in the USA, but the pest status of this insect has not been determined (Kring, Yearian and Steward, 1987). In the absence of effective natural enemies, earhead caterpillars may be better controlled by insecticides combined with adjustment of cultural practices. Meanwhile, work on economic importance, damage thresholds, population ecology and practical control methods needs to be undertaken due to the increasing importance of these minor/occasional pests on some high-yielding cultivars.

## EARHEAD BUGS

### *Sorghum earhead bug*

Sorghum earhead bug, *Calocoris angustatus* (Leth.), is an important pest in South India. The bug prefers sorghum among several host plants (Hiremath, 1986). The nymphs and adults suck the sap from developing grains, resulting in distortion and shrinking. Damaged grains have low starch and protein contents and a high level of free amino acids (Natarajan and Sundara Babu, 1987a). There is a loss in seed germination, 1000-seed weight and grain hardiness, and losses in grain yield may attain 43–94% depending upon the cultivar (Rao and Azam, 1987b; Sharma and Lopez, 1989) or 9.7–42.0

g/panicle (Natarajan and Sundara Babu, 1988d). Adults consume three times as much food as fourth-instar nymphs. The economic threshold has been worked out as 8–15 nymphs, 5–10 feeding adults or 0.06–0.12 ovipositing adults per panicle (Natarajan and Sundara Babu, 1988c). Thus, based on a cost : benefit ratio of 1 : 1, the economic injury level ranged from 0.2 to 1.4 insects/panicle (Sharma and Lopez, 1989).

Females prefer 1–3-day-old panicles for oviposition (Natarajan and Sundara Babu, 1988d). In the laboratory, the pre-oviposition and oviposition periods are 2.7–3.2 and 5.2–9.1 days, respectively (Natarajan and Sundara Babu, 1987b, 1988a). A female lays 99–129 eggs, and nymphs hatch in 6 days. The life cycle is completed in 24–33 days. The adult is a yellowish-green, 1 cm long bug with a sex-ratio of 1M : 1.5–2.7 F in laboratory-reared colonies. The cannibalistic and predatory habit of the bug has been noted by Hiremath, Thontadarya and Jairao (1984) but this does not appear to hamper pest multiplication. The highest pest numbers are reached during flowering and the milky/dough grain stages (Hiremath and Thontadarya, 1984; Natarajan, Sundara Babu and Chelliah, 1989). An increase in bug population is associated positively with relative humidity and negatively with temperature (Hiremath and Thontadarya, 1984; Prabhakar, Rao and Rao, 1986).

Sorghum planted in early June suffers less from pest attack than later plantings (Hiremath and Thontadarya, 1984; Natarajan, Sundara Babu and Chelliah, 1989). Two generations can attack the crop during each cropping season. Screening of sorghum cultivars indicated some resistant sources, viz. IS-2205, IS-2394, IS-4663 and IS-5604 (Natarajan, Sundara Babu and Sridhar, 1988). One or two applications of HCH, malathion, chlorpyrifos, phosalone, carbaryl, dimethoate or demeton-S-methyl control this pest (Rao and Azam, 1987b; Sharma and Leuschner, 1987; Natarajan and Sundara Babu, 1988c) but Sharma and Lopez (1989) applied 2–5 sprays of carbaryl at 500 g a.i./ha between half anthesis and dough stages to obtain significantly higher yields over untreated plots. The use of carbaryl, malathion or endosulfan will continue until alternative pest-management strategies are devised.

### *Other bugs*

Outbreaks of the following bugs have been reported: lygaeids, *Nysius graminicolus* Kol. in Italy (Bin and Colazza, 1986) and *N. plebejus* Dist. in China (Hong, 1986); pyrrhocorid, *Dysdercus voelkeri* Schmidt in Nigeria (Daramola, 1986); mirids, *Eurystylus bellevoyi* Reut. in India (Sharma and Leuschner, 1987); *E. marginatus* Odh. and *Creontiades pallidus* Ramb. in West Africa (Gahukar, Doumbia and Bonzi, 1989) and a pentatomid, *Thyanta preditor* (Fb.), in Brazil (Busoli *et al.*, 1984a). In Nigeria, the local cultivars, BE-5 and HP-3, showed reduced susceptibility to the attack of *D. voelkeri* (Daramola, 1986). No control work has been carried out on these pests.

EARHEAD BEETLES

Beetles feeding on flowers inhibit seed formation, and scarabaeids form a major group. In the Sahelian zone of West Africa, blister beetles, *Psalydolytta fusca* Oliv. and *P. vestita* Duf., have become important pests in certain seasons (Gahukar *et al.*, 1989) and some new beetles have also been reported (Gahukar, 1984b). In India, significant yield losses occurred due to the feeding of *Chiloloba acuta* Wied. (Kundu and Srivastava, 1984). Other beetles of frequent occurrence belong to the genera *Mylabris, Cyanolylta, Anomala, Cylindrothorax, Rhinyptia* and *Schizonycha*. Abundance of blister beetles generally coincides with sorghum flowering and, when an epidemic occurs, insecticides are dusted over the earheads.

## Stored-grain pests

Many Lepidoptera and Coleoptera pests attack sorghum grains stored in traditional and commercial storage structures. Common pests are *Sitophilus oryzae* (L.), *S. zeamais* Motsch., *Sitotroga cerealella* (Oliv.), *Corcyra cephalonica* (St.), *Ephestia cautella* (Wlk.), *Oryzaephilus surinamensis* (L.), *Rhizopertha dominica* (F.), *Tribolium castaneum* (Herbst.) and *T. confusum* (de Duval). *Sitophilus oryzae, S. zeamais* and *S. cerealella* attack exposed grains even in fields. The presence of *Prostephanus truncatus* (Horn) in sorghum shipments has recently been noted for the first time in India (Verma and Lal, 1987). Most of the storage pests are cosmopolitan and damage is caused by larvae feeding and developing within the grains. Primary pests are capable of attacking whole healthy grains, while secondary feeders prefer cracked or damaged grains or flour. Insect feeding results in a reduction in quality, viability and quantity of grains. Pest attack intensity and the losses incurred vary according to the pest complex, the duration and method of storage (Pawar *et al.*, 1984; Hoppe, 1986), and also upon the size and hardness of sorghum grain (Shazali, 1986). A larva of *C. cephalonica* consumes 0.43–0.61 g grain during its lifetime of about 2 weeks (Parameshwar and Jairao, 1987).

The biology of storage pests has been studied in controlled and natural environments (Sattigi, Kulkarni and Thontadarya, 1985a; Shazali and Smith, 1985). Weevils deposit eggs within the grains whereas moths lay eggs on grain surfaces. Life cycles are completed in about a month. The development period and larval mortality are not affected by grain size in *S. oryzae* and *S. cerealella* but larger grains give larger adults of *S. cerealella*, and progeny production is greater in small grains (Shazali, 1987b). *Sitophilus oryzae*, on the other hand, produces more progeny on larger grains which generally favour oviposition (Shazali, 1986). In laboratory culture, these two insects compete for food, and progeny of females decreases as parental density increases (Shazali, 1985). A temperature range of 25–30°C and relative humidity of 60–80% favour insect development and multiplication (Shazali and Smith, 1985; Zafar and Quayyum, 1985), although *S. cerealella* can tolerate temperatures higher than the optimum range (Shazali and Smith,

1985). Low grain moisture content (11%) delayed insect development and a high moisture content (14%) increased the survival rate of *C. cephalonica* (Osman, 1984). Further biological and ecological studies on storage pests are required.

The storage of well-dried grains (less than 12% moisture) under hygienic conditions is the most effective method of pest management. In fact, sun drying is a common practice in most areas. Recently, sunlight has been used successfully to increase the temperature of grain up to 56–75°C for 60–180 seconds in order to kill the insects in infested stocks (Mohan *et al.*, 1987a). Although modern storage facilities are sometimes available through government agencies, metal bins and silos are popular in villages. Trials on the relative susceptibility of sorghum cultivars have identified resistant sources, such as Dabar, to *S. cerealella* and *S. oryzae* (Shazali, 1987a); FD-1 and 303 to *S. oryzae* (Adetunji, 1988); Combine shallu, Rio and Wray to *S. zeamais* (Rego *et al.*, 1984; Shiota and Lara, 1985); CSV-2 to *T. castaneum* (Jhala, Patel and Shah, 1984); and SB-1079, SB-2509, CS-3541, Makadandi and Maguti to *C. cephalonica* (Hugar and Jairao, 1986). Hard grains with a flinty endosperm resist pest attack due both to antixenosis for oviposition and to an antibiosis effect (Shazali, 1987a; Adetunji, 1988). Insecticides are usually used as dusts, sprays, or fumigants, viz. carbaryl, isofenphos, pirimiphos-methyl, fenitrothion, phosphine, methyl bromide (Bengston *et al.*, 1984; Sattigi, Kulkarni and Thontadarya, 1985c; Kavadia, Pareek and Sharma, 1986b; Maranhao and Oliveira, 1987). Tablets of aluminium phosphide are available in the markets but, when fumigation is used, facilities for aeration must be available in the storage structures. The use of ash, neem seed powder, road dust, etc. is also being tested (Lakhani and Patel, 1985). Further study of resistant cultivars and insecticides may lead to better management of storage pests; cultivars resistant to both field and storage pests would be of great help to farmers.

### Insecticide residues

The residues of insecticides applied to the crop can be detected in straw and grains. Factors that determine the persistence and degradation of insecticides in soil or plants include: the ecological conditions, insecticide formulations, number and rate of applications, stage and organ of the plant treated (Naitam and Sukhani, 1984; Bauernfeind, 1987; Gardner and Hornby, 1987; Singh, Srivastava and Agnihotri, 1987; Singh, Srivastava and Singh, 1988). Insecticides used during grain storage may leave residues in or on the grain. These residues have been found to lie within the tolerance limits fixed by FAO (Rajukannu *et al.*, 1985; Sattigi, Kulkarni and Thontadarya, 1985b; Kavadia, Pareek and Sharma, 1986a, b; Maranhao and Oliveira, 1987). Therefore, the use of the safer insecticides as protectants for long-term storage of sorghum grains may be continued.

### Perspectives

Measures undertaken against several pests at a time reduce the cost of control operations for individual pests. Early planting is effective against shoot fly and sorghum midge. Removal and destruction of stubble and crop residues proved to be effective against stem borers. In general, cultural practices are useful in reducing pest populations and plant damage. Varietal resistance is particularly suited to control programmes and offers advantages over other methods (Gebrekidan, 1985; Leuschner, Taneja and Sharma, 1985; Sharma, 1985; Teetes, 1985). Chemical applications are often costly and problems of phytotoxicity, residues, application techniques, etc. may arise. Thus, integrated pest management against a particular pest or a group of pests can be designed to make the control operation most practical for farmers (Sharma, 1985; Sukhani, 1986). The socio-economic situation of farmers and suitable farming systems in a region are important considerations for long-term or permanent control. International and regional team efforts for multidisciplinary approaches are being explored by research and development institutes and agencies (ICIPE, ICRISAT, IITA, IRAT, SAFGRAD, INTSORMIL) in order to formulate interdisciplinary and appropriate pest-management strategies for sorghum.

## References

The survey of literature (1984–89) was concluded March 1990. The review of literature in other than English is based on translations or summaries published in English. Mimeographs, dissertations, project reports, symposia/workshop proceedings and newsletters are not included.

ADESIYUN, A. A. (1986). Control of the stem borer, *Busseola fusca*, on sorghum with granular insecticides. *Samaru Journal of Agricultural Research* **4**, 35–43.

ADETUNJI, J. F. (1988). A study of the resistance of some sorghum seed cultivars to *Sitophilus oryzae* (L). (Coleoptera: Curculionidae). *Journal of Stored Products Research* **24**, 67–71.

ADEYEYE, O. A. AND BLUM, M. S. (1988). A cowpea artificial diet for noctuid larvae. *Insect Science and its Application* **9**, 609–611.

AGRAWAL, B. L., ABRAHAM, C. V. AND HOUSE, L. R. (1988). Inheritance of resistance to midge, *Contarinia sorghicola* Coq. in sorghum. *Insect Science and its Application* **9**, 43–45.

AGRAWAL, B. L., SHARMA, H.C. AND LEUSCHNER, K. (1987). Registration of ICSV-197: a midge resistant sorghum cultivar. *Crop Science* **27**, 1312–1313.

AHMED, T. R., KINDLER, S. D. AND PRUESS, K. P. (1984). Recovery of sorghum varieties from sublethal infestation of chinch bug, *Blissus leucopterus leucopterus* Say (Hemiptera: Lygaeidae). *Journal of Economic Entomology* **77**, 151–152.

ALGHALI, A. M. (1984). Studies on the biology, damage and crop loss assessment of the sorghum midge, *Contarinia sorghicola* Coq. (Diptera: Cecidomyiidae). *Insect Science and its Application* **5**, 253–258.

ALGHALI, A. M. (1985a). Larval instars determination for *Chilo partellus* Swinhoe (Lepidoptera: Pyralidae) reared on sorghum. *Insect Science and its Application* **6**, 677–679.

ALGHALI, A. M. (1985b). Insect-host plant relationships: the spotted stalk borer, *Chilo partellus* (Swinhoe) (Lepidoptera: Pyralidae) and its principal host, sorghum. *Insect Science and its Application* **6**, 315–322.

ALGHALI, A. M. (1986). Effects of cultivar, time and amount of *Chilo partellus*

Swinhoe (Lepidoptera: Pyralidae) infestation on sorghum yield components in Kenya. *Tropical Pest Management* **32**, 126–129.

ALGHALI, A. M. (1987). Effect of time of *Chilo partellus* Swinhoe (Lepidoptera: Pyralidae) infestation on yield loss and compensatory ability in sorghum cultivars. *Tropical Agriculture (Trinidad)* **64**, 144–148.

ALGHALI, A. M. AND SAXENA, K. N. (1988). Larval movement, feeding and development of *Chilo partellus* Swinhoe (Lepidoptera: Pyralidae) on two sorghum cultivars. *Insect Science and its Application* **9**, 7–11.

ALL, J. N. (1988). Fall armyworm (Lepidoptera: Noctuidae) infestation in no tillage cropping systems. *Florida Entomologist* **71**, 268–272.

ALLSOPP, P. G. AND RADFORD, B. J. (1987). Sowing techniques for sorghum on soil infested with black field earwigs and sugarcane wireworms. *Queensland Journal of Agricultural and Animal Sciences* **44**, 15–19.

AMATOBI, C. I. (1985). Oviposition preference and the influence of cultivation on egg hatching of *Oedaleus senegalensis* (Orthoptera: Acrididae). *Samaru Journal of Agricultural Research* **5**, 81–85.

ANAND, M., PANT, J. C. AND SRIVASTAVA, B. G. (1986). Qualitative analysis of amino acids of diapausing and non diapausing larvae of *Chilo partellus* (Swinhoe) and their host plant materials. *Indian Journal of Entomology* **48**, 244–247.

ARCHER, T. L. AND BYNUM, E. D., JR (1986a). Greenbug (Homoptera: Aphididae) and Banks grass mite (Acari: Tetranychidae) distribution on sorghum plants. *Journal of Economic Entomology* **79**, 116–121.

ARCHER, T. L. AND BYNUM, E. D., JR (1986b). Influence of sorghum leaf area on abundance of and damage by the greenbug, *Schizaphis graminum* (Rond.) (Homoptera: Aphididae). *Journal of Kansas Entomological Society* **59**, 653–657.

ARCHER, T.L., BYNUM, E.D., JR AND PETERSON, G. C. (1986a). Importance of plant senescence on grain sorghum resistance to the Banks grass mite *(Acari: Tetranychidae). Journal of Economic Entomology* **79**, 15–17.

ARCHER, T.L., BYNUM, E.D., JR AND PETERSON, G. C. (1986b). Influence of sorghum maturity on abundance and damage potential of Banks grass mite, *Oligonychus pratensis* (Banks). *Experimental and Applied Acaralogy* **2**, 217–222.

ARCHER, T.L., BYNUM, E.D., JR AND ONKAN, A. B. (1988). Abundance of Banks grass mites (Acari: Tetranychidae) on corn and sorghum fertilized with different rates of nitrogen and phosphorus. *Journal of Economic Entomology* **81**, 300–303.

BA-ANGOOD, S. A. AND HUBAISHAN, M. A. (1985). Effect of sowing date and cultivars of sorghum on infestation by *Atherigona yorki, Sesamia cretica* and *Chilo partellus* in Yemen. *Journal of Agricultural Sciences (UK)* **104**, 61–66.

BALASUBRAMANIAN, G., BALASUBRAMANIAN, M., SANKARAR, K. AND MANICKAM, T. S. (1986). Effect of fertilization on the incidence of shoot fly and stem borer in rainfed sorghum. *Madras Agricultural Journal* **73**, 471–473.

BALASUBRAMANIAN, G., BALASUBRAMANIAN, M., GOPALAN, M. AND SIVAPRAKASAM, N. (1987). Seed treatment of sorghum for the control of shoot fly, *Atherigona soccata* Rondani. *Madras Agricultural Journal* **74**, 237–239.

BARBULESCU, A. (1985). Aspects of the control of the cereal greenbug, *Schizaphis graminum*, on sorghum through chemical seed treatment. *Probleme de Protectia Plantelor* **13**, 157–171 (in Rumanian).

BAUERNFEIND, R. J. (1987). Residual effectiveness of insecticides for the control of chinch bugs (Heteroptera: Lygaeidae) in sorghum. *Journal of the Kansas Entomological Society* **60**, 336–339.

BAXENDALE, F. P., TEETES, G. L. AND SHARPE, P. J. H. (1984). Temperature dependent model for sorghum midge (Diptera: Cecidomyiidae) spring emergence. *Environmental Entomology* **13**, 1566–1571.

BAXENDALE, F. P., TEETES, G. L. SHARPE, P. J. H. AND WU, H. (1984). Temperature dependent model for development of non-diapausing sorghum midge (Diptera: Cecidomyiidae). *Environmental Entomology* **13**, 1572–1576.

BENE, G. (1986). Notes on the biology of *Atherigona soccata* Rondani (Diptera:

Muscidae) in Tuscany and Latium. *Redia* **69**, 47–63 (in Italian).

BENGSTON, M., DAVIES, R. A. H., DESMARCHELLER, J. M., PHILLIPS, M. P. AND SIMPSON, B. W. (1984). Organophosphorous and synergised pyrethroid insecticides as grain protectants for stored sorghum. *Pesticide Science* **15**, 500–508.

BEREGOVOY, V. H. AND STARKS, K. J. (1986). Enzyme patterns in biotypes of the greenbug, *Schizaphis graminum* (Rondani) (Homoptera: Aphididae). *Journal of the Kansas Entomological Society* **59**, 517–523.

BEREGOVOY, V. H., STARKS, K. J. AND JANARDAN, K. G. (1988). Fecundity characteristics of the greenbug biotypes C and E cultured on different host plants. *Environmental Entomology* **17**, 59–62.

BERNAYS, E., WOODHEAD, S. AND HAINES, L. (1985). Climbing by new hatched larvae of the spotted stalk borer, *Chilo partellus*, to the top of sorghum plants. *Entomologia Experimentalis et Applicata* **37**, 73–79.

BIN, F. AND COLAZZA, S. (1986). Population explosion of *Nysius graminicolus* Kol. on sorghum in Central Italy (Rhynchota: Lygaeidae). *Informatore Agrario* **42**, 75–77 (in Italian).

BINDRA, O. S. AND AMATOBI, C. I. (1985). Bionomics of *Kraussaria angulifera* Krauss (Orthoptera: Acrididae) in northern Nigeria. *Samaru Journal of Agricultural Research* **1**, 59–67.

BIRADAR, S. G. AND BORIKAR, S. T. (1985). Genetic analysis of shoot fly resistance in relation to growth stages in sorghum. *Zeitschrift für Pflanzenzüchtung* **95**, 173–178.

BOOZAYA-ANGOON, D., STARKS, K. J., WEIBEL, D. E. AND TEETES, G. L. (1984). Inheritance of resistance in sorghum, *Sorghum bicolor*, to the sorghum midge, *Contarinia sorghicola* (Diptera: Cecidomyiidae). *Environmental Entomology* **13**, 1531–1534.

BOTHE, N. N. AND POKHARKAR, P. N. (1985). Role of silica content in sorghum for reaction to shoot fly. *Journal of Maharashtra Agricultural Universities* **10**, 338–339.

BREEN, J. P. AND TEETES, G. L. (1986a). Relationship of yellow sugarcane aphid (Homoptera: Aphididae) density to sorghum damage. *Journal of Economic Entomology* **79**, 1106–1110.

BREEN, J. P. AND TEETES, G. L. (1986b). Yellow sugarcane aphid (Homoptera: Aphididae) infestation duration effect on sorghum. *Journal of Economic Entomology* **79**, 1603–1605.

BROOKS, G. W. AND GILSTRAP, F. E. (1985). Parasitism of sorghum midge in Johnson grass in major vegetational areas of Texas. *Southwestern Entomologist* **10**, 171–175.

BUDHARAJA, K., SINGH, O. P., MISRA, U. S., DHAMDHERE, S. V. AND DEOLE, J. Y. (1984). Seasonal incidence, host plants and efficacy of some insecticides against *Myllocerus maculosus* Desb. infesting hybrid sorghum. *Indian Journal of Agricultural Sciences* **54**, 418–421.

BURTON, R. L., JONES, O. R., BURD, J. D., WICKS, G. A. AND KRENZER, E. G., JR (1987). Damage by greenbug (Homoptera: Aphididae) to grain sorghum as affected by tillage, surface residues and canopy. *Journal of Economic Entomology* **80**, 792–798.

BUSOLI, A., LARA, F. M., GRAZIA, J. AND FERNANDES, D. A. (1984a). Occurrence of *Thyanta preditor* (Fabricius 1791) (Heteroptera: Pentatomidae) damaging sorghum crop in Jaboticabal, Sao Paulo, Brazil. *Anais da Sociedade Entomologica do Brasil* **13**, 179–181 (in Portuguese).

BUSOLI, A., LARA, F. M., GRAVENA, S. AND MALHEIROS, E. B. (1984b). Bioecological aspects of the sorghum midge, *Contarinia sorghicola* (Coquillett 1898) (Diptera: Cecidomyiidae) and their parasites in Jaboticabal. *Anais da Sociedade Entomologica do Brasil* **13**, 167–176 (in Portuguese).

BYNUM, E. D., JR, ARCHER, T. L., LYLE, W. M. AND BORDOVSKY, J. P. (1988). Chlorpyrifos applications for greenbug (Homoptera: Aphididae) control with a new multifunction irrigation system. *Journal of Economic Entomology* **81**, 1781–1784.

CAMPBELL, B. C. AND DREYER, D. L. (1985). Host plant resistance of sorghum: differential hydrolysis of sorghum pectic substances by polysaccharases of greenbug biotypes (*Schizaphis graminum*, Homptera: Aphididae). *Archives of Insect Biochemistry and Physiology* **2**, 203–215.

CASTRO, M. AND PITRE, H. (1988). Potential for using maize as a trap crop for the fall armyworm, *Spodoptera frugiperda* (Lepidoptera: Noctuidae) where sorghum and maize are intercropped on subsistence farms. *Florida Entomologist* **71**, 273–278.

CEREMENOL, G. F., GALVAN, V. R. AND LABATON, G. V. (1984). Life cycle and daily seasonal population fluctuations in the sorghum midge, *Contarinia sorghicola* Coquillet (Diptera: Cecidomyiidae). *Revista Colombiana de Entomologia* **10**, 15–19 (in Spanish).

CHAKRAVORTY, B. P., KUNDU, G. G., HEGDE, S. V. AND SHARMA, J. K. (1985). Effect of *Serratia marcescens* Bizio for the control of sorghum stem borer, *Chilo partellus* (Swinhoe). *Indian Journal of Entomology* **45**, 456–458.

CHANG, C. P. AND FANG, M. N. (1984). Studies on the resistance of sorghum varieties to sorghum aphid, *Melanaphis sacchari* (Zehntner). *Chinese Journal of Entomology* **4**, 97–105 (in Chinese).

CHAUDHARY, J. P., KAUSHIK, S. K., SINGH, R. AND MRIG, K. K. (1987). Role of natural enemies in the suppression of sugarcane leafhopper, *Pyrilla perpusilla* (Walker). *FAO Plant Protection Bulletin* **35**, 15–19.

CHAUDHARY, R. N. AND SHARMA, V. K. (1987). Parasitization in diapausing larvae of *Chilo partellus* (Swinhoe) by *Apanteles flavipes* (Cameron). *Indian Journal of Ecology* **14**, 155–157.

CHAUDHARY, R. N. AND SHARMA, V. K. (1988). Induction and termination of diapause in maize stalk borer, *Chilo partellus* (Swinhoe). *Indian Journal of Entomology* **50**, 523–529.

CRUZ, I. AND VENDRAMIM, J. D. (1988). Effect of different levels of infestation by the greenbug, *Schizaphis graminum* (Rondani, 1952) on susceptible and resistant sorghum. *Pesquisa Agropecuaria Brasileira* **23**, 111–118 (in Portuguese).

DARAMOLA, A. M. (1986). The development of *Dysdercus voelkeri* on three new sorghum varieties and assessment of the damage to grains. *Insect Science and its Application* **7**, 659–662.

DELFINO, M. A. (1985). Discovery of the sugarcane aphid, *Melanaphis sacchari* (Zehntner 1897) in Argentina and Uruguay. *Revista Investigation CIRPON* **2**, 57–64 (in Spanish).

DELOBEL, A. G. L. (1984). A method for evaluating crop losses caused by the sorghum shoot fly, *Atherigona soccata. Agronomie Tropicale* **39**, 350–354 (in French).

DELOBEL, A. G. L. AND LUBEGA, M. C. (1984). Rainfall as a mortality factor in the sorghum shoot fly, *Atherigona soccata* Rond. (Diptera: Muscidae). *Zeitschrift für Angewande Entomologie* **97**, 510–516.

DEPEW, L. J. AND HOOKER, M. L. (1987). Effect of insecticide placement at planting for control of greenbug (Homoptera: Aphididae) on grain sorghum. *Journal of Economic Entomology* **80**, 490–493.

DHALIWAL, J. S., MEHNDIRATTA, P. D. AND ARORA, B. S. (1988). Reaction of some promising sorghum fodder varieties to insect pests. *Journal of Research, Punjab Agricultural University* **25**, 280–286.

DHARMALINGHAM, S., GUTHRIE, W. D., JARVIS, J. L., KINDLER, D., ATKINS, R. E., TSENG, C. T. AND ZHOU, D. (1984). European corn borer (Lepidoptera: Pyralidae): rate of first generation larval mortality in sorghum hybrids compared with inbred lines of maize during whorl stage of plant development. *Journal of Economic Entomology* **77**, 929–931.

DIAZ, S. M. (1988). Taxonomy of the main microhymenopterous parasitoids of *Contarinia sorghicola* Coquillett 1898 (Diptera: Cecidomyiidae) and preliminary observations on some characteristics of their populations in Manfredi (Cordoba). *Revista de Ciencias Agropecuarias* **6**, 55–72 (in Spanish).

DOUMBIA, Y. U. AND GAHUKAR, R. T. (1986). *Atherigona soccata* Rondani and other flies harmful to sorghum in Mali. *Agronomie Tropicale* **41**, 170–172 (in French).

DREYER, D. L. AND CAMPBELL, B. C. (1984). Association of the degree of methylation of intercellular pectin with plant resistance to aphids and with induction of aphid biotypes. *Experientia* **40**, 224–226.

DREYER, D. L., CAMPBELL, B. C. AND JONES, K. C. (1984). Effect of bioregulator treated sorghum on greenbug fecundity and feeding behaviour: implications for host plant resistance. *Phytochemistry* **23**, 1593–1596.

EISENBACH, J. AND MITTLER, T. E. (1987). Extra-nuclear inheritance in a sexually produced aphid: the ability to overcome host plant resistance by biotype hybrids of the greenbug, *Schizaphis graminum*. *Experientia* **43**, 332–334.

ELAMIN, E. M. (1988). Some aspects of the ecology of the stem borer (*Sesamia cretica* Led.) in sugarcane and sorghum in the Sudan. *Beitrage zur Tropischen Landwirtschaft und Veterinaermedizin* **26**, 33–37.

FAO (1979). *Elements of integrated control of sorghum pests.* Plant Production and Protection Division, Document No. 79, Food and Agriculture Organization, Rome, Italy.

FAO (1987). *Production yearbook*, volume 40. Food and Agriculture Organization, Rome, Italy.

FORRESTER, N. W. (1987). Diapause development in the sorghum midge, *Contarinia sorghicola* (Coquillett) (Diptera: Cecidomyiidae). *Bulletin of Entomological Research* **77**, 581–587.

FULLER, B. W. AND REAGAN, T. E. (1988). Comparative predation of the sugarcane borer (Lepidoptera: Pyralidae) on sweet sorghum and sugarcane. *Journal of Economic Entomology* **81**, 713–717.

FULLER, B. W. AND REAGAN, T. E. (1989). The relationship of sweet sorghum plant fiber and survival of the sugarcane borer, *Diatraea saccharalis* (F.) (Lepidoptera: Pyralidae). *Journal of Agricultural Entomology* **6**, 113–118.

FULLER, B. W., REAGAN, T. E. AND FLYNN, J. L. (1988). Economic injury level of the sugarcane borer (Lepidoptera: Pyralidae) on sweet sorghum, *Sorghum bicolor* (L.) Moench. *Journal of Economic Entomology* **81**, 349–353.

GAHUKAR, R. T. (1984a). Seasonal distribution of sorghum midge, *Contarinia sorghicola* (Coq.), and its hymenopterous parasites in Senegal. *Agronomie* **4**, 393–397.

GAHUKAR, R. T. (1984b). Senegal: new insect pests of sorghum. *FAO Plant Protection Bulletin* **32**, 31–33.

GAHUKAR, R. T. (1985). Some species of *Atherigona* (Diptera: Muscidae) reared from *Gramineae* in Senegal. *Annals of Applied Biology* **106**, 399–403.

GAHUKAR, R. T. (1987). Population dynamics of sorghum shoot fly, *Atherigona soccata* (Diptera: Muscidae) in Senegal. *Environmental Entomology* **16**, 910–916.

GAHUKAR, R. T., DOUMBIA, Y. O. AND BONZI, S. M. (1989). *Eurystylus marginatus* Odh., a new pest of sorghum in the Sahel. *Tropical Pest Management* **35**, 212–213.

GAHUKAR, R. T., DOUMBIA, Y. O., BAL, A. B. AND BHATNAGAR, V. S. (1989). Geographic distribution, host plants and seasonal fluctuations of blister beetles in Mali and Senegal. *Insect Science and its Application* **10**, 99–106 (in French).

GANDHALE, D. N., PATIL, B. D. AND POKHARKAR, D. N. (1984). Effective control of sorghum shoot fly (*Atherigona soccata* Rond.) with seed dressers. *Madras Agricultural Journal* **71**, 482–483.

GARDNER, W. A. AND HORNBY, J. A. (1987). Bio-assay of the persistence of *Bacillus thuringiensis* on sorghum panicles. *Journal of Entomological Science* **22**, 210–211.

GARDNER, W. A., NOBLET, R. AND SCHWEHR, R. D. (1984). The potential of microbial agents in managing populations of the fall armyworm (Lepidoptera: Noctuidae). *Forida Entomologist* **67**, 325–332.

GEBRE-AMLAK, A. (1988). Development of maize stalk borer, *Busseola fusca* (Fuller) in wild host plants in Ethiopia. *Journal of Applied Entomology* **106**, 390–395.

GEBREKIDAN, B. (1985). Breeding sorghum for resistance to insects in eastern Africa. *Insect Science and its Application* **6**, 351–357.

GILSTRAP, F. E., KRING, T. J. AND BROOKS, G. W. (1984). Parasitism of aphids (Homoptera: Aphididae) associated with Texas sorghum. *Environmental Entomology* **13**, 1613–1617.

GIRADDI, R. S. AND KULKARNI, K. A. (1987). Efficacy of insecticidal baits and granules against armyworm, *Mythimna separata* (Walker) in sorghum. *Pesticides* **21**, 27–35.

GOPALAN, M. (1988). Chemical control of the mite, *Oligonychus indicus*, on sorghum. *Indian Journal of Plant Protection* **16**, 51–52.

GOUR, M. N. AND DABI, R. K. (1988). Biological control of white grub using *Verticillium lecanii* (Zimmerm) Viegas. *Current Science* **57**, 620–621.

GUJRATI, J. P., THAKUR, R. C. AND SINGH, O. P. (1988). Biology and seasonal incidence of sorghum shoot fly, *Atherigona soccata* Rond. in Madhya Pradesh. *Journal of Insect Science* **1**, 173–177.

GUTHRIE, W. D., HASKINS, F. A. AND GORZ, H. J. (1988). Relationship of European corn borer resistance in sorghum to HCN-p and DIMBOA content in leaf and stalk-collar tissue. *Journal of Agricultural Entomology* **5**, 21–28.

GUTHRIE, W. D., DHARMALINGAM, S., JARVIS, J. L., KINDLER, D., ATKINS, R. E., TSENG, C. T. AND ZHOU, D. (1984a). European corn borer: rate of second generation larval mortality in sorghum hybrids compared with inbred lines of maize during anthesis. *Journal of Agricultural Entomology* **1**, 273–281.

GUTHRIE, W. D., DHARMALINGAM, S., JARVIS, J. L., KINDLER, D., ATKINS, R. E., TSENG, C. T. AND ZHOU, D. (1984b). European corn borer: rate of second generation larval mortality in sorghum hybrids compared with inbred lines of maize during the whorl stage of plant development. *Journal of Economic Entomology* **72**, 929–931.

GUTHRIE, W. D., DHARMALINGAM, S., JARVIS, J. L., KINDLER, D., ATKINS, R. E., TSENG, C. T. AND ZHOU, D. (1985). Resistance of sorghum genotypes to leaf feeding by first generation European corn borer larvae compared with maize genotypes. *Journal of Agricultural Entomology* **2**, 175–184.

HALALLI, M. S. (1985). Studies on inheritance of resistance to shoot fly (*Atherigona soccata* Rondani) in sorghum (*Sorghum bicolor* (L.) Moench). *Mysore Journal of Agricultural Sciences* **19**, 287–288.

HALLMAN, G. J., TEETES, G. L. AND JOHNSON, J. W. (1984a). Relationships of sorghum midge (Diptera: Cecidomyiidae) density to damage to resistant and susceptible sorghum hybrids. *Journal of Economic Entomology* **77**, 83–87.

HALLMAN, G. J., TEETES, G. L. AND JOHNSON, J. W. (1984b). Weight compensation of undamaged kernels in response to damage by sorghum midge (Diptera: Cecidomyiidae). *Journal of Economic Entomology* **77**, 1033–1036.

HANNA, W. W., WISEMAN, B. R., DUNCAN, R. R. AND SCHERTZ, K. R. (1989). Registration of TIFT-MR-88 midge resistant sorghum germplasm. *Crop Science* **29**, 245–246.

HARVEY, T. L. AND THOMPSON, C. A. (1988). Effect of sorghum density and resistance on infestation of greenbug *Schizaphis graminum* (Homoptera: Aphididae). *Journal of Kansas Entomological Society* **61**, 68–71.

HASABE, B. M., PATIL, B. R. AND AJRI, D. S. (1984). Effects of insecticides and interculturing on the infestation of white grubs in sorghum. *Journal of Maharashtra Agricultural Universities* **9**, 216.

HAYES, J. L. (1988). A comparative study of adult emergence phenologies of *Heliothis virescens* (F.) and *H. zea* (Boddie) (Lepidoptera: Noctuidae) on various hosts in field cages. *Environmental Entomology* **17**, 344–349.

HE, F. G., LIEU, L. J. AND ZHENG, G. X. (1987). A preliminary study on the occurrence of *Scymnus babai* Sassji. *Insect Knowledge* **24**, 10–13 (in Chinese).

HIREMATH, I. G. (1986). Host preference of sorghum earhead bug, *Calocoris angustatus* Lethierry (Hemiptera: Miridae). *Entomon* **11**, 121–125.

HIREMATH, I. G. AND THONTADARYA, T. S. (1984). Seasonal incidence of the sorghum earhead bug, *Calocoris angustatus* Lethierry (Hemiptera: Miridae). *Insect Science and its Application* **5**, 469–474.

HIREMATH, I. G., THONTADARYA, T. S. AND JAIRAO, K. (1984). Report on cannibalistic and predatory behaviour of sorghum earhead bug, *Calocoris angustatus* Lethierry (Hemiptera: Miridae). *Agricultural Science Digest (India)* **4**, 67–68.

HONG, S. C. (1986). Damage to sorghum by a lygaeid bug, *Nysius plebejus* Distant. *Chinese Journal of Entomology* **6**, 119–125 (in Chinese).

HONG, S. C. (1987). Relationship of sorghum midge density to the damage of sorghum. *Chinese Journal of Entomology* **7**, 25–29 (in Chinese).

HOPPE, T. (1986). Storage insects of basic food grains in Honduras. *Tropical Science* **26**, 25–38.

HUGAR, P. AND JAIRAO, K. (1986). Preference of sorghum varieties/hybrids for feeding of rice moth, *Corcyra cephalonica* Stainton (Galleridae: Lepidoptera). *Science and Culture* **52**, 134–136.

ICRISAT (1985). *Proceedings of the international sorghum entomology workshop, College Station, USA, 15–21 July 1984.* International Crops Research Institute for the Semi-Arid Tropics, India.

ICRISAT (1989). *Proceedings of the international workshop on sorghum stem borers, Patancheru, India, 17–20 November 1987.* International Crops Research Institute for the Semi-Arid Tropics, India.

INAYATULLAH, C., WEBSTER, J. A. AND FARGO, W. S. (1987). Morphometric variation in alates of greenbug (Homoptera: Aphididae) *Annals of the Entomological Society of America* **80**, 306–311.

JADHAV, S. S., MOTE, U. N. AND BAPAT, D. R. (1988). Performance of sweet stalked sorghum genotypes to shoot fly reaction. *Journal of Maharashtra Agricultural Universities* **13**, 232–234.

JAGTAP, A. B., GHULE, B. D. AND DEOKAR, A. B. (1985). Incidence of *Pyrilla perpusilla* on ratoon sorghum and sugarcane. *Journal of Maharashtra Agricultural Universities* **10**, 101–102.

JHALA, R. C., PATEL, C. B. AND SHAH, A. H. (1984). Studies on preference of *Tribolium castaneum* Herbst. (Tenebrionidae: Coleoptera) for four of different sorghum varieties. *Gujarat Agricultural University Research Journal* **10**, 59–61.

KARLEN, D. L., SADLER, E. D. AND ROACH, S. H. (1989). Grain sorghum yield and susceptibility to insects as affected by early season growth. *Agronomy Journal* **81**, 320–321.

KATTI, G. AND VERMA, S. (1988). Efficacy of different formulations of quinalphos and endosulfan against sorghum stem borer, *Chilo partellus* Swinhoe and midge, *Contarinia sorghicola* Coquillett. *Indian Journal of Plant Protection* **16**, 211–215.

KAVADIA, V. S., PAREEK, B. L. AND SHARMA, K. P. (1986a). Phosphine residues in sorghum stored in mud kothi and metal drum. *Bulletin of Grain Technology* **24**, 19–23.

KAVADIA, V. S., PAREEK, B. L. AND SHARMA, K. P. (1986b). Dispersal of fumigants in empty and sorghum filled mud kothi. *Bulletin of Grain Technology* **24**, 126–131.

KERNS, D. L., PETERS, D. C. AND PUTERKA, G. J. (1987). Greenbug biotype and grain sorghum seed rate surveys in Oklahoma 1986. *Southwestern Entomologist* **12**, 237–242.

KFIR, R. (1988). Hibernation by the lepidopteran stalk borers, *Busseola fusca* and *Chilo partellus* in grain sorghum. *Entomologia Experimentalis et Applicata* **48**, 31–36.

KFIR, R., VAN HAMBURG, H. AND VAN VUUREN, R.(1989). Effect of stubble treatment on the post diapause emergence of the grain sorghum stalk borer, *Chilo partellus* (Swinhoe) (Lepidoptera: Pyralidae). *Crop Protection* **8**, 289–292.

KHURANA, A. D. AND VERMA, A. N. (1985). Some physical plant characters in relation to stem borer and shoot fly resistance in sorghum. *Indian Journal of Entomology* **47**, 14–19.

KHURANA, A. D. AND VERMA, A. N. (1988a). Screening of forage sorghum and dual purpose lines for susceptibility to shoot fly, *Atherigona soccata* (Rondani). *Indian Journal of Entomology* **50**, 422–428.

KHURANA, A. D. AND VERMA, A. N. (1988b). Screening of forage sorghums for susceptibility to stem borer, *Chilo partellus* (Swinhoe). *Indian Journal of Entomology* **50**, 283–284.

KINDLER, S. D., GUTHRIE, W. D. AND JARVIS, J. L. (1987). Damage from infesting sorghum plants with European corn borer egg masses and larvae. *Maydica* **32**, 273–285.

KINDLER, S. D., SPOMER, S. M., HARVEY, T. L., BURTON, R. L. AND STARKS, K. J. (1984). Status of biotype E greenbugs (Homoptera: Aphididae) in Kansas, Nebraska, Oklahoma and northern Texas during 1980–1981. *Journal of the Kansas Entomological Society* **57**, 155–158.

KISHORE, P. (1984). Development of stable resistant cultivars for stem borer, *Chilo partellus* (Swinhoe), in sorghum. *Insect Science and its Application* **5**, 475–479.

KISHORE, P. (1987a). A new sorghum variety resistant to the stem borer, *Chilo partellus* (Swinhoe). *Journal of Entomological Research* **11**, 115–118.

KISHORE, P. (1987b). New sources of resistance to stem borer, *Chilo partellus* (Swinhoe) in sorghum. *Journal of Entomological Research* **11**, 155–157.

KISHORE, P. (1988). Efficacy of fenvalerate dust and its comparison with recommended endosulfan dust against stem borer on sorghum. *Pesticides* **22**, 38–41.

KISHORE, P., RANA, B. S. AND AGARWAL, K. N. (1985). Selection for the shoot fly *Atherigona soccata* Rondani resistance in high yielding varieties of sorghum. *Journal of Entomological Research* **9**, 188–194.

KISHORE, P., RANA, B. S. AND SHARMA, G. C. (1984). Selection for the stem borer *Chilo partellus* (Swinhoe) resistance in segregating generations in sorghum. *Journal of Entomological Research* **8**, 20–24.

KRING, T. J. AND GILSTRAP, F. E. (1984). Efficacy of the natural enemies of grain sorghum aphids (Homoptera: Aphididae). *Journal of the Kansas Entomological Society* **57**, 460–467.

KRING, T. J. AND GILSTRAP, F. E. (1986). Beneficial role of corn leaf aphid, *Rhopalosiphum maidis* (Fitch) (Homoptera: Aphididae) in maintaining *Hippodamia* spp. (Coleoptera: Coccinellidae) in grain sorghum. *Crop Protection* **5**, 125–128.

KRING, T. J., YEARIAN, W. C. AND STEWARD, V. B. (1987). *Pleuroprucha insulsaria* (Guenee) (Lepidoptera: Geometridae): a potential pest of grain sorghum. *Journal of the Kansas Entomological Society* **60**, 476–477.

KRING, T. J., YEARIAN, W. C. AND STEWARD, V. B. (1989). Within field and within panicle distribution of *Heliothis zea* (Lepidoptera: Noctuidae) and *Celama sorghiella* (Lepidoptera: Noctuidae) eggs in grain sorghum. *Environmental Entomology* **18**, 150–156.

KUNDU, G. G. (1985). E-304: a high yielding sorghum derivative resistant to stem borer, *Chilo partellus* Swinhoe. *Bulletin of Entomology* **26**, 25–27.

KUNDU, G. G. AND SRIVASTAVA, K. P. (1984). Estimation of loss in grain yield of sorghum by beetle, *Chiloloba acuta* Wied. *Journal of Entomological Research* **8**, 223–224.

LAKHANI, G. K. AND PATEL, N. G. (1985). New insecticidal control of *Ephestia cautella* Walker. *Pesticides* **19**, 46–48.

LARA, F. M. AND PERUSSI, E. M. (1984). Resistance of sweet sorghum, *Sorghum bicolor* (L.) Moench, to attack by *Diatraea saccharalis* (Fabricius 1794). *Ciencia e Cultura* **36**, 280–286 (in Portuguese).

LEUSCHNER, K., TANEJA, S. L. AND SHARMA, H. C. (1985). The role of host plant resistance in pest management in sorghum in India. *Insect Science and its Application* **6**, 453–460.

MAHADEVAN, N. R. AND CHELLIAH, S. (1986a). Influence of season and weather factors on the occurrence of the sorghum stem borer, *Chilo partellus* (Swinhoe) in Tamil Nadu. *Tropical Pest Management* **32**, 212–214.

MAHADEVAN, N. R. AND CHELLIAH, S. (1986b). Influence of intercropping legumes with sorghum on the infestation of the stem borer, *Chilo partellus* (Swinhoe) in Tamil Nadu, India. *Tropical Pest Management* **32**, 162–163.

MANJUNATHA, M. AND PUTTASWAMY, R. (1989). A new record of spiders as predators of sorghum spider mite *Oligonychus indicus* (Hirst.) (Acari: Tetranychidae). *Current Science* **58**, 272.

MARANHAO, E. A. A. AND OLIVEIRA, J. V. (1987). Residual effects of insecticides against *Sitophilus zeamais* Mots. 1855, in corn (*Zea mays* L.) and sorghum (*Sorghum bicolor* (L.) Moench) grains. *Anais da Sociedade Entomologica do Brasil* **16**, 61–67 (in Portuguese).

MATE, S. N., PHADNANWIS, B. A. AND MEHETRE, S. S. (1988). Studies on growth and physiological factors in relation to the shoot fly attack in sorghum. *Indian Journal of Agricultural Sciences* **22**, 81–84.

MAYO, Z. B., BANKS, D. J., STARKS, K. J. AND VEAL, R. A. (1987). Influence of host plant and insecticide resistance on greenbug (Homoptera: Aphididae) biotype variation measured by fluorescent microscopy. *Environmental Entomology* **16**, 676–679.

MELTON, K. D. AND TEETES, G. L. (1984). Effects of resistant sorghum hybrids on sorghum midge (Diptera: Cecidomyiidae) biology. *Journal of Economic Entomology* **77**, 626–631.

MEN, U. B., MUNDIWALE, S. K. AND BORLE, M. N. (1986). Efficacy of certain insecticides against shoot fly. *Pesticides* **20**, 59–60.

MICHELS, G. J., JR (1987). Development of greenbug (Homoptera: Aphididae) on corn: geographic variations in host plant range of biotype E. *Journal of Economic Entomology* **80**, 394–397.

MICHELS, G. J., JR AND UNDERSANDER, D. J. (1986). Temporal and spatial distribution of the greenbug (Homoptera: Aphididae) on sorghum in relation to water stress. *Journal of Economic Entomology* **79**, 1221–1225.

MIZE, T. W. AND WILDE, G. (1986a). Reproduction of chinch bug (Heteroptera: Lygaeidae) on new resistant sources of grain sorghum. *Journal of Economic Entomology* **79**, 664–667.

MIZE, T. W. AND WILDE, G. (1986b). New resistant germplasm to the chinch bug (Heteroptera: Lygaeidae) in grain sorghum: contribution of tolerance and antixenosis as resistance mechanism. *Journal of Economic Entomology* **79**, 42–45.

MIZE, T. W. AND WILDE, G. (1986c). New grain sorghum sources of antibiosis to the chinch bug (Heteroptera: Lygaeidae). *Journal of Economic Entomology* **79**, 176–180.

MODINI, M. P., PAGE, F. D. AND FRANZMANN, B. A. (1987). Diurnal oviposition activity in grain sorghum by *Contarinia sorghicola* (Coquillett) (Diptera: Cecidomyiidae). *Journal of the Australian Entomological Society* **26**, 293–294.

MOGAL, B. H., MALI, A. R. AND RAJPUT, S. G. (1984). Effects of relative humidity and storage period on viability of carbofuran treated sorghum seeds and effectivensss against shoot fly. *Journal of Maharashtra Agricultural Universities* **9**, 149–151.

MOHAN, S., JAYARAJ, S. AND RANGARAJAN, A. V. (1988). A reliable egg count method to fix economic threshold level for sorghum shoot fly, *Atherigona soccata* Rond. *Entomon* **13**, 169–172.

MOHAN, S. AND KARUPPUCHAMY, P. (1987). A potential predator for sorghum mite,

*Oligonychus indicus* (Hirst.). *Current Science (India)* **56**, 845–846.

MOHAN, S. AND TAJUDDIN, A. (1988). Efficacy of a battery operated suction trap for sampling insects. *Indian Journal of Plant Protection* **16**, 71–73.

MOHAN, S., BALASUBRAMANIAN, G., GOPALAN, M. AND JAYARAJ, S. (1987a). Solar heat treatment: a novel method to check rice weevil and red flour beetle infestation in sorghum during storage. *Madras Agricultural Journal* **74**, 235–236.

MOHAN, S., JAYARAJ, S., PURUSHOTHAMAN, D. AND RANGARAJAN, A. V. (1987b). Can the use of *Azospirillum* biofertilizer control sorghum shoot fly. *Current Science (India)* **56**, 723–725.

MOTE, U. N. (1984a). Performance of carbofuran treated and untreated sorghum seed mixtures on shoot fly incidence and yield of sorghum hybrid CSH-8R. *Indian Journal of Plant Protection* **12**, 49–52.

MOTE, U. N. (1984b). Occurrence of armyworm on rabi sorghum. *Journal of Maharashtra Agricultural Universities* **9**, 230.

MOTE, U. N. (1984c). Unusual occurrence of pyrilla on rabi sorghum. *Journal of Maharashtra Agricultural Universities* **9**, 231.

MOTE, U. N. (1984d). Sorghum species resistant to shoot fly. *Indian Journal of Entomology* **46**, 241–242.

MOTE, U. N. (1985a). Relation between the shoot fly damage and sorghum yields during rainy season. *Indian Journal of Plant Protection* **11**, 145–147.

MOTE, U. N. (1985b). Effective and economical insecticides against sorghum midge. *Plant Protection Bulletin (India)* **37**, 13–15.

MOTE, U. N. (1986a). Comparison of carbofuran granule slurry treatment with other chemical treatments against shoot fly. *Indian Journal of Plant Protection* **14**, 83–86.

MOTE, U. N. (1986b). Correlation between the dead hearts caused by shoot fly, *Atherigona soccata* Rondani, and the yield of sorghum hybrids. *Indian Journal of Entomology* **48**, 356–357.

MOTE, U. N. (1986c). Correlation between the degree of damage due to stem borer, *Chilo partellus* (Swinhoe) and the yield of sorghum grain. *Indian Journal of Entomology* **48**, 357–358.

MOTE, U. N. AND BAPAT, D. R. (1985). Screening of high yielding sorghum genotypes against shoot fly. *Journal of Maharashtra Agricultural Universities* **10**, 91–92.

MOTE, U. N. AND KADAM, J. R. (1984). Efficacy of new insecticides and their formulations against shoot fly on rabi sorghum hybrids. *Plant Protection Bulletin (India)* **36**, 5–6.

MOTE, U. N. AND KADAM, J. R. (1985). Relative efficacy of different insecticides against sorghum earhead caterpillar, *Heliothis armigera* Hubner. *Pesticides* **19**, 51–52.

MOTE, U. N. AND RAMSHE, D. G. (1987). Effects of fertilizers and plant population on the incidence of shoot fly (*Atherigona soccata* Rondani) on some sorghum cultivars. *Indian Journal of Entomology* **49**, 490–495.

MOTE, U. N., KADAM, J. R. AND BAPAT, D. R. (1985). Recovery resistance to shoot fly in some sorghum hybrids. *Journal of Maharashtra Agricultural Universities* **10**, 190–193.

MOTE, U. N., KADAM, J. R. AND BAPAT, D. R. (1986). Antibiosis mechanism of resistance to sorghum shoot fly. *Journal of Maharashtra Agricultural Universities* **11**, 43–46.

NAITAM, N. R. AND SUKHANI, T. R. (1984). Effect of ecological conditions on the persistence of carbofuran and isofenphos in soil and sorghum seedlings. *Indian Journal of Entomology* **46**, 452–459.

NAITAM, N. R. AND SUKHANI, T. R. (1985). Ovipositional behaviour of sorghum shoot fly, *Atherigona soccata* Rondani, under different soil, plant and weather parameters. *Indian Journal of Entomology* **47**, 195–200.

NATARAJAN, K. (1985). Chemical control of shoot fly and stem borer in sorghum. *Indian Journal of Plant Protection* **13**, 41–42.

NATARAJAN, K. AND CHELLIAH, S. (1986a). Studies on the mechanism of resistance in sorghum accessions to the grain midge, *Contarinia sorghicola* Coquillett. *Insect Science and its Application* **7**, 751–755.

NATARAJAN, K. AND CHELLIAH, S. (1986b). Chemical control of shoot fly and stem borer of sorghum. *Pesticides* **20**, 23–24.

NATARAJAN, K. AND CHELLIAH, S. (1986c). Management of sorghum grain midge, *Contarinia sorghicola* Coq., using plant resistance and insecticides. *Indian Journal of Plant Protection* **14**, 77–79.

NATARAJAN, N. AND SUNDARA BABU, P. C. (1987a). Effect of earhead bug feeding on nutritional content of sorghum seeds. *Madras Agricultural Journal* **74**, 139–141.

NATARAJAN, N. AND SUNDARA BABU, P. C. (1987b). Bionomics of sorghum earhead bug, *Calocoris angustatus* Lethierry. *Madras Agricultural Journal* **74**, 186–192.

NATARAJAN, N. AND SUNDARA BABU, P. C. (1988a). Culture of sorghum earhead bug, *Calocoris angustatus* Lethierry, in the laboratory. *Tropical Pest Management* **34**, 356.

NATARAJAN, N. AND SUNDARA BABU, P. C. (1988b). Seasonal occurrence of *Heliothis armigera* Hb. infesting sorghum panicles. *Tropical Pest Management* **34**, 338–342.

NATARAJAN, N. AND SUNDARA BABU, P. C. (1988c). Economic injury level for sorghum earhead bug, *Calocoris angustatus* Lethierry, in southern India. *Insect Science and its Application* **9**, 395–398.

NATARAJAN, N. AND SUNDARA BABU, P. C. (1988d). Damage potential of earhead bug on sorghum panicle. *Insect Science and its Application* **9**, 411–414.

NATARAJAN, N., SUNDARA BABU, P. C. AND CHELLIAH, S. (1989). Seasonal occurrence of sorghum earhead bug, *Calocoris angustatus* Lethierry. (Hemiptera: Miridae) in southern India. *Tropical Pest Management* **35**, 70–77.

NATARAJAN, N., SUNDARA BABU, P. C. AND SRIDHAR, R. P. (1988). Grading sorghum entries for their resistance to earhead bug. *Madras Agricultural Journal* **75**, 216–217.

NIMBALKAR, V. S. AND BAPAT, D. R. (1987). Genetic analysis of shoot fly resistance under high level of shoot fly infestation in sorghum. *Journal of Maharashtra Agricultural Universities* **12**, 331–334.

NIU, Z. M. (1987). Study of bioprediction of occurrence of *Aphis sacchari. Shanxi Agricultural Science* **4**, 9–11 (in Chinese).

NWANZE, K. F. (1988). Distribution and seasonal incidence of some major insect pests of sorghum in Burkina Faso. *Insect Science and its Application* **9**, 313–321.

OKUDA, T. (1988). Effect of artificial wetting and rainfall on the larval diapause of a stem borer, *Busseola fusca* in Western Kenya. *Entomologia Experimentalis et Applicata* **48**, 263–267.

OSMAN, N. (1984). Assessment of damage by the rice moth, *Corcyra cephalonica* (St.) on different grains at four levels of moisture content. *Pertanika* **7**, 53–58 (in Malaysian).

PAIR, S. D. AND GROSS, H. R. (1984). Field mortality of pupae of the fall armyworm, *Spodoptera frugiperda* (J. E. Smith), by predators and a newly discovered parasitoid, *Diapetimorpha introita* (Cresson). *Journal of the Georgia Entomological Society* **19**, 22–26.

PAIR, S. D., RAULSTON, J. R., STARKS, K. J. AND MARTIN, P. B. (1986). Fall armyworm (Lepidoptera: Noctuidae) parasitoids: differential spring distribution and incidence on corn and sorghum in the Southern United States and Northeastern Mexico. *Environmental Entomology* **15**, 342–348.

PARAMESHWAR, H. AND JAIRAO, K. (1987). Food consumption by the larvae of rice moth, *Corcyra cephalonica* (Stainton). *Agricultural Science Digest (India)* **7**, 190–198.

PATEL, J. R. AND JOTWANI, M. G. (1986a). Effect of different populations of sorghum shoot fly on efficacy of carbofuran seed treatment. *Indian Journal of Entomology* **48**, 187–191.

PATEL, J. R. AND JOTWANI, M. G. (1986b). Effect of ecological factors on incidence and damage by sorghum midge, *Contarinia sorghicola. Indian Journal of Entomology* **48**, 220–222.

PATEL, R. H., DESAI, K. B., DESAI, M. S. AND KUKADIA, M. V. (1986). Combining ability for resistance to earhead worms in sorghum. *Gujarat Agricultural University Research Journal* **11**, 79–81.

PATEL, V. N., SHEKHAWAT, P. S. AND BHALANI, P. A. (1987). Insecticidal control schedule for stem borer and armyworm pests of hybrid sorghum under rainfed conditions in Saurashtra region. *Pesticides* **21**, 12–15.

PATHAK, R. A. (1985). Genetic variation of stem borer resistance and tolerance in three sorghum crosses. *Insect Science and its Application* **6**, 359–364.

PATIL, R. B. AND JARHAD, A. L. (1988). Storability of seeds of sorghum hybrids under ambient conditions. *Seed Research* **19**, 88–91.

PAUL, M. D., KULKARNI, N. AND MURTHY, K. N. (1984). Genetic variability in the F3 progenies for shoot fly resistance in grain sorghum. *Indian Journal of Agricultural Sciences* **54**, 253–255.

PAWAR, C. S., SITHANANTHAM, S., BHATNAGAR, V. S., SRIVASTAVA, C. P. AND REED, W. (1988). The development of sex pheromone trapping of *Heliothis armigera* at ICRISAT, India. *Tropical Pest Management* **34**, 39–43.

PAWAR, D. B., AMBEKAR, J. S., PATIL, B. D. AND POKHARKAR, P. N. (1984). Studies on the estimation of losses in storage due to insect pests in sorghum hybrid. *Bulletin of Grain Technology* **22**, 73–76.

PAWAR, V. M., JADHAV, G. D. AND KADAM, B. S. (1984a). Compatibility of oncol 50SP with different fungicides on sorghum (CS-3541) against shoot fly (*Atherigona soccata* Rondani). *Pesticides* **18**, 9–10.

PAWAR, V. M., JADHAV, G. D. AND KADAM, B. S. (1984b). Bioefficiency of oncol 40EC, carbofuran 3G, and carbofuran 50SP against sorghum shoot fly (*Atherigona soccata* Rondani). *Pesticides* **18**, 44–46.

PAWAR, V. M., SHIRSHIKAR, S. P. AND SAWALE, D. A. (1985). Bioefficiency and compatibility of oncol formulations against shoot fly. *Asian Farm Chemicals* **1**, 21–24.

PEREIRA, R. P., DACOSTA, R. A. M. AND FERREIRA, J. M. (1987). Incidence of *Diatraea* in genotypes of sorghum in various planting dates. *Pesquisa Agropecuaria Brasileira* **22**, 23–25 (in Portuguese).

PETERSON, G. C., ALI, A. E. B., TEETES, G. L., JONES, J. W. AND SCHAEFER, K. (1989). Grain sorghum resistance to midge by yield loss vs. visual scores. *Crop Science* **29**, 1136–1140.

PITRE, H. N. (1988). Relationship of fall armyworm (Lepidoptera: Noctuidae) from Florida, Honduras, Jamaica and Mississippi: susceptibility to insecticides with reference to migration. *Florida Entomologist* **71**, 56–61.

PRABHAKAR, B., RAO, P. K. AND RAO, B. H. K. (1986). Studies on the seasonal prevalence of certain hemiptera occurring on sorghum. *Entomon* **11**, 95–99.

PUTERKA, G. J., PETERS, D. C., KERNS, D. L., SLOSSER, J. E., BUSH, L., WORRALL, D. W. AND MCNEW, R. W. (1988). Designations of two new greenbug (Homoptera: Aphididae) biotypes G and H. *Journal of Economic Entomology* **81**, 1754–1759.

QUICKE, D. L. J. AND WHARTON, R. A. (1989). *Myosoma nyanzaensis* sp. n. (Hymenoptera: Braconidae) parasitic on the stem borer pest, *Chilo partellus* (Swinhoe) (Lepidoptera: Pyralidae) in Kenya. *Bulletin of Entomological Research* **79**, 1–5.

RADFORD, B. J. AND ALLSOPP, P. G. (1986). Soil insect pests: control under conservation cropping. *Queensland Agricultural Journal* **112**, 189.

RADFORD, B. J. AND ALLSOPP, P. G. (1987). Use of insecticides and a press wheel to control soil insects affecting sorghum and sunflower establishment in southern Queensland. *Journal of the Australian Entomological Society* **26**, 161–167.

RAINA, A. K., THINDWA, H. Z., OTHIENO, S. M. AND DOUGLASS, L. W. (1984). Resistance in sorghum to sorghum shoot fly (Diptera: Muscidae): oviposition on selected cultivars. *Journal of Economic Entomology* **77**, 648–651.

RAJA, S., THAKUR, S. S., RAO, B. K. AND KAUR, A. (1986a). Quantitative changes in the glycogen content during growth and development of *Chilo partellus* (Lepidoptera: Pyralidae). *Current Science (India)* **55**, 423–425.

RAJA, S., THAKUR, S. S., RAO, B. K. AND KAUR, A. (1986b). Variations in the lipid content of the fat body during metamorphosis of *Chilo partellus* Swinhoe (Lepidoptera: Pyralidae). *Current Science (India)* **55**, 630–632.

RAJA, S., THAKUR, S. S., RAO, B. K. AND KAUR, A. (1986c). Protein changes in the haemolymph of *Chilo partellus* (Swinhoe) (Lepidoptera: Pyralidae) during ontogenesis. *Philippine Journal of Science* **115**, 253–257.

RAJUKKANNU, K., HABEEBULLAH, B., DORAISAMY, P. AND BALASUBRAMANIAN, M. (1985). Insecticides residues in sorghum: effect of different formulations. *Madras Agricultural Journal* **72**, 241–243.

RAMASWAMY, S. B., MA, P. N. K. AND PITRE, H. N. (1988). Calling rhythm and pheromone titers in *Spodoptera frugiperda* (J. E. Smith) (Lep., Noctuidae) from Mississippi and Hondurus. *Journal of Applied Entomology* **106**, 90–96.

RAMOSKA, W. A. AND TODD, T. (1985). Variation in efficacy and viability of *Beauveria bassiana* in the chinch bug (Hemiptera: Lygaeidae) as a result of feeding activity on selected host plants. *Environmental Entomology* **14**, 146–148.

RANA, B. S., SINGH, B. U., RAO, V. J. M., REDDY, B. B. AND RAO, N. G. P. (1984). Inheritance of stem borer resistance in sorghum. *Indian Journal of Genetics and Plant Breeding* **44**, 7–14.

RAO, D. V. S. AND AZAM, K. M. (1987a). Relative efficacy of some new insecticides as soil and seed dressers in the control of sorghum shoot fly. *Indian Journal of Plant Protection* **15**, 180–183.

RAO, D. V. S. AND AZAM, K. M. (1987b). Economical control of sorghum earhead bug. *Indian Journal of Plant Protection* **15**, 194–197.

RAO, P. K., BHUSHAN, V. S. AND AZAM, K. M. (1986). Biology of sorghum shoot bug, *Peregrinus maidis* (Ashmead). *Andhra Agricultural Journal* **33**, 94–96.

REGO, A. M., VEIGA, A. F. S. L., PIMENTEL, M. L., RODRIGUES, Z. A. AND OLIVEIRA, M. L. (1984). Response of cultivars of *Sorghum bicolor* (L.) Moench, to attack by *Sitophilus zeamais* Motschutsky 1855 (Coleoptera: Curculionidae). *Ciencia e Cultura* **36**, 988–992 (in Portuguese).

RICE, M. E. AND WILDE, G. E. (1988). Experimental evaluation of predators and parasitoids in suppressing greenbugs (Homoptera: Aphididae) in sorghum and wheat. *Environmental Entomology* **17**, 836–841.

RICHTER, A. R., FUXA, J. R., ABDEL-FATTAH, M. (1987). Effect of host plant on the susceptibility of *Spodoptera frugiperda* (Lepidoptera: Noctuidae) to a nuclear polyhedrosis virus. *Environmental Entomology* **16**, 1004–1006.

ROSSETTO, C. J., NAGAI, V. AND OVERMAN, J. (1984). Mechanism of resistance in sorghum variety AF-28 to *Contarinia sorghicola* (Diptera: Cecidomyiidae). *Journal of Economic Entomology* **77**, 1439–1440.

SANDHU, G. S., DHALIWAL, G. S. AND SIDHU, B. S. (1986). Resistance of forage sorghum to shoot fly (*Atherigona soccata* Rond.). *Indian Journal of Agricultural Sciences* **56**, 753–756.

SATTIGI, H. N., KULKARNI, K. A. AND THONTADARYA, T. S. (1985a). Studies on the biology of lesser grain borer on sorghum under laboratory conditions. *Journal of Maharashtra Agricultural Universities* **10**, 252–254.

SATTIGI, H. N., KULKARNI, K. N. AND THONTADARYA, T. S. (1985b). Residues of insecticides used for the control of stored grain beetle pests of sorghum. *Pesticides* **19**, 61–62.

SATTIGI, H. N., KULKARNI, K. N. AND THONTADARYA, T. S. (1985c). Chemical control of *Sitophilus oryzae* L. infesting sorghum seeds during storage. *Seed Research* **13**, 102–106.

SAXENA, K. N. AND ONYANGO, J. D. (1987). Attraction of the stem borer, *Chilo partellus* (Swinhoe) larvae (Lepidoptera: Pyralidae) by certain coloured surfaces for trapping. *Applied Entomology and Zoology* **22**, 493–498.

SCHWAGER, B., PITRE, H. AND GOURLEY, L. (1984). Field evaluation of sorghum characteristics to fall armyworm. *Journal of the Georgia Entomological Society* **19**, 333–339.

SEGARRA-CARMONA, A. E. AND BARBOSA, P. (1988). The sorghum midge, *Contarinia sorghicola*, a sorghum pest newly introduced to Puerto Rico. *Journal of Agriculture of the University of Puerto Rico* **72**, 161–163.

SESHUREDDY, K. V. (1985). Relative susceptibility and resistance of some sorghum lines to stem borers in Western Kenya. *Insect Science and its Application* **6**, 401–404.

SHAH, N. K. AND GARG, D. K. (1988). Exposure of root grubs during ploughing and their predation by birds. *Indian Journal of Entomology* **55**, 131–132.

SHARMA, G. C. AND RANA, B. S. (1985). Genetics of ovipositional non preference and dead heart formation governing shoot fly resistance in sorghum. *Journal of Entomological Research* **9**, 104 105.

SHARMA, H. C. (1985). Strategies for pest control in sorghum in India. *Tropical Pest Management* **31**, 167–185.

SHARMA, H. C. AND LEUSCHNER, K. (1987). Chemical control of sorghum head bugs (Hemiptera: Miridae). *Crop Protection* **6**, 334–340.

SHARMA, H. C. AND LOPEZ, V. F. (1989). Assessment of avoidable losses and economic injury levels for the sorghum head bug, *Calocoris angustatus* Leth. (Hemiptera: Miridae) in India. *Crop Protection* **8**, 429–435.

SHARMA, H. C., VIDYASAGAR, P. AND LEUSCHNER, K. (1988a). Field screening of sorghum for resistance to sorghum midge (Diptera: Cecidomyiidae). *Journal of Economic Entomology* **81**, 327–334.

SHARMA, H. C., VIDYASAGAR, P. AND LEUSCHNER, K. (1988b). No-choice cage technique to screen for resistance to sorghum midge (Diptera: Cecidomyiidae). *Journal of Economic Entomology* **81**, 415–422.

SHAZALI, M. E. H. (1985). Intraspecific competition and progeny production in *Sitophilus oryzae* (L.) (Coleopt.) and *Sitotroga cerealella* (Oliv.) (Lepid.) in sorghum grain. *Anzeiger für Schädlingskunde Pflanzenschutz Umweltschutz* **58**, 121–123.

SHAZALI, M. E. H. (1986). Effect of sorghum grain size on development ecology of *Sitophilus oryzae*. *Zeitschrift für Angewandte Entomologie* **73**, 293–300.

SHAZALI, M. E. H. (1987a). Relative susceptibility of stored sorghum varieties to *Sitophilus oryzae* (L.) and *Sitotroga cerealella* (Oliv.). *Beitrage zur Tropischen Landwirtschaft und Veterinarmedizin* **25**, 61–67.

SHAZALI, M. E. H. (1987b). The influence of sorghum grain size on the development of *Sitotroga cerealella* (Olivier) (Lepidoptera: Gelechiidae). *Zeitschrift für Angewandte Zoologie* **74**, 367–375.

SHAZALI, M. E. H. AND SMITH, R. H. (1985). Life history studies of internally feeding pests of stored sorghum: *Sitotroga cerealella* (Oliv.) and *Sitophilus oryzae* (L.). *Journal of Stored Products Research* **21**, 171–178.

SHIOTA, C. M. AND LARA, F. M. (1985). Resistance of sorghum genotypes to attack by the maize weevil under laboratory conditions. *Fitossan* **6–9**, 1–10 (in Portuguese).

SILVAIN, J. F. (1986). Use of pheromone traps as warning system against attacks of *Spodoptera frugiperda* larvae in French Guyana. *Florida Entomologist* **69**, 139–147.

SINGH, B. U. (1987). Varietal resistance in sorghum to midge, *Contarinia sorghicola* Coquillet (Diptera: Cecidomyiidae). *Insect Science and its Application* **8**, 129–144.

SINGH, B. U. AND RANA, B. S. (1984). Influence of varietal resistance on oviposition and larval development of stalk borer, *Chilo partellus* Swinhoe, and its relationship to field resistance in sorghum. *Insect Science and its Application* **5**, 287–296.

SINGH, B. U. AND RANA, B. S. (1986). Resistance in sorghum to the shoot fly, *Atherigona soccata* Rondani. *Insect Science and its Application* **7**, 577–587.

SINGH, B. U. AND RANA, B. S. (1989). Varietal resistance in sorghum to spotted stem borer, *Chilo partellus* (Swinhoe). *Insect Science and its Application* **10**, 3–27.

SINGH, B. U., RANA, B. S. AND RAO, N. G. P. (1987). Stability of resistance to stem borer in sorghum. *Indian Journal of Entomology* **49**, 358–362.

SINGH, G. AND BALAN, J. S. (1986). Host plants and natural enemies of *Heliothis armigera* (Hubner) in Haryana. *Indian Journal of Entomology* **13**, 175–178.

SINGH, R. AND CHAUDHARY, J. P. (1987). Relative suitability of some graminaceous host plants to *Mythimna loreyi* (Dup.). *Zeitschrift für Angewandte Zoologie* **74**, 367–375.

SINGH, R., MRIG, K. K. AND CHAUDHARY, J. P. (1987a). Feeding and ovipositional preference of *Mythimna separata* (Wlk.) and effect of temperature and humidity on its reproductive biology. *Indian Journal of Entomology* **14**, 102–110.

SINGH, R., MRIG, K. K. AND CHAUDHARY, J. P. (1987b). Incidence and survival of *Mythimna separata* on cereal crops in Hisar. *Indian Journal of Agricultural Sciences* **57**, 59–60.

SINGH, R. P., SRIVASTAVA, K. P. AND AGNIHOTRI, N. P. (1987). Effect of seasonal variations on the persistence of some systemic insecticides in soil and sorghum crop and their relative efficacy against shoot fly, *Atherigona soccata* (Rond.). *Journal of Entomological Research* **11**, 107–114.

SINGH, S. P. AND VERMA, A. N. (1988a). Monitoring of shoot fly, *Atherigona soccata* (Rondani) in traps and their periodic incidence in sorghum. *Crop Research (India)* **1**, 76–83.

SINGH, S. P. AND VERMA, A. N. (1988b). Inheritance of resistance to shoot fly, *Atherigona soccata* (Rondani) in forage sorghum. *Journal of Insect Science* **1**, 49–52.

SINGH, S. P. AND VERMA, A. N. (1988c). Antibiosis mechanism of resistance to stem borer, *Chilo partellus* (Swinhoe) in sorghum. *Insect Science and its Application* **9**, 579–582.

SINGH, S. P. AND VERMA, A. N. (1988d). Combining ability for stem borer, *Chilo partellus* (Swinhoe) resistance in sorghum. *Insect Science and its Application* **9**, 665–668.

SINGH, U. C., MISHRA, U. S., DHAMDHERE, S. V. AND DWIVEDI, U. S. (1985). Carry over of the sorghum stem borer, *Chilo partellus* (Swinhoe) in off season in different crops. *Journal of Entomological Research* **9**, 170–173.

SINGH, Y. P., SRIVASTAVA, A. S. AND SINGH, S. V. (1988). Residues of phosphamidon, endosulfan and monocrotophos in/on sorghum grains. *Indian Journal of Entomology* **50**, 17–23.

SITHOLE, S. Z. (1985). The shoot fly (*Atherigona soccata*). *Zimbabwe Agricultural Journal* **82**, 9–10.

SKOROSZEWSKI, R. W. AND VAN HAMBURG, H. (1987). The release of *Apanteles flavipes* (Cameron) (Hymenoptera: Braconidae) against stalk borers of maize and grain sorghum in South Africa. *Journal of the Entomological Society of South Africa* **50**, 249–255.

SLOSSER, J. E., WITZ, J. A., PUTERKA, G. J., PRICE, J. R. AND HARTSTACK, A. W. (1987). Seasonal changes in bollworm (Lepidoptera: Noctuidae) moth catches in pheromone traps in a large area. *Environmental Entomology* **16**, 1296–1301.

STEWARD, V. B., KRING, T. J. AND YEARIAN, W. C. (1989). Evaluation of survey methods for estimating sorghum midge (Diptera: Cecidomyiidae) densities in grain sorghum. *Journal of Entomological Science* **24**, 47–52.

SUKHANI, T. R. (1986). Insect pest management in sorghum. *Plant Protection Bulletin (India)* **38**, 57–61.

TAN, W. O., LI, S. M., GUI, H. P. AND GAO, R. P. (1985). A study of the inheritance of aphid resistance in sorghum. *Shanxi Agricultural Science* **8**, 12–14 (in Chinese).

TANEJA, S. L. AND LEUSCHNER, K. (1986). Simple trap for monitoring sorghum shoot fly. *Indian Journal of Plant Protection* **14**, 83–86.

TANEJA, S. L., SESHUREDDY, K. V. AND LEUSCHNER, K. (1986). Monitoring of shoot fly population in sorghum. *Indian Journal of Plant Protection* **14**, 29–36.

TEAKLE, R. E. AND BYRNE, V. S. (1989). Food selection by larvae of *Heliothis armigera* (Hubner) (Lepidoptera: Noctuidae) on grain sorghum. *Journal of the Australian Entomological Society* **27**, 293–296.

TEAKLE, R. E., JENSEN, J. M. AND GILES, J. E. (1985). Susceptibility of *Heliothis armigera* to a commercial nuclear polyhedrosis virus. *Journal of Invertebrate Pathology* **46**, 166–173.

TEETES, G. L. (1985). Insect resistant sorghums in pest management. *Insect Science and its Application* **6**, 443–451.

TEETES, G. L., BECERRA, M. T. AND PETERSON, G. C. (1986). Sorghum midge (Diptera: Cecidomyiidae) management with resistant sorghum and insecticides. *Journal of Economic Entomology* **79**, 1091–1095.

THOMSON, M. S. AND ALL, J. N. (1984). The use of oviposition on artificial substrates as a survey tool for the fall armyworm. *Florida Entomologist* **67**, 349–357.

THONTADARYA, T. S. AND RAO, K. J. (1987). Biology of *Orius maxidentex* Ghauri (Hemiptera: Anthocoridae), a predator of the sorghum earhead midge, *Contarinia sorghicola* (Coquillett). *Mysore Journal of Agricultural Sciences* **21**, 27–31.

TREACY, M. F., PARKER, R. C., ANDERSON, R. M., SCHMIDT, K. M. AND BENEDICT, J. H. (1986). Soybean and cottonseed oils as adjuvants and diluents for insecticides used to control sorghum midge. *Southwestern Entomologist* **11**, 39–43.

UNNITHAN, G. C. (1987). Development and reproductive biology of the maize stem borer, *Busseola fusca* Fuller (Lepid., Noctuidae). *Journal of Applied Entomology* **104**, 172–179.

UNNITHAN, G. C. AND DELOBEL, A. G. L. (1985). Effects of temperature on longevity, fecundity and fertility of the sorghum shoot fly, *Atherigona soccata* Rondani. *Insect Science and its Application* **6**, 513–516.

UNNITHAN, G. C. AND SESHUREDDY, K. V. (1985). Oviposition and infestation of the sorghum shoot fly, *Atherigona soccata* Rondani, on certain sorghum cultivars in relation to their relative resistance and susceptibility. *Insect Science and its Application* **6**, 409–412.

UNNITHAN, G. C., SAXENA, K. N., BENTLEY, M. D. AND HASSANALI, A. (1987). Role of sorghum extract in eliciting oviposition on a non host by the sorghum shoot fly, *Atherigona soccata* Rondani (Diptera: Muscidae). *Environmental Entomology* **16**, 967–970.

VERMA, B. R. AND LAL, B. (1987). India: review of *Prostephanus truncatus* (Horn) from India with report of a new host. *FAO Plant Protection Bulletin* **35**, 100.

VERMA, K. K. AND JOTWANI, M. G. (1985). Life processes of the sorghum stem borer, *Chilo partellus* Swinhoe, of different agro-climatic areas. *Indian Journal of Entomology* **45**, 477–478.

VILLATA, C. A. (1986). Effect of dimethoate used as a seed treatment for the control of the cereal greenbug (*Schizaphis graminum* (Rond.)). *Revista Agronomica de Manfredi* **2**, 23–33 (in Spanish).

WAQUIL, J. M., TEETES, G. L. AND PETERSON, G. C. (1986a). Adult sorghum midge (Diptera: Cecidomyiidae) non preference for a resistant hybrid sorghum. *Journal of Economic Entomology* **79**, 455–458.

WAQUIL, J. M., TEETES, G. L. AND PETERSON, G. C. (1986b). Sorghum midge (Diptera: Cecidomyiidae) adult ovipositional behaviour on resistant and susceptible sorghum hybrids. *Journal of Economic Entomology* **79**, 530–532.

WAQUIL, J. M., TEETES, G. L. AND PETERSON, G. C. (1986c). Comparison of immature sorghum midge (Diptera: Cecidomyiidae) development on resistant and susceptible sorghums. *Journal of Economic Entomology* **79**, 833–837.

WEIBEL, D. E. (1986). Registration of eight bloomless parental lines of sorghum. *Crop Science* **26**, 842.

WEIBEL, D. E. AND STARKS, K. J. (1986). Greenbug non preference for bloomless sorghum. *Crop Science* **26**, 1151–1153.

WIDSTROM, N. W., WISEMAN, B. R. AND MCMILLAN, W. W. (1984). Patterns of resistance in sorghum to the sorghum midge. *Crop Science* **24**, 791–793.

WILDE, G. AND MIZE, T. (1984). Enhanced microbial degradation of systemic pesticides in soil and its effect on chinch bug, *Blissus leucopterus leucopterus* (Say) (Heteroptera: Lygaeidae) and greenbug, *Schizaphis graminum* Rondani (Homoptera: Aphididae) control in seedling sorghum. *Environmental Entomology* **13**, 1079–1082.

WILDE, G., RUSS, O. AND MIZE, T. (1986). Tillage, cropping patterns and insecticide use practice: effects and efficacy of planting time treatment for controlling greenbug (Homoptera: Aphididae) and chinch bug (Heteroptera: Lygaeide) in seedling sorghum. *Journal of Economic Entomology* **79**, 1364–1365.

WILDE, G., STUART, J. AND HATCHETT, J. H. (1986). Chinch bug (Heteroptera: Lygaeidae) reproduction on selected small grains and genetic sources. *Journal of the Kansas Entomological Society* **59**, 550–551.

WILDE, G., MIZE, T., STUART, J., WHITWORTH, J. AND KINSINGER, R. (1984). Comparison of planting time applications of granular or liquid insecticides and liquid fertilizer plus insecticides combinations for control of chinch bugs (Heteroptera: Lygaeide) and greenbugs (Homoptera: Aphididae) on seedling sorghum. *Journal of Economic Entomology* **77**, 706–708.

WISEMAN, B. R., DUNCAN, R. R. AND WIDSTROM, N. W. (1988). Registration of SGIRL-MR-3 and SGIRL-MR-4 midge resistant sorghum germplasms. *Crop Science* **28**, 202–203.

WISEMAN, B. R. AND LOVELL, G. R. (1988). Resistance to the fall armyworm in sorghum seedlings from Ethiopia and Yemen. *Journal of Agricultural Entomology* **5**, 17–20.

WISEMAN, B. R., PITRE, H. N., GOURLEY, L. AND FALES, S. L. (1984). Differential growth responses of fall armyworm larvae on developing sorghum seeds incorporated into a meridic diet. *Florida Entomologist* **67**, 357–367.

WISEMAN, B. R., PITRE, H. N., FALES, S. L. AND DUNCAN, R. R. (1986). Biological effects of developing sorghum panicles in meridic diet on fall armyworm (Lepidoptera: Noctuidae) development. *Journal of Economic Entomology* **79**, 1637–1640.

WOOD, T. G., BEDNARZIK, M. AND ADEN, H. (1987). Damage to crops by *Microtermes nadjensis* (Isoptera: Macrotermitinae) in irrigated semi desert areas of the Red Sea coast. 1. The Tihama region of the Yemen Arab Republic. *Tropical Pest Management* **33**, 142–150.

WOODHEAD, S. AND TANEJA, S. L. (1987). The importance of the behaviour of young larvae in sorghum resistance to *Chilo partellus*. *Entomologia Experimentalis et Applicata* **45**, 47–54.

YOUM, O., BROWNING, H. W. AND GILSTRAP, F. E. (1988). Population dynamics of stalk borers attacking corn and sorghum in the Texas Rio Grande valley. *Southwestern Entomologist* **13**, 199–204.

YOUNG, W. R. AND TEETES, G. L. (1977). Sorghum entomology. *Annual Review of Entomology* **22**, 193–218.

ZAFAR, M. A. AND QUAYYUM, H. A. (1985). Effect of temperature on the development of rice weevil, *Sitophilus oryzae* (Linn.) (Curculionidae: Coleoptera) on four food grains. *Journal of Agricultural Research (Pakistan)* **23**, 197–201.

# 3
# Stored-Products Entomology in the Tropics

ATAUR RAHMAN KHAN AND ABDUL MANNAN

*Department of Zoology, Rajshahi University, Rajshahi-6205, Bangladesh*

Introduction
Storage pests
Taxonomy of storage pests
Damage by storage pests
Direct damage—Indirect damage—Extent of damage
Control of storage pests
Mechanical control—Physical control—Chemical control—Biological control—Legislative control
Guidelines for future research
Conclusions
Acknowledgements
References

## Introduction

The intensification of food production has led to several problems in the postharvest phase, including the major concern of pest infestation during storage. This is further aggravated by the increased attention paid to maintenance of buffer stocks to provide food security for a country. Pest problems have increased side by side with the increase in the amount of food stockpiled and the longer duration of storage. Such pest problems are more acute in the tropics than in temperate zones because the environment in the former is more conducive to the growth and development of pests.

Insects are the dominant group of storage pests, but a second group of arthropods, the Acari or mites, also contains a large number of important pests of stored food and other commodities. Other animal groups also contain pest species, so pest control experts must often deal with simultaneous infestations by insects, mites, rodents and birds.

Among the insect pests of stored commodities, the beetles (Coleoptera) are by far the most numerous group, followed by the moths (Lepidoptera). Many other groups are also important and these include: ants (Hymenoptera) (Smith, 1965; Mallis, 1982), cockroaches (Dictyoptera) (Cochran, 1982; Mallis, 1982), flies (Diptera) (Greenberg, 1971, 1973; Mallis, 1982), psocids

*Agricultural Zoology Reviews* — Vol. 4, February 1991
0269–0543 $20.00 + $0.00 © Intercept Ltd, PO Box 716, Andover, Hampshire SP10 1YG, UK

(Psocoptera) (Turner, 1975; Mallis, 1982), silverfish (Thysanura) (Wygodzinsky, 1972; Mallis, 1982) and springtails (Collembola) (Mallis, 1982). However, some of them are only briefly mentioned here.

Gorham (1985) gave a fairly detailed account of the insect pests and their control in the Latin American and Caribbean regions. The aim of this article is to present important aspects of insect and mite pests of stored products, to discuss their control and to suggest guidelines for future research. The literature on stored-products pests is enormous and it is impossible to gather all of the information in a single review. In this review only selected works have been mentioned and the authors apologize for any unintentional omissions.

## Storage pests

Pests attacking stored commodities have been described in a number of works, e.g. Hinton (1945), Pruthi and Singh (1950), Cotton (1956), Metcalf, Metcalf and Flint (1962), Davey (1965), Busvine (1966), Munro (1966), Alam (1971), Benson (1973), Atwal (1976), United States Department of Agriculture (1979), Freeman (1980), Davis (1981), Hill (1983), Wilbur and Mills (1985), Davidson and Lyon (1987), Gorham (1990a,b) and those mentioned in *Table 1*. Insects infesting various stored products are listed in *Table 2*.

**Table 1.** Selected references on some storage pests

| Species | References |
|---|---|
| *Cadra cautella* (Walker) | Kamel, Saleh and Badawi (1976) |
| *Callosobruchus* spp. | Rahman and Sapra (1942); Arora and Pajni (1957); Rajak and Pandey (1965); Raina (1970); Begum and Rahman (1974); Pajni and Jit (1979); Southgate (1979); Tauthong and Wanleelag (1981) |
| *Corcyra cephalonica* Staint. | Kamel and Hassenein (1967); Pajni, Rose and Gill (1976); Hodges (1979) |
| *Cryptolestes pussiloides* (Steel and Howe) | Lefkovitch (1964) |
| *Doloessa viridis* (Zell.) | Baldos (1979) |
| *Laemophloeus minutus* Oliv. | Davies (1949) |
| *Lasioderma serricorne* (F.) | Howe (1957) |
| *Necrobia rufipes* (DeGeer) | Simmons and Ellington (1925) |
| *Oryzaephilus* spp. | Back and Cotton (1926); Howe (1956); Joshi (1974) |
| *Pachymerus chinensis* Linn. | Singh (1965) |
| *Plodia interpunctella* (Hub.) | Williams (1964) |
| *Prostephanus truncatus* (Horn) | Wright and Spilman (1983); Hodges (1986) |
| *Ptinus* spp. | Howe and Burges (1953) |
| *Rhizopertha dominica* (F.) | Potter (1935) |
| *Sitophilus* spp. | Khan (1949); Howe (1952); Arida (1974); Longstaff (1981); Bhuiyan and Quiniones (1988) |
| *Sitotroga cerealella* (Oliv.) | Alam and Shahjahan (1964) |
| *Tinea pellionella* L. | Cheema (1957) |
| *Tribolium* spp. | Good (1933); Park (1934); Stanley (1965); Sokoloff (1974); Dawson (1977); Pajni and Virk (1982a,b); Pajni and Bhookul (1984) |
| *Trogoderma* spp. | Rahman, Sohi and Sapra (1945); Lindgren, Vincent and Krohne (1951); Beal (1954); Hadaway (1956); Lindgren and Vincent (1959); Rajagopal and Balasubramanian (1977) |

**Table 2.** Insects associated with stored products

| Species | Common names | Products infested |
|---|---|---|
| Coleoptera | | |
| Anobiidae | | |
| *Lasioderma serricorne* (F.) | Cigarette beetle/tobacco beetle | Dried tobacco, foodstuffs, turmeric |
| *L. testaceum* L. | Cheroot beetle | Dried tobacco, foodstuffs, turmeric |
| *Stegobium paniceum* (L.) | Drug store beetle | Foodstuffs, stored turmeric, ginger, chilli, coriander |
| *Gastrallus indicus* L. | Book worm | Printed matter |
| Bostrichidae | | |
| *Rhizopertha dominica* (F.) | Lesser grain beetle | Rice, wheat, maize, flour |
| *Dinoderus ocellaris* (F.) | Ghoon beetle | Dried bamboo, furniture, rice, wheat |
| *Prostephanus truncatus* (Horn) | Larger grain borer | Corn, soft wheat, dried cassava |
| Brenthidae | | |
| *Araeocerus fasciculatus* (Deg.) | Coffee bean weevil | Coffee beans, seeds |
| Bruchidae | | |
| *Callosobruchus chinensis* (L.) | Pulse beetle/Oriental cowpea bruchid | Pulses |
| *C. maculatus* (F.) | Spotted cowpea bruchid | Pulses |
| *Bruchus pisorum* (L.) | Pea weevil | Pulses |
| *Caryedon serratus* (Oliv.) | Groundnut borer | Pulses, groundnuts |
| *Acanthoscelides obtectus* (Say) | Bean bruchid | Pulses |
| Cerylonidae | | |
| *Murmidius ovalis* (Beck) | — | Paddy, rice, corn, spices |
| Cleridae | | |
| *Necrobia rufipes* (De Geer) | Red-legged ham beetle/copra beetle | Copra, oilseeds, dried fish, rice, wheat, mixed feed |
| Cucujidae | | |
| *Cryptolestes ferrugineus* (Stephens) | Rust-red grain beetle | Grains |
| *C. pusillus* (Schönherr) | Flat grain beetle | Grains |
| *Laemophloeus minutus* (Oliv.) | Flat grain beetle | Rice, wheat, maize |
| Curculionidae | | |
| *Sitophilus oryzae* (L.) | Rice weevil | Rice, maize, foodstuffs |
| *S. zeamais* Motsch. | Maize weevil | Maize, rice, sorghum, mung-bean |
| Dermestidae | | |
| *Trogoderma granarium* Everts | Khapra beetle | Grains, groundnut |
| *Dermestes lardarius* L. | Lardar beetle | Dried animal matter |
| *D. maculatus* (Deg.) | Hide beetle | Dried animal matter |

*cont'd*

**Table 2.** *cont'd*

| Species | Common names | Products infested |
|---|---|---|
| *Attagenus fasciatus* L. | Carpet beetle | Dried animal matter |
| *A. piceus* Oliv. | Black carpet beetle | Dried animal matter |
| *Lophocateres pusillus* (Klug) | Siamese grain beetle | Grains, turmeric |
| Mycetophagidae | | |
| *Typhaea stercorea* (L.) | Hairy fungus beetle | Maize |
| Nitidulidae | | |
| *Carpophilus dimidiatus* (F.) | Corn sap beetle | Rice, corn, flour |
| *C. hemipterus* (L.) | Dried fruit beetle | Dried fruits |
| Ptinidae | | |
| *Gibbium psylloides* (Czenp.) | — | Scavengers, bite holes in textiles |
| *Ptinus tectus* Boield. | Australian spider beetle | Cereals, cereal products and spices, often found as scavengers of miscellaneous debris |
| *P. fur* (L.) | White-marked spider beetle | Cereals, cereal products and spices, often found as scavengers of miscellaneous debris |
| Silvanidae | | |
| *Oryzaephilus surinamensis* (L.) | Saw-toothed beetle | Rice, wheat, peas, flour |
| *O. mercator* (Fauvel) | Merchant grain beetle | Wheat, mixed feed |
| *Ahasverus advena* (Waltl) | Foreign grain beetle | Rice, maize |
| Tenebrionidae | | |
| *Tribolium castaneum* (Herbst) | Red flour beetle | Grains, flour, mixed feed |
| *T. confusum* Duval. | Confused flour beetle | Grains, flour, mixed feed |
| *Tenebrio molitor* L. | Yellow mealworm | Flour, mixed feed |
| *Latheticus oryzae* (Waterh.) | Long-headed flour beetle | Flour, mixed feed |
| *Palorus subdepressus* (Wall.) | Depressed flour beetle | Grains, flour |
| *P. ratzeburgii* (Weissmann) | Small-eyed flour beetle | Cereal products |
| *Alphitobius diaperinus* (Panzer) | Lesser mealworm | Rice, wheat |
| *A. laevigatus* (F.) | Black fungus beetle | Whole grains, wheat bran, etc. |
| *Gnathocerus cornutus* (F.) | Broad-horned flour beetle | Grains, flours |
| *G. maxillosus* (F.) | Slender-horned flour beetle | Grains, flours |
| Trogositidae | | |
| *Tenebroides mauritanicus* (L.) | Cadelle beetle | Grains, mixed feed |

*cont'd*

**Table 2.** *cont'd*

| Species | Common names | Products infested |
|---|---|---|
| Lepidoptera | | |
| Galechiidae | | |
| *Sitotroga cerealella* (Oliv.) | Rice moth/Angoumois moth | Rice, wheat, maize flour |
| Oecophoridae | | |
| *Hoffmannophila pseudospretella* (Staint.) | Brown house moth | Grain and grain products; cause damage to carpets |
| *Endrosis sarcitrella* | White-shouldered house moth | Grain and grain products; cause damage to carpets |
| Phycitidae | | |
| *Cadra cautella* (Walker) | Almond moth | Grains, dried fruits, almonds |
| *Ephestia elutella* (Hub.) | Tobacco moth | Tobacco, dried fruits, cocoa beans |
| *E. (Anagasta) kuehniella* (Zell.) | Mediterranean flour moth | Flour |
| *Plodia interpunctella* (Hübner) | Indian meal moth | Dried fruits, meals, flour |
| *Corcyra cephalonica* Staint. | Rice moth | Cereals, pulses, dried fruits and fishes |
| *Hypsopygia costalis* (F.) | Clover hayworm | Clover |
| *Pyralis farinalis* | Meal snout moth | Maize |
| Pyraustidae | | |
| *Doloessa viridis* (Zeller) | Green rice moth | Milled rice, maize, sorghum |
| Tineidae | | |
| *Setomorpha tineoides* Wallsingham | — | Dried tobacco |
| *Tinea pellionella* L. | Cloth moth | Woollen-cloths, carpets, skin, feathers |
| Psocoptera | | |
| Liposclelidae | | |
| *Embidopsocus* sp. | — | Rice, bean |
| *Liposclelis entomophilus* | — | Rice, maize, mung-bean |
| *L. botrychophilus* | — | Rice, maize, cassava |

**Table 3.** Mites associated with stored commodities

| Species | Host |
|---|---|
| *Acarus siro* L. | Flour, all kinds of farinaceous products, cheese, grain, hay |
| *A. farris* (Oudemans) | Cereals and legumes |
| *A. immobilis* Griffiths | Cereals |
| *Aleuroglyphus ovatus* | Cereals |
| *Blomia freemani* | Cereals, fruit, vegetables |
| *Caloglyphus barlesei* (Michael) | Damp grain |
| *C. mycophagus* | Fruit, vegetables, fungi |
| *Carpoglyphus lactis* (L.) | Dried fruits and jam |
| *Chortoglyphus arcuatus* | Cereals, fruit, vegetables |
| *Dermatophagoides farinae* Hughes | Cereals |
| *Gohieria fusca* | Flour, rice, corn, bran |
| *G. domesticus* | Cereals, legumes, seed, proteinaceous food |
| *Lardoglyphus konoi* (Sana and Asanuma) | Proteinaceous and high-fat foods |
| *Lepidoglyphus destructor* (De Geer) | Many kinds of stored food |
| *Rhizoglyphus echinopus* | Damp grain, nuts, bulbs |
| *Tyrophagus casei* (L.) | Cheese, damp flour, old honeycomb |
| *T. longior* (Gervais) | Cereals, legumes, seed, fruit, vegetables |
| *T. palmarum* | Cereals, legumes, seed, fruit, vegetables |
| *T. putrescentiae* (Schrank) | Cereals, legumes, seed, fruit, vegetables |
| *T. similis* | Cereals, legumes, seed, fruit, vegetables |

A large number of mites infest stored produce, and various workers have listed mites found on stored products in different parts of the world (Griffiths, 1960; Krantz, 1961; Sinha, 1963; Champ, 1966; Zdarkova, 1967; Cusack, Evans and Brennan, 1975; Griffiths *et al*, 1976; Hughes, 1976;

**Table 4.** Ants associated with households

| Ant species | Common name |
|---|---|
| *Acanthomyops claviger* (Roger) | Small yellow ant |
| *A. interjectus* (Mayr) | Large yellow ant |
| *Atta texana* (Buckley) | Texas leaf-cutting ant |
| *Camponotus pennsylvanicus* De Geer | Black carpenter ant |
| *C. ferrugineus* (Fabr.) | Red carpenter ant |
| *C. abdominalis floridanus* (Buckley) | Florida carpenter ant |
| *Conomyrma insana* (Buckley) | Pyramid ant |
| *Iridomyrmex humilis* (Mayr) | Argentine ant |
| *I. pruinosus* (Roger) | — |
| *Labidus coecus* (Latreille) | — |
| *Monomorium destructor* (Jerdon) | — |
| *M. floricola* (Jerdon) | — |
| *M. minimum* (Buckley) | Little black ant |
| *M. pharaonis* (L.) | Pharaoh ant |
| *Neivamyrmex nigrescens* (Cresson) | — |
| *N. opacithorax* (Emery) | — |
| *Ochetomyrmex auropunctatus* (Roger) | Little fire ant |
| *Paratrechina longicornis* (Latreille) | Crazy ant |
| *Pheidole megacephala* Fabricius | — |
| *Solenopsis geminata* Fabricius | Hormiga brava, fire ant |
| *S. invicta* Buren | Red imported fire ant |
| *S. molesta* (Say) | Thief ant |
| *S. richteri* Forel | Black imported fire ant |
| *S. xyloni* McCook | Southern fire ant |
| *Tapinoma melanocephalum* Fabricius | — |
| *T. sessile* (Say) | Odorous house ant |
| *Tetramorium caespitum* (L.) | Pavement ant |

**Table 5.** Cockroaches associated with stored food

| Species | Common name |
|---|---|
| *Blatella germanica* (L.) | German cockroach |
| *Blatta orientalis* L. | Oriental cockroach |
| *Nauphoeta cinerea* (Oliver) | Lobster cockroach |
| *Neostylopyga rhombifolia* (Stoll) | — |
| *Periplaneta americana* (L.) | American cockroach |
| *P. australasiae* (F.) | Australian cockroach |
| *P. brunnea* (Burmeister) | Large brown cockroach |
| *Phaetalia pallida* (Brunner) | — |
| *Pycnoscletus surinamensis* (L.) | Surinam cockroach |
| *Shelfordella tartara* Adel. | — |
| *Supella longipalpa* (F.) | Brown-banded cockroach |
| *S. supellectilium* (Serville) | Brown-banded cockroach |

Jeffrey, 1976; Pagliarini, 1979; Tseng, 1979). Some of the mites associated with stored commodities are listed in *Table 3*, and household ants and cockroaches are listed in *Tables 4* and *5*.

Wilbur and Mills (1985) classified the insects that, to varying degrees, have become adapted to farm and commercially stored grains into the following four categories based on the nature and extent of their association with the grain:

1. Major pests. These comprise those few species that are particularly well adapted for living in stored grain, e.g. *Anagasta kuehniella* (Zell.), *Plodia interpunctella* (Hub.), *Rhizopertha dominica* (F.), *Tribolium castaneum* (Herbst), etc.
2. Minor pests. This group consists of a large number of insects and mites that may become abundant locally and are occasionally damaging to grain and milled products, e.g. *Stegobium paniceum* (L.), *Gnathocerus cornutus* (F.), *Tenebrio molitor* L., etc. However, some of these species approach major pest status under favourable conditions.
3. Incidental pests. This category consists of approximately 150 species of insects and mites. Some of these are occasionally present in grain, and others occur frequently in grain and in large numbers along with major and minor pests. They rarely damage grain except by contamination from their presence. The food of many of them consists of fungi and micro-organisms rather than the grain itself. Various families, viz. Blattidae, Cryptophagidae, Nitidulidae, Ptinidae, etc. contain the majority of the insect pests of this category.
4. Parasites and predators. These include insects and mites that prey upon the other three groups.

## Taxonomy of storage pests

The taxonomy of storage pests is more or less well documented, and some of the more important taxonomic works on storage insects are: Hinton (1945, general; 1948, *Tribolium* spp.); Beal (1954, *Trogoderma* sp.); Crowson (1956, Coleoptera); Southgate (1958, *Callosobruchus* spp.); Metcalf, Metcalf and Flint (1962, general); Halstead (1964, *Sitophilus* spp.; 1973, *Silvanus* spp.;

1980, *Oryzaephilus* spp.; 1986, Coleoptera); Morrison (1964, *Sitophilus* spp.); Busvine (1966, general); Britton (1970, general); Hinton and Corbet (1972, general); Arnett (1973, Coleoptera); Brendell (1975, Coleoptera); Richards and Davies (1977, general); and Gorham (1990a,b, general).

The most important general taxonomic works on mites are those of Hughes (1976) and Krantz (1978).

## Damage by storage pests

### DIRECT DAMAGE

#### *Kernel damage*

Much of the damage done by grain pests is directly to the kernels. The degree of damage to germ and endosperm and its accompanying loss of weight is determined by such factors as pest species and their numbers, type of grain, temperature, moisture content and length of storage.

#### *Contamination*

Grain pests contaminate more than they consume. This contamination results from:

1. the presence of living or dead insects or insect parts;
2. cast exuviae, egg shells, pupal cases and cocoons;
3. faecal matter;
4. noxious and persistent odours; and
5. webbing of food.

#### *Grain dust*

The activity of pests inevitably results in dust consisting of flour, excreta and broken kernels. Certain insects, e.g. lesser grain borers and cadelles, may produce so much dust that fumigation becomes unsuccessful unless the dust is removed.

#### *Damage to wooden structures and to containers*

Many insects, e.g. cadelle larvae, lesser grain borers and dermestid larvae, tunnel into the wooden parts of granaries, warehouses, boxcars and ships and weaken these structures.

### INDIRECT DAMAGE

#### *Dry heating and moisture migration in storage*

The metabolic activities of grain-infesting insects within a grain mass may produce temperatures up to 106°F (41°C) even during winter. As a result, moisture is deposited on grain surfaces, and this encourages growth of moulds and stimulates germination. Consequently, caking and spoiling occur. Dry grain heating with no moisture deposition occurs in grains stored at below 15% moisture content.

### *Lowered germination of seed*

The germination of grains stored as seeds is impaired or destroyed by heating or very moist conditions, or by damage to either germ or endosperm.

### *Distribution of fungi and other micro-organisms*

Insects infesting stored grains carry grain-damaging moulds to containers and may distribute them throughout the grain mass. Various pathogens, e.g. *Salmonella* and the fungi that produce toxins such as aflatoxins, likewise may be distributed by pests.

### *Insect parts in food products*

Very often insects inside grain kernels are processed into flour and other cereal products along with the germ and endosperm. The occurrence of insects or of their faeces, odours or silks is extremely repulsive to all consumers and causes a loss in value in terms of both acceptability and nutritional quality.

## EXTENT OF DAMAGE

Substantial losses occur every year in various stored products due to infestations by insects and other organisms in the developing countries. The quantitative assessment of losses has been a very difficult task because of the extremely variable and complex traditions that influence storage practices, and the lack of proper 'loss-survey' techniques. However, reports indicate serious depreciations in both quality and quantity of different stored commodities due to inadequate pest control practices.

Post-harvest losses of stored produce are expressed in terms of direct weight and nutrient loss (Howe, 1965; Krishnamurthy, 1975; Watters and Shuyler, 1977). According to Alam (1971), 5–8% of food grain seeds and different stored commodities are lost annually due to storage pests. Depadua (1974) estimated a loss of unhusked rice of 2–6% for South-East Asia and Girst (1975) reported an average of 5% loss in village stores. Shahjahan (1974) observed an average loss of 8% of rice due to *Sitotroga cerealella* (Oliv.) infestations only. Huq (1980) reported a 2.6% average farm-level storage loss in the weight of unhusked rice in 3–6 months due to insect infestation and Pandya, Ali and Alam (1980) estimated losses of 6.0, 6.5, 5.6, 8.6 and 6.5% for stored unhusked rice, wheat, maize, grain and pulses, respectively, in India. Greeley (1980) reported yearly losses of 3.3, 2.4 and 2.5% for stored dry raw paddy, parboiled paddy and parboiled rice, respectively, by insect pests in Bangladesh. Harris (1984) estimated a loss of 3–10% in the developing world. Bhuiyan, Alam and Karim (1988) recorded average weight losses of stored rice of 9.56 and 3.31% due to infestations of *Sitophilus oryzae* (L.) and *S. cerealella* Oliv., respectively, in Bangladesh. Malek and Parveen (1989) reported average annual losses in weight and viability of 6.77 and 48.01%, respectively, in stored rice due to insect pests.

Huysmans (1982) gave an average estimated weight loss of 12% for staple cereals for all developing countries and Semple, Caliboso and Satio (1983) gave estimated weight losses of various stored commodities in the ASEAN (Association of Southeast Asian Nations) members in detail.

The accuracy and comparability of loss assessment methods are very difficult to ascertain. According to Tyler (1982), estimates are extremely variable from commodity to commodity and from country to country or even different parts of the same country, due to lack of standardization of methodology and the variable climatic conditions under which the products are grown, harvested, stored, processed and handled.

Schulten (1982) listed several constraints to the effective implementation of post-harvest loss reduction tactics in tropical Africa and these appear, in principle, to be applicable to other tropical countries. Schulten's constraints are:

1. lack of co-ordination among the various national institutes/agencies involved in loss prevention;
2. lack of trained personnel in research, warehouse management, quality control and extension;
3. lack of information on post-harvest technologies that have proved effective elsewhere;
4. lack of accurate information on the magnitude of losses in different operations within the post-harvest system;
5. lack of appropriate loss assessment methods;
6. lack of storage capacity;
7. lack of an effective transport and distribution system;
8. lack of grades and standards that can be applied in the field for assessing quality; and
9. lack of differential pricing of the various grades to create incentives for farmers to deliver better quality grain and for investment in improved facilities.

## Control of storage pests

For the protection of stored commodities against pest infestations, the maxim that prevention is better than cure is highly relevant. Much of this damage can be avoided by adopting preventive measures, which include the elimination of the source of infestation in the warehouses (such as old bags and sacks and the mixing of infested old grains with new) and avoiding the use of transport carriages that harbour insects. It is good practice to store grain as dry as possible and not to allow dampness to spoil it. This theme has been echoed very recently by Prevett (1990).

MECHANICAL CONTROL

Mechanical control of storage insects can be achieved in the following ways:

1. Sieving and winnowing. These are commonly practised against stored grain pests. A good number of these insects are removed by these operations, particularly the larvae of *Tribolium castaneum* (Herbst) and *Trogoderma granarium* Everts.
2. Using mechanical traps. Different types of traps have been devised for collecting and killing insects. These are:
   (i) air suction traps, which are fixed in warehouses against stored grain pests;
   (ii) electric traps, i.e. metal screens on which insects or other pests are electrocuted.

PHYSICAL CONTROL

There is a renewed interest in the application of physical control methods against storage pests because consumers are becoming more aware of the consequences of immoderate use of chemical pesticides in the food chain. Various physical control practices have been reviewed by a number of researchers (Watters, 1972; Shejbal, 1980; Armitage, 1987; Bell, 1987; Lessard, 1987; Navarro and Jay, 1987; Wilkin and Nelson, 1987). Physical control methods give immediate tangible results and are generally popular with, and convincing to, farmers. Several physical control measures are available and these are:

1. High-temperature fluidized beds have reached a pilot scale for the disinfestation of grains and their products. It has been confirmed that short exposures to temperatures above 60°C are generally effective for the disinfestation of grains (Evans, 1981, 1987a). Evans' method is based on fluidized-bed drying techniques (Dermott and Evans, 1978) and permits rapid heating of individual grains. In tropical countries solar dryers provide an opportunity for disinfestation by the heat of the sun.
2. Stored dried fruits are protected by keeping them in refrigerated warehouses in different parts of the world. The multiplication of pests in grains is slowed or stopped by using mobile refrigeration units for grain elevators or by chilled-air ventilation of stores. Cooling and drying methods in storage pest management are discussed by Evans (1987b).
3. Controlled and modified atmosphere storage practices for protecting grains have developed greatly during recent years. Much information on these subjects is now available (Hyde *et al.*, 1973; Anonymous, 1984; Evans, 1987b).
4. Dry foodstuffs can be disinfested by exposing them to non-ionizing electromagnetic energy, microwaves or high-frequency dielectric heating. The same procedures can be applied to packaged goods.
5. Ionizing radiations such as X-rays and $\gamma$-rays allow sterilization at lower dosages but are lethal at higher dosages. The application of ionizing

radiations in controlling storage pests, particularly in grain, is currently under extensive study in various countries. For this, large irradiation centres with ionizing radiation sources, radio-isotopes or accelerated beams of electrons have been constructed, principally for grain-insect sterilization.

6. Mechanical handling or physical force have some applications in flour milling or semolina processing, e.g. entoleters, impactors, fluidized-bed separators, etc.
7. A number of marginal means of insect management also have a physical effect, e.g. mineral salt admixture in grain or sound disturbance.

However, it must be stressed that much research is still needed before any of these physical techniques for pest control can be used widely.

### CHEMICAL CONTROL

The use of chemical pesticides is still the chief method of control for insect and mite pests of stored products. Chemical control of storage pests with conventional pesticides is practised in the following ways:

1. by fumigation of stored commodities either under airtight sheets or in gastight storage structures;
2. by spraying or admixture with pesticides of low mammalian toxicity;
3. by treatment of the surfaces of storage structures with persistent pesticides;
4. by space treatment of air in a store; or
5. by impregnation of packaging materials.

Comprehensive accounts of the use of chemicals have been given by Bond (1973), Storey, Speirs and Henderson (1979), Quinlan (1982), Anonymous (1985), Champ and Highley (1986), Edwards (1986), Diekmann (1988), etc. The use of various additives is described in McFarlane (1989). Fumigants are the most used of the chemical pesticides and, of the many available, phosphine and methyl bromide are the best proven and the most commonly available. Among synthetic insecticides pirimiphos-methyl, phoxim, malathion, fenitrothion, chlorpyrifos, chlorpyrifos-methyl, methacrifos, dichlorvos, etc. (organophosphates), bendiocarb, propoxur, etc. (carbamates) and bioresmethrin, permethrin, cypermethrin, deltamethrin, etc. (synthetic pyrethroids) are all used.

The use of insect growth regulators as an alternative method of pest control has received much attention recently. Growth regulators are of three kinds: juvenile hormones, which affect metamorphosis and prevent transformation into adults; ecdysone hormones, which regulate moulting; and antichitin compounds, which prevent chitin formation during moulting. These compounds are considered quite safe for higher animals. The use of juvenile hormone analogues in pest control has been reviewed recently by Edwards and Menn (1981) and Edwards (1986). The growth regulators dimilin, methoprene, Bowers JH and Attacus JH isolated by Paguia and Morallo-

Rejesus (1977) were found to be effective against *Corcyra cephalonica* Staint., *Plodia interpunctella* (Hub.), *Sitophilus* spp. and *Rhizopertha dominica* (F.) (Fajardo and Morallo-Rejesus, 1980; Morallo-Rejesus and Javier, 1978).

Natural plant products (botanicals) represent a very promising group of pesticides because of their apparent safety to mammals. Several publications on botanicals as pest control agents have been made (Jacobson and Crosby, 1967; Teotia and Tewari, 1971; Agarwal, Deshpande and Tipnis, 1973; Pandey, Singh and Tewari, 1976; Pandey and Verma, 1977; Saradamma, Dale and Nair, 1977; Tilak, 1977; Schoonhoven, 1978; Stenmark, 1978; Verma and Pandey, 1978; Cox, 1981; Schmutterer, Ascher and Rembold, 1981; Prakash, Pasalu and Mathur, 1982; Rajasekaran and Kumaraswami, 1985).

Mites can be controlled by similar methods to those used for insects. However, control of mites in tropical stored produce is, at present, mainly achieved as an incidental side-effect of insect control procedures. Haines (1982) comments, 'If specific control efforts are directed at tropical storage mites in the future, they will need to take account of the fact that many insecticides used at current recommended dosages are not efficient as acaricides.'

The problem of resistance to pesticides provides a major reason for the development of improved pest management strategies that are less completely dependent upon pesticide use. There is a number of reviews on resistance to pesticides, including Champ and Dyte (1976), Watson and Brown (1977), Georghiou and Saito (1983), Champ (1986a,b) and Muggleton (1987).

Resistance to pesticides within a pest population may be innate, behavioural, biochemical and/or physiological. But it becomes more pronounced with pesticide usage because the weak or susceptible members are eliminated from the population, leaving the tolerant individuals to reproduce. Generally, pests with short life cycles develop resistance faster because the selection process proceeds more rapidly. Detoxification mechanisms in the body enable them to survive, and these survival factors can be passed on to subsequent generations through genetic material. Cross- or multiple-resistance can also occur, and this magnifies the problem. In addition, there is evidence that a relationship exists between the food and the degree of resistance. However, this phenomenon is not clearly understood. The problem of resistance build-up can be lessened in the following ways:

1. Pesticides should be applied only when absolutely necessary, using compounds that break down quickly and to which cross- or multiple-resistance has not developed.
2. Pesticide mixtures and continuous-release formulations should be avoided.
3. Proper dosages should be applied, and when there is evidence of resistance unrelated compounds should be used in a rotation programme.
4. Pesticide applications should be kept to a minimum and, where feasible,

these should be substituted by other methods of pest control.

## BIOLOGICAL CONTROL

Biological control of pests has received much recent attention due to the obvious hazards produced by chemical pesticides.

### *Use of resistant varieties*

In recent years several investigations have been directed at screening varieties of grain and seed crops for resistance to pests, with the objective of minimizing requirements for pesticide treatments. It has been suggested that cereals resistant to storage insects would become major components of pest management programmes (Dobie, 1974, 1977; Cogburn *et al.*, 1980; Rathore, Bhattacharya and Sachan, 1980). A resistant variety would reduce the rate of increase of a pest population in the following ways:

1. by causing a reduction in the rate of egg-laying; and/or
2. by extending the developmental period; and/or
3. by causing high mortality of the developing insects.

Some maize and sorghum varieties have been observed to be resistant to *Sitophilus* spp. (Bernardo, 1971, 1972; Bernabe-Adalla and Bernardo, 1976a,b). Also reported are instances of resistance to *Trogoderma granarium* in different varieties of wheat (Atwal and Dhaliwal, 1971), to *Sitotroga cerealella* in 14 varieties of paddy grains (Abraham *et al.*, 1972) and to *Sitophilus oryzae* in various high-yielding varieties of wheat (Chakravorty and Mathew, 1972; Singh, Agarwal and Girish, 1972).

Further examples of resistance are those to *Callosobruchus analis, C. chinensis* and *C. maculatus* in chick peas (Raina, 1971), to *C. chinensis* in peas (Singh and Talluri, 1972), to *C. maculatus* in various pulses (Gokhle, 1973) and to *C. chinensis* in varieties of *Dolichos lablab* (Janarthanan, 1973). Malit (1973) and Epino (1980) examined several accessions of mung bean and found resistance to *C. chinensis*. The current status of research in this area has been reviewed by Dobie (1984).

### *The use of pheromones*

A pheromone is a chemical secreted to the exterior by one individual causing a specific reaction in receiving individuals of the same species. Many species of storage beetles, moths and mites produce pheromones which stimulate specific responses, viz. alarm, dispersal, aggregation or attraction of the opposite sex. There have been several reviews of pheromones in stored-product insects and their application in pest management (Burkholder, 1981, 1982; Sifner *et al.*, 1983; Hodges, 1984a,b; Burkholder and Ma, 1985; Campion, Hall and Prevett, 1987).

Pheromones may be used in two ways for storage pets control:

1. For pest monitoring. Conventional insect traps, such as refuge traps or sticky traps, may be baited with sex-attractants or aggregation pheromones to increase trap efficiency and thereby afford an improved means of monitoring pest populations.
2. In pest control. Pheromones may be applied in three ways for controlling pests:
   (i) Traps baited with sex-attractants may be used for mass trapping of insects. However, aggregation pheromones attracting both sexes may prove more effective.
   (ii) Sex-attractants or aggregation pheromones may be used to lure pests to a source of a pathogen or chemosterilant.
   (iii) The air of a storage structure may be perfused with sex pheromone to disrupt normal courtship behaviour and prevent mating among pests.

With further development, pest monitoring or control by pheromones will provide opportunities for improved pest-management strategies in tropical stored commodities and the techniques could be integrated with other control measures to keep pest populations at acceptable levels (Hodges, 1984a,b).

### *The use of parasites and predators*

Most animals are subject to attack by natural enemies, which include organisms that kill them or impair their health by acting as parasites, predators or pathogens. Haines (1984) has reviewed the current status of research on the role of parasites and predators in the management of storage pests. The list of arthropod parasites and predators in *Table 6* includes the important groups.

Many of the naturally occurring parasites and predators are widely distributed in tropical stores. Before useful control strategies can be developed and implemented, comprehensive observations and studies, both basic and applied, need to be made (Haines, 1984). The major priorities for future research include:

1. observations on the prey/host range of many of the natural enemies for which such basic data are lacking, and surveys of their distribution and occurrence;
2. ecological and behavioural studies, under relevant environmental conditions, on species that appear to have potential as control agents;
3. increased emphasis on the use of field trials to evaluate the efficiency of potential biocontrol agents.

### *The use of insect diseases*

Storage pests are subject to a number of diseases caused by pathogenic viruses, bacteria, fungi, Protozoa, Rickettsiae and Nematoda. These diseases

**Table 6.** Insects and mites as parasites and predators on storage pests

| Insects/mites | Hosts |
|---|---|
| Insecta | |
| Hymenoptera | |
| *Anisopteromalus calandrae* (How.) | *C. chinensis, C. analis, S. oryzae, L. serricorne* and *R. dominica* larvae and pupae |
| *Bracon hebetor* Say | *C. cephalonica* and *S. cerealella* larvae |
| *Choetospila elegans* Westwood | *S. oryzae* and *R. dominica* larvae and pupae |
| *Dinarmus basalis* (Rond.) | *C. chinensis, C. analis* and *C. maculatus* larvae and pupae |
| *Rhabdepyris rhizoperthae* Menon, Chatterjee & Sarup | *R. dominica* larvae |
| *R. zeae* Turner & Waterston | *T. confusum* larvae |
| *Trichogramma* spp. | *S. cerealella* eggs |
| *Uscana mukerjii* (Mani) | *C. chinensis* eggs |
| Hemiptera | |
| *Lyctocoris campestris* | Feed on beetles, moths and mites |
| *Xylocoris flavipes* Reuter | |
| Diptera | |
| *Scenopinus fenestralis* (L.) | Predatory on larvae of moths and beetles |
| Acari (mites) | |
| *Blattisocius tarsalis* (Berlese) | Larvae of moths, e.g. *C. cautella, E. kuehniella, P. interpunctella*, etc. |
| *Cheyletus eruditus* (Schrank) | *A. siro* |
| *Pediculoides ventricocus* | *R. dominica* eggs and larvae |
| *Pyemotes* sp. | *S. oryzae* larvae |

are frequently fatal to the insects, especially the larval stage. Transmission of the pathogens takes place through feeding, by contact, during mating or from the female to its progeny during oviposition. Hodges (1984a) reviewed this subject comprehensively.

Viruses attack larvae of several storage moths (Hunter, 1973; Hunter, Collier and Hoffman, 1975; McGaughey, 1975) and have potential as control agents. It is known that treatment of *Plodia interpunctella* with granulosis virus may be combined with malathion to produce a higher mortality than would be achieved if either was applied separately (Hunter, Collier and Hoffman, 1975).

Among bacteria, *Bacillus thuringiensis* Berliner is the most important pathogen, with several commercial preparations available. This pathogen has been tested on several storage moths, e.g. *Plodia interpunctella, Ephestia kuehniella, Ephestia elutella, Cadra cautella, Sitotroga cerealella* (Burges, 1969; McGaughey, 1976) and *Corcyra cephalonica* (Venkatraman, Mathur and Llander, 1962). The bacterium appears very effective against all of these moths except for *S. cerealella*, with young larvae the most sensitive stages.

Many protozoans are severe pathogens of storage pests (*Table 7*). The potential of protozoans in controlling storage pests has been stressed by a number of workers, e.g. McLaughlin (1971), Brooks (1980), Henry (1981), Khan and Selman (1989). The control of storage beetles through the integration of protozoans with low doses of chemical pesticides has also been reported (Khan, 1981; Khan and Selman, 1984, 1987, 1988).

**Table 7.** A list of Protozoa considered to be potential control agents of storage pests

| Protozoa | Hosts |
|---|---|
| *Adelina tribolii* Bhatia<br>*Farinocystis tribolii* Weiser<br>*Lymphotropha tribolii* Ashford<br>*Nosema whitei* Weiser<br>*N. oryzaephili* Burges, Canning and Hurst | Stored-products beetles, viz. *Tribolium* spp. and *Oryzaephilus surinamensis* |
| *Mattesia trogodermae* | *Trogoderma* spp. |
| *Vairimorpha (Nosema) plodiae* Kellen and Lindgren | Storage moths, e. g. *Plodia interpunctella* |

Insect pathogens offer distinct possibilities for the control of stored product pests, and they may be admixed with bulk grain or disseminated by insects lured to a source of pathogen by an attractant. However, considerable research is still required in this area of work.

### *The use of sterile insects*

This technique involves the release of sterile or cytoplasmically incompatible insects into and/or around storage facilities. Males or both of the sexes may be utilized since both are often sterilized by similar methods. When the sterile and normal insects compete for mates, the result is a reduction in the number of progeny in the $F_1$ generation and, if further sterile adults are available to compete with the $F_1$ individuals, the pest population may be reduced still further.

Insects are sterilized by exposure to ionizing radiation, e.g. $\gamma$-rays and X-rays, and by treatment with chemosterilants. The application of sterile control requires an initial assessment of the size of the pest population to be controlled, the rearing of an appropriate number of cytoplasmically incompatible insects or insects to be sterilized, and the release of sterile insects into the pest population (Hodges, 1984b). However, there is an alternative approach, in which it may be possible to avoid insect release altogether by luring insects to a source of chemosterilant by a pheromone or food attractant.

It is important that some of the more feasible possibilities should be thoroughly tested before it can be seen whether or not the sterile-release method could eventually be used as an alternative method for control of tropical stored-products insects. Bower (1974) and, more recently, Hodges (1984b) have reviewed the status of research in this area.

## LEGISLATIVE CONTROL

For control of certain insect pests, government legislation has been passed to make control measures mandatory. However, these are extreme measures and apply only to certain very serious pests of national importance. The Khapra beetle, *Trogoderma granarium*, is probably the most serious international pest of stored grains, by nature of its biology. Regular introductions of this beetle occur accidentally in the harbour areas of Dar es Salaam and Mombasa from cargo ships. For this situation, very strict control measures are practised to ensure that there is no spread of infestations beyond the dockside areas. Eradication measures are taken against the pest in the dockside warehouses and infested ships are usually fumigated.

Import quarantine procedures alone are of limited value for control of storage pests because of the widespread distribution of most stored-products pests (Haines, 1982). There are three situations, however, in which quarantine techniques are valuable:

1. to prevent importation to a previously unaffected area of those few species that have a limited distribution but can be serious pests, e.g. *T. granarium*;
2. to prevent importation of pesticide-resistant strains to countries or areas where resistance has not yet developed; and
3. to facilitate immediate disinfestation of important commodities, irrespective of the pests involved, in order to prevent cross-infestation and, of course, further damage to the important consignment.

## Guidelines for future research

Although the information in the literature on stored-products entomology is extensive, much more research is still required:

1. Detection, extraction and evaluation of natural pesticides from locally available plants should be carried out.
2. A continuous search for alternative pesticides, in conjunction with periodical diagnosis of the resistance status of various economically important pest species, should be made.
3. There is a need for exploring innovative pesticide formulations that provide improved dispensing systems, e.g. controlled-release formulations.
4. Crop resistance is probably one of the most promising ways of managing pest and disease attack, and some progress in the development of resistant varieties has been made by geneticists and plant breeders. However, much research is yet needed in this line.
5. Extensive biological and ecological studies of pests and their environment are essential to improve understanding of insect behaviour and to establish treatment thresholds.
6. Development of detection techniques for early or hidden infestations of pests, particularly those involving species that are subject to quarantine and are difficult to detect.
7. Improved techniques for pesticide application to improve coverage and which are adaptable to varied formulations and make for ease of application are required.
8. Reliable methods for monitoring and assessment of pest populations and pest damage with the determination of economic thresholds are needed. This was first stressed about 35 years ago by Parkin (1956).

9. Theoretical models of the effects of various pest management methods, singly and in combination, which can be tested in reality and altered according to findings, should be drafted. Development of integrated pest-management programmes will result ultimately in greater pest control efficiency and reduce dependence on chemical pesticides.
10. Farmers should be trained in pest-management methodology through efficient extension services.
11. Appropriate warehouse design is a basic requirement for the safe storage of grains and other commodities. To this end, the modification and improvement of existing warehouse designs to make them more suitable for holding products must be pursued.
12. Studies should be undertaken to identify and establish the degree of influence of various socio-economic and technical factors on the adoption or non-adoption of pest-control technologies.
13. There is a need for a better understanding of the costs and benefits of pest management strategies to ensure that the technology recommended is economically viable.

## Conclusions

It is envisaged that the management of storage pests will continue to depend on the use of chemical pesticides, given the need for a fast, reliable and cost-effective method of pest control. The judicial use of pesticides is likely to remain essential in the foreseeable future, in the context of overall pest-control strategies, i.e. within the framework of an integrated pest management (IPM) system. However, other methods of control, viz. some physical and mechanical methods, the development of resistant varieties and use of pheromones are expected to play an important role. The potential for biological control of pests is obviously great. The utilization of biocontrol agents in pest-management strategies has not received so much enthusiasm because, in general, they cannot produce the spectacular results achievable with chemical pesticides. Clearly, however, biocontrol agents have potential for keeping pest populations below economic threshold levels and so should constitute a part of the overall pest-management system.

## Acknowledgements

The authors would like to extend sincere thanks to the following scientists for kindly supplying literature: Dr B. R. Champ, Research Programme Coordinator, ACIAR, Canberra, Australia; Dr J. R. Gorham, Scientific Director, Food and Drug Administration, Washington, DC, USA; and Dr P. F. Prevett, ODNRI, Storage Department, Slough, UK.

## References

ABRAHAM, C. C., THOMAS, B., KARUNAKARAN, K. AND GOPALAKRISHNAN, R. (1972). Relative susceptibility of different varieties of paddy to infestation by

Angoumois grain moth, *Sitotroga cerealella* Oliver, as influenced by the amylose content of endosperm. *Bulletin of Grain Technology* **10**, 263–266.

AGARWAL, D. C., DESHPANDE, R. B. AND TIPNIS, H. P. (1973). Insecticidal activity of *Acarus calasnus* on stored grain insects. *Pesticides* **7**, 21–23.

ALAM, M. Z. (1971). *Pests of stored grains and other stored products and their control.* Agricultural Information Service, Dhaka, Bangladesh.

ALAM, M. Z. AND SHAHJAHAN, M. (1964). On the biology of rice moth, *Sitotroga cerealella* Oliv. (Galechiidae, Lepidoptera) in East Pakistan. In *A Review of Research: Division of Entomology (1947–1964)* (M. Z. Alam, A. Ahmed, S. Alam and A. Islam, Eds), pp. 128–137. Agricultural Information Service and East Pakistan Agricultural Research Institute, Dacca, Bangladesh.

ANONYMOUS (1984). Controlled atmospheres and fumigation in grain storages. *Proceeding of the International Symposium on Practical Aspects of CA and Fumigation in Grain Storages, Perth, 11–22 April 1983*. Elsevier, Amsterdam.

ANONYMOUS (1985). Manual of pest control for food security of grain stocks. *FAO Plant Production and Protection Paper*, No. 63.

ARIDA, G. (1974). *Biology of the rice weevil,* Sitophilus oryzae (L.) *on and preference for 2 corn varieties*. BS Thesis, University of the Philippines, Los Banos.

ARMITAGE, D. M. (1987). Controlling insects by cooling grain. In *Stored Products Pest Control* (T.J. Lawson, Ed.), pp. 219–227. British Crop Protection Council Monograph No. 37. The British Crop Protection Council, Thornton Heath.

ARNETT, R. H., JR (1973). *The Beetles of the United States (A Manual for Identification)*, 4th edition. The American Entomological Institute, Ann Arbor.

ARORA, G. L. AND PAJNI, H. R. (1957). Some observations on the biology and oviposition of *Bruchus analis* (Bruchidae: Coleoptera). *Research Bulletin of the Panjab University* **128**, 453–470.

ATWAL, A. S. (1976). *Agricultural Pests of India*. Kalyani Publishers, Delhi and Ludhiana.

ATWAL, A. S. AND DHALIWAL, G. S. (1971). The rate of increase of the population of *Trogoderma granarium* Evertson on different varieties of wheat. *Indian Journal of Entomology* **33**, 172–175.

BACK, E. A. AND COTTON, R. T. (1926). Biology of the saw-toothed grain beetle, *Oryzaephilus surinamensis*. *Journal of Agricultural Research* **33**, 435–452.

BALDOS, E. P. (1979). *The biology and external anatomy of* Doloessa viridis *(Zell.) and its susceptibility to insecticides*. MS Thesis, University of the Philippines, Los Banos.

BEAL, R. S. (1954). Synopsis of the economic species of *Trogoderma* occurring in the United States with description of a new species (Coleoptera: Dermestidae). *Annals of the Entomological Society of America* **49**, 559–566.

BEGUM, A. AND RAHMAN, S. (1974). Biology of *Callosobruchus analis* (F.). *Bangladesh Journal of Zoology* **2**, 171–178.

BELL, C. H. (1987). Effect of grain moisture content on the establishment and maintenance of a low oxygen atmosphere containing carbon dioxide. In *Stored Products Pest Control* (T. J. Lawson, Ed.), British Crop Protection Monograph No. 37, pp. 237–246. The British Crop Protection Council, Thornton Heath.

BENSON, J. F. (1973). The biology of Lepidoptera infesting stored products, with special reference to population dynamics. *Biological Review* **48**, 1–26.

BERNABE-ADALLA, C. AND BERNARDO, E. N. (1976a). Mechanism of resistance of ten maize varieties to the rice weevil complex. *Philippine Agriculturist* **69**, 112–120.

BERNABE-ADALLA, C. AND BERNARDO, E. N. (1976b). Correlation between husk characters and weevil infestation of 51 varieties and lines of maize in the field. *Philippine Agriculturist* **69**, 121–129.

BERNARDO, E. N. (1971). Varietal resistance of corn and sorghum to rice weevil, *Sitophilus oryzae* L. Presented at the 2nd annual conference on intensified corn production program. Los Banos College, Laguna, University of the Philippines, March 1971. Mimeograph.

BERNARDO, E. N. (1972). Varietal resistance of corn to the rice weevil, *Sitophilus oryzae* (L.). *Philippine Entomologist* **2**, 195–204.

BHUIYAN, I. M. AND QUINIONES, A. C. (1988). Biology of the corn weevil, *Sitophilus zeamais* Motsch. *University Journal of Zoology,* Rajshahi University **7**, 27–31.

BHUIYAN, I. M., ALAM, S. AND KARIM, A. N. M. R. (1988). Studies on the nature and extent of damage to stored rice caused by Rice Weevil *Sitophilus oryzae* (Linnaeus) and Angoumois Grain Moth *Sitotroga cerealella* and their control in Bangladesh. *Bangladesh Journal of Zoology* **16**, 99–109.

BOND, E. J. (1973). Chemical control of stored grain insects and mites. In *Grain Storage: Part of a System* (R. N. Sinha and W. E. Muir, Eds), pp. 137–179. The Anvil Publishing Company.

BOWER, J. H. (1974). Potential for genetic control of stored-product insect populations. *Proceedings of the 1st International Working Conference on Stored Products Entomology, Savannah, Georgia, USA*, pp. 167–180.

BRENDELL, M. J. D. (1975). *Handbook for the Identification of British Insects: Coleoptera (Tenebrionidae)*, volume 5, part 10. Royal Entomological Society of London.

BRITTON, E. B. (1970). Coleoptera. In *The Insects of Australia, A Textbook for Students and Research Workers* (I. M. Mackerras, Ed.). Melbourne University Press, Australia.

BROOKS, W. M. (1980). Production and efficacy of Protozoa. *Biotechnology and Bioengineering* **22**, 1415–1440.

BURGES, H. D. (1969). Control of insects by *Bacillus thuringiensis*. *Proceedings of the 5th British Insecticide and Fungicide Conference, Brighton, November*, pp. 405–411.

BURKHOLDER, W. E. (1981). Biomonitoring of stored-product insects. In *Management of Insect Pests with Semiochemicals* (R. Mitchell, Ed.), pp. 29–40. Plenum, New York.

BURKHOLDER, W. E. (1982). Reproductive biology and communication among grain storage and warehouse beetles. *Journal of the Georgia Entomological Society* **17** (Suppl. 2), 1–10.

BURKHOLDER, W. E. AND MA, M. (1985). Pheromones for monitoring and control of stored-product insects. *Annual Review of Entomology* **30**, 257–272.

BUSVINE, J. R. (1966). *Insects and Hygiene*, 2nd edition. Methuen, London.

CAMPION, D. G., HALL, D. R. AND PREVETT, P. F. (1987). Use of pheromones in crop and stored products pest management: control and monitoring. *Insect Science and its Application* **8**, 737–741.

CHAKRAVORTY, D. P. AND MATHEW, G. (1972). The effect of host plant nutrition on the susceptibility of seeds of several exotic wheat varieties to *Sitophilus oryzae* Lin. *Bulletin of Grain Technology* **10**, 116–119.

CHAMP, B. R. (1966). Insects and mites associated with stored products in Queensland. 4. Acarina and Pseudo-scorpions. *Queensland Journal of Agriculture and Animal Science* **23**, 197–210.

CHAMP, B. R. (1986a). Occurrence of resistance to pesticides in grain storage pests. In *Pesticides and Humid Tropical Grain Storage Systems* (B. R. Champ and E. Highley, Eds), pp. 229–255. Australian Centre for International Agricultural Research, Canberra.

CHAMP, B. R. (1986b). Pest resistance to pesticides. In *Ecology and Management of Food Industry Pests*. US Food and Drug Administration, Technical Bulletin No. 4.

CHAMP, B. R. AND DYTE, C. E. (1976). *Report of the FAO Global Survey of Pesticide Susceptibility of Stored Grain Pests*. FAO Plant Production and Protection Series, No. 5. FAO, Rome.

CHAMP, B. R. AND HIGHLEY, E. (1986). *Pesticides and Humid Tropical Grain Storage*

*Systems: ACIAR Proceedings No. 14.* Australian Centre for International Agricultural Research, Canberra.

CHEEMA, P. S. (1957). Studies on the bionomics of the case-bearing clothes moth, *Tinea pellionella* L. *Bulletin of Entomological Research* **47**, 167–182.

COCHRAN, D. G. (1982). *Cockroaches. Biology and Control.* WHO/VBC/82. **856**, 1–53.

COGBURN, R. R. BOLLICH, C. N., JOHNSTON, T. H. AND MCILARATH, W. O. (1980). Environmental influences on resistance to *Sitotroga cerealella* in varieties of rough rice. *Environmental Entomology* **9**, 689–693.

COTTON, R. T. (1956). *Pests of Stored Grain and Grain Products*, 2nd edition. Burgess, Minneapolis.

COX, A. (1981). Neem—A pesticide potential. *International Pest Control* **3**, 68–70; 78.

CROWSON, R. A. (1956). Coleoptera: Introduction and key to families *Handbook for the Identification of British Insects* **4**, 1–59.

CUSACK, P. D., EVANS, G. O. AND BRENNAN, P. A. (1975). A survey of mites of stored grain and grain products in the Republic of Ireland. *Scientific Proceedings of the Royal Dublin Society Series B* **3**, 273–329.

DAVEY, P. M. (1965). Insect pests of stored products in the tropics and the commodities and conditions in which they occur. *Tropical Stored Products Information* **10**, 377–386.

DAVIDSON, R. H. AND LYON, W. F. (1987). *Insect Pests of Farm, Garden, and Orchard*, 8th edition. John Wiley, New York.

DAVIES, R. G. (1949). The biology of *Laemophloeus minutus* Oliv. *Bulletin of Entomological Research* **40**, 63–82.

DAVIS, R. (1981). Stored-grain insects and their control. In *CRC Handbook of Transportation and Marketing in Agriculture. Vol. 2, Field Crops* (E. E. Finney, Jr, Ed.) CRC, Boca Raton.

DAWSON, P. S. (1977). Life history strategy and evolutionary history of *Tribolium* flour beetles. *Evolution* **31**, 226–229.

DEPADUA, D. H. (1974). *Post-harvest Rice Technology in Indonesia, Malaysia, The Philippines and Thailand.* International Development Research Centre. Ottawa, Canada.

DERMOTT, T. AND EVANS, D. E. (1978). An evaluation of fluidized-bed heating as a means of disinfesting wheat. *Journal of Stored Products Research* **14**, 1–12.

DIEKMANN, M. (1988). Control of storage pests. In *Quality Seed Production* (A. J. G. van Gastel and J. Kerley, Eds), pp. 73–81. International Centre for Agricultural Research in the Dry Areas, Aleppo, Syria.

DOBIE, P. (1974). The laboratory assessment of the inherent susceptibility of maize varieties to post-harvest infestation by *Sitophilus zeamais* Motsch. (Coleoptera, Curculionidae). *Journal of Stored Products Research* **10**, 183–197.

DOBIE, P. (1977). The contribution of the Tropical Stored Products Centre to the study of insect resistance in stored maize. *Tropical Stored Products Information* **34**, 7–22.

DOBIE, P. (1984). Biological methods for integrated control of insects and mites in tropical stored products. I. The use of resistant varieties. *Tropical Stored Products Information* **48**, 4–8.

EDWARDS, J. P. (1986). Practical developments in the use of juvenile hormone analogues for pest control. *Proceedings of the 7th British Pest Control Conference, 1986,* Session 3, Paper 1.

EDWARDS, J. P. AND MENN, J. J. (1981). The use of juvenoids in insect pest management. In *Chemie der Pflanzenschutz und Schadlingsbekampfungsmittel* (R. Wegler, Ed.), pp. 185–214. Springer Verlag, Berlin.

EDWARDS, J. P., SHORT, J. E. AND ABRAHAM, L. (1987). New approaches to the protection of stored products using an insect juvenile hormone analogue. In *Stored Products Pest Control* (T. J. Lawson, Ed.), British Crop Protection

Council Monograph No. 37, pp. 197–205. British Crop Protection Council, Thornton Heath.

EPINO, P. B. (1980). *Mechanism of resistance in mungbean,* Vigna radiata *(L) Witczek to* Callosobruchus chinensis *(L.).* MS Thesis, University of the Philippines, Los Banos, Laguna.

EVANS, D. E. (1981). Thermal acclimation in several species of stored grain beetles. *Australian Journal of Zoology* **29**, 483–492.

EVANS, D. E. (1987a). The influence of heating on the mortality of *Rhyzopertha dominica* (F.) (COLEOPTERA: BOSTRICHIDAE). *Journal of Stored Products Research* **23**, 73–77.

EVANS, D. E. (1987b). In *Integrated Pest Management* (A. J. Burner *et al.*, Eds), pp. 425–461. Academic Press, London.

FAJARDO, V. G. AND MORALLO-REJESUS, B. (1980). Effect of ten insect growth regulators on *Plodia interpunctella* Hubn. and *Corcyra cephalonica* Staint. *Philippine Entomologist* **4**, 169–176.

FREEMAN, J. A. (Ed.) (1980). *Common Insect Pests of Stored Food Products*, 6th edition. British Museum (Natural History), London.

GEORGHIOU, G. P. AND SAITO, T. (1983). *Pest Resistance to Pesticides.* Plenum, New York.

GIRST, D. H. (1975). *Rice*, 5th edition. Longman, London.

GOKHLE, V. C. (1973). Developmental compatibility of several pulses in Bruchidae. I. Growth and development of *Callosobruchus maculatus* (Fab.) on host-seeds. *Bulletin of Grain Technology* **11**, 28–31.

GOOD, N. E. (1933). Biology of the flour beetles, *Tribolium confusum* Duv. and *T. ferrugineum* Fab. *Journal of Agricultural Research* **46**, 327–334.

GORHAM, J. R. (1985). *Insect Pests of Stored Food in Latin America and the Caribbean Region.* Armed Forces Pest Management Board, Washington, DC. (Mimeographed)

GORHAM, J. R. (Ed.) (1990a). *Ecology and Management of Food-industry Pests.* US Food and Drug Administration, Technical Bulletin No. 4. Association of Official Analytical Chemists, USA, in press.

GORHAM, J. R. (Ed.) (1990b). *Insect and Mite Pests in Food: An Illustrated Key.* Association of Official Analytical Chemists, USA, in press.

GREELEY, M. (1980). Farm level post-harvest losses: the myth of the soft third option. In *Proceedings of the Post-production Workshop on Food Grains.* Bangladesh Council of Scientific and Industrial Research, Dacca.

GREENBERG, B. (1971). *Flies and Diseases, Vol. 1. Ecology, Classification and Biotic Associations.* Princeton University Press, Princeton.

GREENBERG, B. (1973). *Flies and Diseases, Vol. 2. Biology and Disease Transmission.* Princeton University Press, Princeton.

GRIFFITHS, D. A. (1960). Somc ficld habitats of mites of stored food products. *Annals of Applied Biology* **48**, 134–144.

GRIFFITHS, D. A., WILKIN, D. R., SOUTHGATE, B. J. AND LYNCH S. M. (1976). A survey of mites in bulk grain stored on farms in England and Wales. *Annals of Applied Biology* **82**, 180–184.

HADAWAY, A. B. (1956). The biology of the Dermestid beetle *Trogoderma granarium* Everts and *Trogoderma versicolor* (Creutz). *Bulletin of Entomological Research* **46**, 781–796.

HAINES, C. P. (1982). Pest management in stored products. *Protection Ecology* **4**, 321–330.

HAINES, C. P. (1984). Biological methods for integrated control of insects and mites in tropical stored products. III. The use of predators and parasites. *Tropical Stored Products Information* **48**, 17–24.

HALSTEAD, D. G. H. (1964). The separation of *Sitophilus oryzae* (L.) and *S. zeamais* Motschulsky (Col., Curculionidae), with a summary of their distribution. *Entomologist's Monthly Magazine* **99**, 72–74.

HALSTEAD, D. G. H. (1973). *A revision of the genus* Silvanus *Latreille (s.l.) (Coleoptera: Silvanidae)*. Bulletin of the British Museum (Natural History): Entomology, volume 2, No. 9.

HALSTEAD, D. G. H. (1980). A revision of the genus *Oryzaephilus* Ganglabauer, including descriptions of related genera (Coleoptera: Sylvanidae). *Zoological Journal of the Linnean Society* **69**, 271–374.

HALSTEAD, D. G. H. (1986). Keys for the identification of beetles associated with stored products. I. Introduction and key to families. *Journal of Stored Products Research* **22**, 163–203.

HARRIS, K. L. (1984). Postharvest grain losses in the developing world. *Cereal Foods World* **29**, 456–458.

HENRY, J. E. (1981). Natural and applied control of insects by protozoa. *Annual Review of Entomology* **26**, 49–73.

HILL, D. S. (1983). *Agricultural insect pests of the tropics and their control*, 2nd edition. Cambridge University Press, Cambridge.

HINTON, H. E. (1945). *A monograph of beetles associated with stored products*, volume 1. British Museum (Natural History), London.

HINTON, H. E. (1948). A synopsis of the genus *Tribolium* Macley with some remarks on the evolution of the species groups (Coleoptera, Tenebrionidae). *Bulletin of Entomological Research* **39**, 13–56.

HINTON, H. E. AND CORBET, A. S. (1972). *Common insect pests of stored food products, a guide to their identification*. Economic Series No. 15, 5th edition. British Museum (Natural History), London.

HODGES, R. J. (1979). *A review of the biology and control of the rice moth* Corcyra cephalonica *Stainton (Lepidoptera: Galleriidae)*. Report of the Tropical Products Institute, London, G125.

HODGES, R. J. (1984a). Biological methods for integrated control of insects and mites in tropical stored products. IV. The use of insect diseases. *Tropical Stored Products Information* **48**, 27–31.

HODGES, R. J. (1984b). Biological methods for the integrated control of insects and mites in tropical stored products. V. The use of sterile insects. *Tropical Stored Products Information* **48**, 33–36.

HODGES, R. J. (1986). The biology and control of *Prostephanus truncatus* (Horn) (Coleoptera: Bostrichidae): a destructive storage pest with an increasing range. *Journal of Stored Products Research* **22**, 1–14.

HOWE, R. W. (1952). The biology of the rice weevil, *Calandra oryzae* (L.). *Annals of Applied Biology* **39**, 168–180.

HOWE, R. W. (1956). The biology of two common storage species of *Oryzaephilus* (Coleoptera: Cucujidae). *Annals of Applied Biology* **44**, 341–355.

HOWE, R. W. (1957). A laboratory study of the cigarette beetle, *Lasioderma serricorne* (F.) (Col., Anobiidae) with a critical review of the literature on its biology. *Bulletin of Entomological Research* **48**, 9–58.

HOWE, R. W. (1965). Losses caused by insects and mites in stored foods and feeding stuffs. *Nutrition Abstracts Review* **35**, 285–293.

HOWE, R. W. AND BURGES, H. D. (1953). Studies on beetles of the family Ptinidae. IX – A laboratory study of the biology of *Ptinus tectus* Boield. *Bulletin of Entomological Research* **44**, 461–516.

HUGHES, A. M. (1976). *The Mites of Stored Food and Houses*. Ministry of Agriculture, Fisheries and Food Technical Bulletin 9. HMSO, London.

HUNTER, D. K. (1973). Viruses of stored-products insects and their potential as control agents. In *Some recent advances in insect pathology (Symposium at the Annual Meeting of the Entomological Society of America, Los Angeles, 1971). Miscellaneous Publications of the Entomological Society of America* **9**, 62–65.

HUNTER, D. K., COLLIER, S. J. AND HOFFMAN, D. F. (1975). Compatibility of malathion and the granulosis virus of the Indian Meal Moth. *Journal of Invertebrate Pathology* **25**, 389–390.

HUQ, A. K. F. (1980). Rice in Bangladesh: estimation of food losses in farm level storage. Paper presented at the Post-production Workshop on Food Grains. A BCSIR-IDS Collaborative Research Project, Dacca, Bangladesh.

HUYSMANS, G. (1982). Paper presented at the meeting of the Technical Committee on Agricultural Cooperation of the PTA for Eastern and Southern African States, Lusaka, Zambia.

HYDE, M. B., BAKER, A. A., ROSS, A. C. AND LOPEZ CESAR, O. (1973). *Airtight grain storage*. FAO Agricultural Services Bulletin No. 17.

JACOBSON, M. AND CROSBY, D. G. (1967). *Naturally Occurring Insecticides*. Marcel Dekker, New York.

JANARTHANAN, R. (1973). A note on the relative susceptibility of seeds of *Dolichos lablab* varieties to the pulse beetle, *Callosobruchus chinensis* L. *Bulletin of Grain Technology* **11**, 70–72.

JEFFREY, I. G. (1976). A survey of the mite fauna of Scottish farms. *Journal of Stored Products Research* **12**, 149–156.

JOSHI, G. P. (1974). The biology of *Oryzaephilus mercator* F. on different varieties of wheat. *Zeitschrift für Angewandte Entomologie* **77**, 121–126.

KAMEL, A. H. AND HASSENEIN, M. H. (1967). Biological studies on *Corcyra cephalonica* Stainton (Lepidoptera: Galleriidae). *Bulletin de la Société Entomologique d'Egypte* **51**, 175–198.

KAMEL, A. H., SALEH, M. R. A. AND BADAWI, A. (1976). The biology of *Cadra cautella* Walker, a pest of date fruits in the New Valley (Egypt) (Lepidoptera-Pyralidae). *Agricultural Research Review* **54**, 127–131.

KHAN, A. R. (1981). *The combined action of organophosphorus insecticides and Microsporidians on* Tribolium castaneum. Ph.D. Thesis, University of Newcastle upon Tyne.

KHAN, A. R. AND SELMAN, B. J. (1984). Effect of insecticide, microsporidian, and insecticide–microsporidian doses on the growth of *Tribolium castaneum* larvae. *Journal of Invertebrate Pathology* **44**, 230–232.

KHAN, A. R. AND SELMAN, B. J. (1987). Effect of pirimiphos methyl, *Nosema whitei* and pirimiphos methyl–*N. whitei* doses on the growth of *Tribolium castaneum* adults. *Journal of Invertebrate Pathology* **49**, 336–338.

KHAN, A. R. AND SELMAN, B. J. (1988). On the mortality of *Tribolium castaneum* adults from larvae sub-lethally treated with pirimiphos methyl, *Nosema whitei* and pirimiphos methyl–*N. whitei* doses. *Entomophaga* **33**, 377–380.

KHAN, A. R. AND SELMAN, B. J. (1989). *Nosema* spp. (Microspora: Microsporida: Nosematidae) of stored-product Coleoptera and their potential as microbial control agents. *Agricultural Zoology Reviews* **3**, 193–221.

KHAN, M. Q. (1949). A contribution to further knowledge of structure and biology of weevils *Sitophilus oryzae* and *Sitophilus granaria*. *Indian Journal of Entomology* **11**, 143–201.

KRANTZ, G. W. (1961). The biology and ecology of granary mites of the Pacific northwest. I. Ecological considerations. *Annals of the Entomological Society of America* **54**, 160–174.

KRANTZ, G. W. (1978). *A Manual of Acarology*, 2nd edition. Oregon State University Press, Corvallis.

KRISHNAMURTHY, K. (1975). Post harvest losses in food grains. *Bulletin of Grain Technology* **13**, 33–49.

LEFKOVITCH, L. P. (1964). The biology of *Cryptolestes pussiloides* (Steel & Howe) (Coleoptera, Cucujidae), a pest of stored cereals in the Southern Hemisphere. *Bulletin of Entomological Research* **54**, 649–656.

LESSARD, F. E. (1987). Control of storage insects by physical means and modified environmental conditions. Feasibility and Applications. In *Stored Products Pest Control* (T. J. Lawson, Ed.), British Crop Protection Council Monograph No. 37, pp. 209–218. The British Crop Protection Council, Thornton Heath.

LINDGREN, D. L. AND VINCENT, L. E. (1959). Biology and control of *Trogoderma granarium* Everts. *Journal of Economic Entomology* **52**, 312–319.

LINDGREN, D. L., VINCENT, L. E. AND KROHNE, H. E. (1951). The Khapra beetle, *Trogoderma granarium* Everts. *Hilgardia* **24**, 1–36.

LONGSTAFF, B. C. (1981). Biology of the grain pest species of the genus *Sitophilus* (Coleoptera: Curculionidae): a critical review. *Protection Ecology* **2**, 83–130.

MCFARLANE, J. A. (1989). *Guidelines for pest management research to reduce stored food losses caused by insects and mites.* Overseas Development and Natural Resources Institute Bulletin No. 22.

MCGAUGHEY, W. H. (1975). Compatibility of *Bacillus thuringiensis* and granulosis virus treatments of stored grain with four grain fumigants. *Journal of Invertebrate Pathology* **26**, 247–250.

MCGAUGHEY, W. H. (1976). *Bacillus thuringiensis* for controlling three species of moths in stored grain. *Canadian Entomologist* **108**, 105–112.

MCLAUGHLIN, R. E. (1971). Use of protozoans for microbial control of insects. In *Microbial Control of Insects and Mites* (H. D. Burges and N. W. Hussey, Eds), pp. 151–172. Academic Press, New York.

MALEK, M. A. AND PARVEEN, B. (1989). Effect of insect infestation on the weight loss and viability of stored B.R. 3 paddy. *Bangladesh Journal of Zoology* **17**, 83–85.

MALIT, F. (1973). *Varietal performance and behaviour of* Callosobruchus chinensis *(L.) on* Phaseolus aureus *Roxb.* Unpublished BS Thesis, University of the Philippines, Los Banos, Laguna.

MALLIS, A. (1982). *Handbook of Pest Control.* Franzak and Foster, Cleveland.

METCALF, C. L., METCALF, R. L. AND FLINT, W. P. (1962). *Destructive and Useful Insects: Their Habits and Control*, TMH edition. Tata McGraw-Hill Publishing, New Delhi.

MORALLO-REJESUS, B. AND JAVIER, P. A. (1978). Effect of ten insect growth regulators in two major coleopterous pests of stored grains. University of the Philippines, Los Banos, Laguna. Unpublished report.

MORRISON, E. O. (1964). Taxonomy of rice weevils, *Sitophilus oryzae* (L.) and *S. zeamais* Mots. and an annotated bibliography relevant to the ecology of the species. *Texas Journal of Science* **16**, 243–253.

MUGGLETON, J. (1987). Insecticide resistance in stored product beetles and its consequences for their control. In *Stored Products Pest Control* (T. J. Lawson, Ed.), British Crop Protection Council Monograph No. 37, pp. 177–186. The British Crop Protection Council, Thornton Heath.

MUNRO, J. W. (1966). *Pests of stored grains.* Hutchinson, London.

NAVARRO, S. AND JAY, E. G. (1987). Application of modified atmospheres for controlling stored grain insects. In *Stored Products Pest Control* (T. J. Lawson, Ed.), British Crop Protection Council Monograph No. 37, pp. 229–236. The British Crop Protection Council, Thornton Heath.

PAGLIARINI, N. (1979). Studies on the mites of stored cereals in Yugoslavia. In *Recent Advances in Acarology*, volume 1, pp. 305–309. Academic Press, New York.

PAGUIA, P. AND MORALLO-REJESUS, B. (1977). Morphogenetic and ovicidal effects of two natural and three synthetic juvenile hormones on corn borer, *Ostrinia furnacalis* Greene. *National Research Council of the Philippines Research Bulletin* **32**, 58–80.

PAJNI, H. R. AND BHOOKUL, N. (1984). Some aspects of the biology of *Trogoderma granarium* Everts (Dermestidae: Coleoptera). *Research Bulletin of the Panjab University, Science* **35**, 19–21.

PAJNI, H. R. AND JIT, S. (1979). Some observations on the biology of *Callosobruchus analis* (Fabr.) (Bruchidae: Coleoptera). *Research Bulletin of the Panjab University, Science* **27**, 141–146.

PAJNI, H. R. AND VIRK, N. (1982a). Observations on some aspects of the biology of *Tribolium castaneum* Herbst. *Research Bulletin of the Panjab University, Science* **33**, 155–158.

PAJNI, H. R. AND VIRK, N. (1982b). A note on the life-cycle of *Tribolium castaneum* Herbst (Coleoptera: Tenebrionidae). *Research Bulletin of the Panjab University, Science* **33**, 159–163.

PAJNI, H. R., ROSE, H. S. AND GILL, K. M. (1976). Some observations on the biology of *Corcyra cephalonica* (Stainton) (Lepidoptera: Galleriidae). *Research Bulletin of the Panjab University, Science* **27**, 223–224.

PANDEY, G. P. AND VERMA, B. K. (1977). Annona (custard apple) seed powder as protectant of mung against pulse beetle, *Callosobruchus maculatus* (Fabr.). *Bulletin of Grain Technology* **15**, 100–194.

PANDEY, N. D., SINGH, S. R. AND TEWARI, G. C. (1976). Use of some plant powders, oils and extracts as protectants against pulse beetle, *Callosobruchus chinensis* Linn. *Indian Journal of Entomology* **38**, 110–113.

PANDYA, A. C., ALI, N. AND ALAM, A. (1980). *Post-harvest technology in India.* Central Institute of Agricultural Engineering, Nabibagh, Bhopal, India.

PARK, T. (1934). Observations on the general biology of the flour beetle *Tribolium confusum. Quarterly Reviews of Biology* **9**, 36–54.

PARKIN, E. A. (1956). Stored product entomology. *Annual Review of Entomology* **1**, 223–240.

POTTER, C. (1935). The biology and distribution of *Rhizopertha dominica* (Fab.). *Transactions of the Entomological Society of London* **83**, 449–482.

PRAKASH, A., PASALU, I. C. AND MATHUR, K. C. (1982). Evaluation of plant products as grain protectants in paddy storage. *International Journal of Entomology, India* **1**, 75–77.

PREVETT, P. F. (1990). Stored-product entomology – the challenge of the next decade. *Bulletin of Entomological Research* **80**, 3–6.

PRUTHI, H. S. AND SINGH, M. (1950). *Pests of Stored Grain and their Control.* Indian Council for Agricultural Research Miscellaneous Bulletin No. 57.

QUINLAN, J. K. (1982). *Grain Protectants for Insect Control.* United States Department of Agriculture. Agricultural Research Service Marketing Bulletin 72.

RAHMAN, K. A. AND SAPRA, A. N. (1942). Studies on the grain pests in the Punjab. II. Biology of *Bruchus analis* Fab. and *Bruchus chinensis* Linn. (Bruchidae: Coleoptera). *Indian Journal of Agricultural Sciences* **12**, 851–864.

RAHMAN, K. A., SOHI, G. S. AND SAPRA, A. N. (1945). Studies on the grain pests in the Punjab *VI.* Biology of *Trogoderma granarium* Everts. *Indian Journal of Agricultural Sciences* **15**, 85–92.

RAINA, A. K. (1970). *Callosobruchus* spp. infesting stored pulses (grain legumes) in India and comparative study of their biology. *Indian Journal of Entomology* **32**, 303–310.

RAINA, A. K. (1971). Comparative resistance to three species of *Callosobruchus* in a strain of chickpea (*Cicer arietinum* L.). *Journal of Stored Products Research* **7**, 213.

RAJAGOPAL, D. AND BALASUBRAMANIAN, R. (1977). Biology of the khapra beetle, *Trogoderma granarium* Everts. *Current Research* **6**, 199–201.

RAJAK, R. L. AND PANDEY, N. D. (1965). A life-history study of the pulse beetle, *Callosobruchus chinensis* Linn. *Labdev. Journal of Science and Technology* **3**, 119–123.

RAJASEKARAN, B. AND KUMARASWAMI, T. (1985). Control of storage insects with seed protectants of plant origin. In *Behavioural and Physiological Approaches in Pest Management* (A. Ragupathy and S. Jayaraj, Eds), pp. 15–17. Tamil Nadu Agricultural University, Coimbatore, Tamil Nadu, India.

RATHORE, Y. S., BHATTACHARYA, A. K. AND SACHAN, G. C. (1980). Use of the concept of distance and group constellation for classifying the susceptibility of sorghum varieties to *Ephestia cautella* (Walker). *Journal of Stored Products Research* **16**, 39–42.

RICHARDS, O. W. AND DAVIES, R. G. (1977). *Imms' General Textbook of Entomol-*

*ogy*, 10th edition, volume 2: *Classification and Biology*. Chapman and Hall, London.

SARADAMMA, K., DALE, D. AND NAIR, M. R. G. K. (1977). On the use of neem seed kernel powder as a protectant for stored paddy. *Agricultural Research Journal of Kerala* **15**, 102–103.

SCHMUTTERER, H., ASCHER, K. R. S. AND REMBOLD, H. (1981). *Natural pesticides from the Neem tree* (Azadirachta indica) *A. Juss*. German Agency for Technical Cooperation, Erchborn, Germany.

SCHOONHOVEN, A. V. (1978). Use of vegetable oils to protect stored beans from bruchid attack. *Journal of Economic Entomology* **71**, 254–256.

SCHULTEN, G. G. M. (1982). Post harvest losses in tropical Africa and their prevention. *Food and Nutrition Bulletin* **4**, 2–9.

SEMPLE, R. L., CALIBOSO, P. M. AND SATIO, G. C. (1983). Reduction of losses with recent advances in pest control techniques. *Proceedings of the Seminar on Recent Advances in Food Protection and Preservation, 21 October 1983*, pp. 63–91. Philtrade Complex, Manila.

SHAHJAHAN, M. (1974). Extent of damage of unhusked stored rice by *Sitotroga cerealella* (Oliv.) (Lepidoptera: Galechiidae) in Bangladesh. *Journal of Stored Products Research* **10**, 23–26.

SHEJBAL, J. (1980). *Controlled Atmosphere Storage of Grains*. Elsevier, Amsterdam.

SIFNER, F., ZDAREK, J., HARDY, I. AND KALVODA, L. (1983). Pheromone traps for the management of phycittid moths. *Crop Protection* **2**, 463–472.

SIMMONS, P. AND ELLINGTON, G. W. (1925). The ham beetle *Necrobia rufipes* de Geer. *Journal of Agricultural Research* **30**, 845–863.

SINGH, H. (1965). *Pachymerus chinensis* Linn. – its biology and extent of damage to grain. *Plant Protection Bulletin* **16**, 23–28.

SINGH, J. AND TALLURI, B. R. (1972). Studies on the relative resistance of some varieties of pea (*Pisum sativum*) to *Callosobruchus chinensis* L. *Bulletin of Grain Technology* **10**, 37–40.

SINGH, K., AGARWAL, N. S. AND GIRISH, G. K. (1972). Studies on the loss in weight caused by *Sitophilus oryzae* Linn. to various high yielding varieties of wheat. *Bulletin of Grain Technology* **10**, 271–275.

SINHA, R. N. (1963). Stored product acarology in Canada. In *Advances in Acarology* (J. R. Naegele, Ed.), volume 1, pp. 70–78. Cornell University Press, Ithaca, New York.

SMITH, M. R. (1965). *House-infesting Ants of the Eastern United States*. United States Department of Agriculture Technical Bulletin No. 1326.

SOKOLOFF, A. (1974). *The biology of Tribolium with special emphasis on genetic aspects*, volume 2. Clarendon Press, Oxford.

SOUTHGATE, B. J. (1958). Systematic notes on species of *Callosobruchus* of economic importance. *Bulletin of Entomological Research* **49**, 591–599.

SOUTHGATE, B. J. (1979). Biology of the Bruchidae. *Annual Review of Entomology* **24**, 449–473.

STANLEY, M. S. M. (1965). *The development of the confused flour beetle* Tribolium confusum *Duval*. Ph.D. Dissertation, University of Utah.

STENMARK, A. (1978). Pyrethroids – an interesting group of insecticides. *Vaxtskydds-notiser* **42**, 87–95.

STOREY, C. L., SPEIRS, R. D. AND HENDERSON, L. S. (1979). *Insect Control in Farm-Stored Grain*. United States Department of Agriculture Farmer's Bulletin 2269.

TAUTHONG, P. AND WANLEELAG, N. (1981). Studies on the life of the southern cowpea weevil, *Callosobruchus chinensis* L. and its control. In *Pests of Stored Products*, BIOTROP Special Publication No. 9, pp. 87–98. Bogor, Indonesia.

TEOTIA, T. R. S. AND TEWARI, G. C. (1971). Dharek drupes and leaves as protectants against *Sitotroga cerealella* Oliv. infesting wheat seeds. *Bulletin of Grain Technology* **9**, 7–12.

TILAK, B. D. (1977). Pest control strategy in India. In *Crop Protection Agents – Their Biological Evaluation* (McFarlane, Ed.), pp. 99–109. Academic Press, London.

TSENG, Y. H. (1979). Studies on the mites infesting stored food products in Taiwan. In *Recent Advances in Acarology* (J.R. Rodriguez, Ed.), volume 1, pp. 311–316. Academic Press, New York.

TURNER, B. D. (1975). The Psocoptera of Jamaica. *Transactions of the Royal Entomological Society of London* **126**, 533–609.

TYLER, P. S. (1982). Misconception of food losses. *Food and Nutrition Bulletin* **4**, 21–24.

UNITED STATES DEPARTMENT OF AGRICULTURE (1979). *Stored-Grain Insects*. Agriculture Handbook 500. Washington, DC, USA.

VENKATRAMAN, T. V., MATHUR, U. K. AND LLANDER, R. (1962). Experiments on the possible use of *Bacillus thuringiensis* in the control of crop-pests. I. Tests with two formulations of *B. thuringiensis* against some insect pests. *Indian Journal of Entomology* **24**, 274–277.

VERMA, B. K. AND PANDEY, G. P. (1978). Treating of stored greengram seed with edible oils for protection from *Callosobruchus maculatus* (Fabr.). *Indian Journal of Agricultural Sciences* **48**, 72–75.

WATSON, D. L. AND BROWN, A. W. A. (1977). *Pesticide Management and Insecticides*. Academic Press, New York.

WATTERS, F. L. (1972). Control of storage insects by physical means. *Tropical Stored Product Information* **23**, 13–28.

WATTERS, F. L. AND SHUYLER, H. R. (1977). Control of post-harvest food losses. *FAO Plant Protection Bulletin* **25**, 184–188.

WILBUR, D. A. AND MILLS, R. B. (1985). Stored Grain Insects. In *Fundamentals of Applied Entomology* (R. E. Pfadt, Ed.), 4th edition, pp. 552–576. MacMillan, New York.

WILKIN, D. K. AND NELSON, G. (1987). Control of insects in confectionary walnuts using microwaves. In *Stored Products Pest Control* (T. J. Lawson, Ed.), British Crop Protection Council Monograph No. 37, pp. 247–254. The British Crop Protection Council, Thornton Heath.

WILLIAMS, G. C. (1964). The life-history of the Indian meal moth, *Plodia interpunctella* (Hübner). (Lepidoptera: Phycitidae) in a warehouse in Britain and on different foods. *Annals of Applied Biology* **53**, 459–475.

WRIGHT, V. F. AND SPILMAN, T. J. (1983). An annotated bibliography of *Prostephanus truncatus* (Horn) (Coleoptera: Bostrichidae): A pest of stored grain. *Tropical Stored Products Information* **46**, 25–30.

WYGODZINSKY, P. (1972). A review of the silverfish (Lepismatidae, Thysanura) of the United States and the Caribbean area. *American Museum Novitates* **2481**, 1–26.

ZDARKOVA, E. (1967). Stored food mites in Czechoslovakia. *Journal of Stored Products Research* **3**, 155–175.

4

# Current Approaches to Cassava Pest Management in Subtropical Africa

R. T. GAHUKAR

*Pest Control Consultant, Plot 220, Reshimbag, Nagpur 440009, India*

Introduction
Cassava mealybug
Bionomics—Natural enemies—Pest management
Other insect pests
Coleoptera—Heteroptera—Homoptera—Isoptera—
Lepidoptera—Orthoptera—Thysanoptera
Cassava green mite
Bionomics—Pest management
Red mites
Limitations to and perspectives of cassava pest management in traditional agriculture
Conclusions
Acknowledgements
References

## Introduction

Cassava, *Manihot esculenta* Crantz (*M. utilissima* Pohl.), is a principal staple food crop in the African subtropics. It is a major source of carbohydrates (tubers), proteins and vitamins (leaves) for about 200 million people (Sylvestre and Arraudeau, 1983). Tubers are harvested 9–15 months (or even 2 years as in Zaire) after planting and are consumed in fresh, fermented or dried forms. Leaves are dried and consumed just after harvest as a vegetable. Flour is used in confectionery, bread, biscuits, etc. In 1986, about 60.8 million tonnes of cassava was produced over an area of 7.68 million hectares (FAO, 1987). Major cassava-growing zones are situated in Cameroon, Ghana, Madagascar, Mozambique, Nigeria, Tanzania, Uganda and Zaire. Tuber yield varies from 3000 to 11 000 kg/ha. Cassava is a popular crop among low-income farmers because:

1. It is easy to manage (less labour and care).
2. Cassava is a secured source of calories since the crop can adapt to diverse climatic conditions and produces moderate yields even in low-fertility soils.

*Agricultural Zoology Reviews* — Vol. 4, February 1991
0269–0543 $20.00 + $0.00 © Intercept Ltd, PO Box 716, Andover, Hampshire SP10 1YG, UK

3. Tubers can be left in the ground for 2–3 years and harvested whenever needed.
4. The crop is used successfully in rotations every 2–5 years.

It may be grown as a monocrop or mixed with yam, corn, plantain, cocoyam, groundnut, banana, sweet potato, cowpea, beans and other vegetables (Okigbo, 1978). The crop is grown in areas receiving an annual rainfall of 750 mm and above, on various types of soil. Warm and moist conditions are most suitable for cassava cultivation. The length of wet and dry seasons varies according to climatic zones, viz. there is no pronounced dry season in equatorial rain forest; a wet season of 9–10 months and dry season of 2–3 months in guinea-savanna; and 7–9 months' wet season in sudano-guinea zones. Planting is done after the first rains. Cultural operations are mostly by hand and inputs such as fertilizers and pesticides are not applied.

Cassava is attacked by a number of pests in the field and during storage of dried chips. At present, the cassava mealybug, *Phenacoccus manihoti* Matile-Ferrero, and the cassava green mite, *Mononychellus tanajoa* (Bondar), are considered key pests, outbreaks of which have been reported frequently. Pest attack results in an average annual loss of 30% in root yield (Herren and Bennett, 1984) but this may reach as much as 50% (Herren, Lema and Neuenschwander, 1983), 65% (Walker, Heydon and Guthrie, 1985) or 70% (Korang-Amoakoh, Cudjoe and Adjaklde, 1987). Also, there is a loss in quality and quantity of leaves harvested, and the quality of planting material derived from the stems of the current crop deteriorates considerably. The variegated grasshopper, *Zonocerus variegatus* L., tobacco whitefly, *Bemisia tabaci* Genn., and the larger grain borer, *Prostephanus truncatus* Horn, are potential pests that are gaining economic importance in the region.

At present, farmers are not taking measures to prevent the spread of pests or to control them in fields or stores. Therefore, research on the development and application of control methods is continuing at national and regional levels so that the full potential of this crop can be realized. The information gathered on biological control agents during the past few years is extensive. The purpose of this paper is to review the literature on pest problems and to discuss the pros and cons of global pest management in traditional cassava production systems. This information may facilitate the exchange of genetic material and views of research workers in the region.

## Cassava mealybug

The cassava mealybug, *Phenacoccus manihoti* Matile-Ferrero (Pseudococcidae: Homoptera) is presumed to be of South American origin and was accidently introduced to Africa in the early 1970s on imported cassava cuttings. It was described as a new species by Matile-Ferrero (1977) and was first recorded in Zaire and the Congo. It spread rapidly to Senegal, Gambia, Nigeria, Benin, Sierra Leone, Gabon, Malawi and Angola and thereafter to south and east Africa (Ezumah and Knight, 1978; Leuschner, 1978b; Matile-Ferrero, 1978; Nwanze, 1978; Nwanze, Leuschner and Ezumah, 1979;

Akinlosotu and Leuschner, 1981; Boussienguet, 1986a; Neuenschwander and Herren, 1988). Pest outbreaks in 1976–77 and 1978–85 were usually associated with droughts (Hennessey and Muaka, 1987) but, by the end of 1986, the pest had spread at rates of up to 300 km/year to nearly 25 countries, covering about 70% of the African cassava belt situated between 15°N and 20°S (Neuenschwander and Herren, 1988).

Cassava mealybug is a key pest in Africa. The loss in tuber yield varies considerably between countries and years, e.g. 84% in Zaire (Nwanze, 1982), 10–100% in Ghana (Walker, Heydon and Guthrie, 1985), 9–75% (Schulthess, 1987) or 96% (Neuenschwander *et al.*, 1989) in the whole African cassava belt. This variation is attributed to the time of planting and tuber harvest, weather and field conditions. Ayanru (1987) compared the yields from insecticide-treated plots with those from untreated plots and estimated losses of 281 and 74 g per plant in the average dry weight of tubers and stems, or 10.6 and 10% of tuber and stem dry matter, respectively. Plants were taller (97.6 cm) in treated than in untreated plots (74.6 cm). When pest infestation level increases, the average harvest index declines (Neuenschwander *et al.*, 1989). However, losses caused by insects alone are difficult to distinguish from those caused by other insect pests and diseases. Losses due to *P. manihoti* attack tend to be higher during drought periods. Data on cassava acreage, production and yield/ha are not available for all countries of the cassava zone and there is no consensus among research and extension agents on loss figures. Therefore, an average loss of 30% in tuber yield has been considered appropriate by the Africa-wide Biological Control Programme (ABCP) and by Walker, Heydon and Guthrie (1985).

In addition to the major loss in tuber yield, there is partial or even complete defoliation. The use of infested cuttings as planting material has led to cassava disappearing from some regions (Neuenschwander and Herren, 1988). In Ghana, the price of food preparations and planting material, after mealybug outbreaks, rose by 9 and 21 times, respectively (Walker, Heydon and Guthrie, 1985). A similar situation exists in other countries of the region. Crop losses are less severe in rain forest areas than in savanna zones (Neuenschwander and Herren, 1988) but, during surveys in eastern Ghana and central Ivory Coast undertaken in 1986, Neuenschwander *et al.* (1989) did not find heavy pest attack. These observations suggest that more intensive surveys should be undertaken and efforts to reduce losses should be concentrated in semi-arid regions where the pest often has a patchy distribution.

## BIONOMICS

### *Host plants*

Several alternative host plants, though of minor importance, have been recorded; viz. a ceara rubber tree, *Manihot glaziovii*, and a tuberless perennial cassava hybrid, *M. esculenta* x *M. glaziovii*, in Zaire (Hennessey and Muaka, 1987); *Acalypha* sp., *Croton hirtus* and *Hibiscus* sp. in Sierra

Leone (James, 1987); okra, pepper, lettuce and wild spinach in Nigeria (Iheagwam, 1981a).

*Nature of damage*

Adults and nymphs feed continuously on the leaves and terminal shoots, resulting in deformed and stunted internodes and growing points (Neuenschwander *et al.*, 1989). Death of the shoot induces the production of new shoots which show bunchy top symptoms. Mealybugs migrate to leaves when shoots are destroyed and plants become weaker and are likely to wither. Sooty moulds often develop on plant parts as a result of honeydew accumulation and may impede photosynthesis. Ants feed on the honeydew and protect the mealybugs from natural enemies, mostly coccinellid beetles (Neuenschwander and Hammond, 1988). In the case of severe pest attack, complete defoliation occurs and tubers are small or not formed at all. Generally, a second-year crop on which the tubers develop is more susceptible to pest attack than a first-year crop (Nenon, 1985). Pest infestation increases when tubers are harvested at intervals and during the first part of the rainy season because of mobilization of carbohydrates from roots for new growth (Neuenschwander *et al.*, 1989).

Mealybug attack and defoliation decrease the content of tuber and stem carbohydrate, phosphorus, potassium, iron, zinc and magnesium, and increase the content of calcium in mature lamina, stem and tuber parenchymatous tissues, magnesium and sodium in roots, and nitrogen in roots and stems (Ayanru and Sharma, 1986; Ayanru, 1987).

*Pest biology and ecology*

About 600 eggs are laid by a mealybug. Egg-laying is more prolific in mealybugs feeding on petioles than in those feeding on leaves. Hatching takes place within a week and nymphs pass through three instars. 'Crawlers' disperse from the egg-laying site, because there are limited numbers of safe high-quality feeding sites, and settle down preferably on the underside of leaves, but later move to veins, stems and growing points (van Alphen *et al.*, 1989). Pinkish mealybugs, measuring 3–4 mm, are always covered with a white waxy secretion. Mealybugs on stem tips grow faster and have a higher lifetime fecundity than mealybugs on leaves (van Alphen *et al.*, 1989). A generation is completed in 3 weeks during the rainy season and 28–33 days in summer (Fabres, 1981a,c; Fabres and Boussienguet, 1981; Iheagwam and Eluwa, 1983; LeRu and Papierok, 1987).

Sampling procedures for determination of population density and infestation level have been outlined by Baumgaertner and Schulthess (1986), Boussienguet (1986b) and Schulthess, Baumgaerther and Herren (1989). Schulthess, Baumgaertner and Herren (1987) studied the effect of temperature, leaf age, crop cultivar and drought stress on net production rate (*R*o), intrinsic rate of increase ($r_m$) and generation time (*G*) of the cassava mealybug. LeRu and Papierok (1987) suggested the innate capacity of increase ($r_c$) instead of ($r_m$). Obviously, both $r_m$ and $r_c$ are based on development of immature stages, age-specific fecundity and survivorship of mealybugs, but $r_c$ is calculated on the basis of fortnightly readings. Therefore, daily observations are not required and a large number of experimental data can be obtained simultaneously.

*Factors affecting population abundance and distribution*

*Temperature.* The temperature threshold for mealybug development is 14.7°C (LeRu and Papierok, 1987). When mealybugs were reared at a range of temperatures between 20° and 35°C, they all died at 35°C before reaching the adult stage (LeRu and Papierok, 1987; Schulthess, Baumgaertner and Herren, 1987). Nsiama-She (1985) reported that $r_m$ values remained more or less constant between 28° and 32.5°C.

*Crop season.* Populations of mealybugs become dense in dry months as the temperature (>30°C) and relative humidity (>75%) are favourable for a rapid succession of pest generations (Iheagwam, 1981b; Iheagwam and Eluwa, 1983; LeRu and Fabres, 1987). Dispersal of immature stages is also faster, often resulting in further pest outbreaks (Fabres, 1981b).

*Rainfall.* During the rainy season, rains wash the insects from plants and kill them (Fabres, 1981c; Fabres and Boussienguet, 1981; Schulthess, Baumgaertner and Herren, 1987)

*Cassava cultivar.* Lema and Herren (1985a) recorded a high *R*o and short *G* values on an improved sweet cultivar TMS-300, whereas a low *R*o and the longest *G* values were noted on the local bitter cultivar Isunikankiyan by Schulthess, Baumgaertner and Herren (1987). The lowest $r_m$ value (0.157) was recorded on an improved bitter cultivar TMS-91934 and the highest $r_m$ (0.170) on local bitter cultivar Egedudu (Schulthess, Baumgaertner and Herren, 1987).

*Crop age.* Fabres and LeRu (1988) noted higher population densities on 2-year-old plants than on a 1-year plantation during the dry season. The reasons for this difference have not been elucidated.

*Plant canopy.* The movement of mealybugs on and between plants and between plant and ground may be important because natural enemies and wind could act differently at different levels on the plants and so change the population distribution (Fabres and LeRu, 1988).

*Leaf stage.* Young leaves are heavily attacked at the time of greatest pest abundancc. Also, an old colony on old leaves produces a colony which migrates to young leaves (Fabres and LeRu, 1988). Schulthess, Baumgaertner and Herren (1987) noted no difference in $r_m$ of mealybugs which were reared on leaves of different ages. However, the $r_c$ value was higher on old leaves than on fresh leaves because of the high fecundity of females and the high *Ro* values (LeRu and Papierok, 1987; Fabres and LeRu, 1988).

*Leaf fall.* The survival of mealybugs on leaves which have fallen to the ground is certainly lower than for those which remain on the plant.

*Water stress.* Water stress has little influence on life-table statistics of cassava mealybug (Schulthess, Baumgaertner and Herren, 1987) and moderate water stress in the plant often favours pest development (Nwanze, Leuschner and Ezumah, 1979). When plants were subjected to water stress (8.4% soil moisture) and infested with mealybugs, the $r_c$ was 0.175, compared to watered plants (11% soil moisture) on which the $r_c$ was 0.158, but water-stressed plants showed a high rate of pest mortality (Fabres and LeRu, 1988). Jones (1959) reported more cyanogenic heterosides in plants under water stress but on which mealybug growth capacity was greater.

*Plant nutrients.* Constituents of plant sap such as minerals, carbon, total nitrogen, amino acids, soluble sugars, etc. may affect insect abundance. Recently, Gutierrez *et al.* (1988a) mentioned the positive association of increased concentrations in the plant of cyanogenic glucosides (which are converted into toxic hydrocyanic acid) with higher levels of fertilization, whereas the effects of nitrogen deprivation were less pronounced than those of water stress; this is probably because the nitrogen content of leaves and other plant parts varies with the nitrogen content of the soil and affects pest dynamics via reduced fecundity, a longer development period and higher intrinsic mortality rates (Gutierrez *et al.*, 1988a).

## NATURAL ENEMIES

Among about 130 predators and parasitoids recorded on cassava mealybug, at least 20 species are common in Africa. Neuenschwander, Hennessey and Herren (1987) gave a list of natural enemies surveyed in the cassava belt, reporting 78.5% of the natural enemies for the first time; 33.1% were collected in west Africa, 36.9% in central Africa and 30% in both regions. The ABCP, based at the International Institute for Tropical Agriculture (IITA), Ibadan, Nigeria, and the Biological Control Programme, based at Cotonou, Benin, are doing further work on natural enemies (Herren, 1987).

### *Parasitoids*

*Indigenous parasitoids.* Few parasitoids have been found attacking cassava mealybug in central and west Africa, the major species being the encyrtids, *Anagyrus bugandaensis* Compere, *A. pseudococci* (Girault), *A. ? pullus* Compere and *Anagyrus* spp. (Fabres, 1981a; Boussienguet, 1986a; Neuenschwander, Hennessey and Herren, 1987; Neuenschwander and Hammond,

1988). These parasitoids occasionally attack the mealybug (Boussienguet, 1986a; Neuenschwander, Hennessey and Herren, 1987). Monitoring of *Anagyrus* spp. was done with the help of sticky traps in the Congo (Fabres, 1981b). Recently, an aphelinid, *Coccophagus* sp., has been recorded in Zaire (Neuenschwander, Hennessey and Herren, 1987) but no information has been gathered on its efficacy and frequency of occurrence.

*Introduced parasitoids.* A wasp, *Epidinocarsis (Apoanagyrus) lopezi* (De Santis) (Hymenoptera: Encyrtidae), which has a neotropical origin, was introduced into Nigeria in 1981 by ABCP. It is a solitary internal parasitoid specific to *P. manihoti*. Recently, Neuenschwander and Herren (1988) described the release, establishment and dispersal of *E. lopezi* in African countries. First releases were made in 1981 at IITA and in south-western Nigeria in 1982 (Lema and Herren, 1985b). By 1984, the parasitoid had established in 70–98% of fields in south-western Nigeria and up to the northern limit of regular mealybug distribution (Herren *et al.*, 1987). Later, about 50 releases in 34 regions in 12 countries were made and by July 1986 the parasitoid's establishment was noted from 16 countries covering an area of 750 000 $km^2$, viz. Angola, Benin, Cameroon, Congo, Gabon, Gambia, Ghana, Guinea Bissau, Ivory Coast, Malawi, Nigeria, Rwanda, Senegal, Togo, Zaire and Zambia (Herren *et al.*, 1987; Neuenschwander, Hammond and Hennessey, 1987b; Neuenschwander and Herren, 1988). By the end of 1986, it had reached about 25 countries (nearly 1.5 million $km^2$) covering 70% of the African cassava belt, which spreads over the sudano-savanna, guinea-savanna, equatorial rain forest and highlands (Neuenschwander and Herren, 1988).

Female parasitoids are attracted to mealybugs and also to infested leaves and uninfested leaves from partly infested cassava plants because of odours (synomones) emitted by infested plants (Nadel and van Alphen, 1986a, b). Males are better able than females to perceive infested plants from a distance (van Alphen *et al.*, 1989). This attraction facilitates the parasitoid's search for mealybug-bearing plants. After arrival, parasitoids are arrested by mealybug wax which acts as a contact kairomone (Langenbach and van Alphen, 1986). Adults mate as soon as they emerge. Females generally prefer older mealybugs for reproduction (Neuenschwander and Sullivan, 1987) and reject those which have failed to complete their development (Loehr *et al.*, 1988). Also, Kraaijeveld and van Alphen (1986) observed a weak rejection of already parasitized mealybugs by *E. lopezi* females when testing the suitability of host insects. The female parasitoid can exploit the second to fourth instars. Host feeding is more prevalent on younger stages (Neuenschwander and Sullivan, 1987) and parasitism is most successful on third and early fourth instars (Kraaijeveld and van Alphen, 1986; Loehr *et al.*, 1988). Early instars of mealybugs are preferred for male eggs, and old ones for female eggs (Kraaijeveld and van Alphen, 1986). Offering a greater number of hosts increases the rate of oviposition but the percentage of hosts parasitized decreases (Odebiyi and Bokonon-Ganta, 1986). At a constant temperature, no daily rhythmicity in egg-laying could be detected in the laboratory. Males do not form leks and no local mating structure exists. The number of eggs in ovarioles/female is positively correlated with the size of the female (van Alphen *et al.*, 1989).

In mealybugs into which the parasitoid lays eggs, black intrusions are formed. They originate first from encapsulation (cellular capsule formed by the host) and second from melanization (deposit of dark pigment). Generally, a single egg is laid at each oviposition attempt (Neuenschwander and Sullivan, 1987; Sullivan and Neuenschwander, 1988). Numerous hosts die without forming a mummy, which is the effect of mutilation and abortive parasitism (particularly at temperatures <30°C) (Neuenschwander and Sullivan, 1987). The stung mealybugs show retarded growth and low fecundity (Neuenschwander and Madojemu, 1986; Loehr *et al.*, 1988), and may die under difficult conditions. Thus, some mealybugs are killed through feeding on the haemolymph and host tissues by an adult wasp and due to wounds received during oviposition (Neuenschwander and Madojemu, 1986) and some mealybugs are killed due to successful development of parasitoid larvae (Neuenschwander and Sullivan, 1987).

After stinging a mealybug, a female parasitoid often turns and feeds on it. Feeding is concentrated on young instars which defend themselves by wriggling, but less vigorously than the older ones which defend by giving jerking movements of the abdomen (Neuenschwander and Sullivan, 1987). The killing power of the parasitoid is 2–3 times greater at high temperature than is apparent from the percentage of mummies formed (Neuenschwander and Sullivan, 1987). Mortality from adult feeding increases with increasing host densities (Loehr *et al.*, 1988). Starved females consume slightly more mealybugs than well-fed parasitoids at 30°C but their feeding is less important on old instars and increases only slightly with increased oviposition.

*Epidinocarsis lopezi* larvae pass through four instars. Development of male larvae is faster than that of female larvae. The immature parasitoid dies in some of the mealybugs having slow growth, and is usually encapsulated within the first 2 weeks of development; only about 12–16% of the total larval population is affected in this way (Neuenschwander and Sullivan, 1987). Live larvae which are encapsulated can often free themselves. During the first days of larval development, complete encapsulation of the parasitoid is not frequent but it is observed frequently in successfully parasitized mealybugs (Neuenschwander and Sullivan, 1987; Sullivan and Neuenschwander, 1988). Most melanization is limited to wounded tissues of the host. Partial encapsulation of the tail end of *E. lopezi* larva may sometimes happen but these larvae escape from encapsulation or from chemically or physically injured host tissues (Neuenschwander and Sullivan, 1987). Complete and fatal encapsulation of parasitoid larvae rarely occurs but supernumerary larvae are killed if the mealybug is stung several times (Neuenschwander and Sullivan, 1987; Nenon, Guyomard and Hemon, 1988). Thus, melanization of parasitoid larvae together with subsequent death or arrestment of development of the mealybug has been observed as late as 2 weeks after the parasitoid's first attack (Neuenschwander and Sullivan, 1987).

The pupal stage of *E. lopezi* is passed inside the mummified mealybugs.

Parasitoid survival is negligible in first-instar mealybug nymphs and the majority of females and males emerge from fourth- and second-instar nymphs, respectively (Kraaijeveld and van Alphen, 1986). Taking an average of 2, 6, 4 and 6 days for the development of egg, larva, prepupa and pupa, respectively, a generation takes 18 days, but this depends on the climatic conditions (Odebiyi and Bokonon-Ganta, 1986). At the optimum temperature of 27°C, the parasitoids pass through a generation in 2 weeks (Odebiyi and Bokonon-Ganta, 1986). The lower temperature threshold is 13.3°C (Lema and Herren, 1985a). Male adults are smaller than females, and big females live longer and reproduce better than small females (Kraaijeveld and van Alphen, 1986). Generally, an adult wasp lives 13–25 days and the sex ratio is 1 male : 3 females (Odebiyi and Bokonon-Ganta, 1986). Adult parasitoid longevity increases substantially at low temperatures (Ganga, 1984; Odebiyi and Bokonon-Ganta, 1986).

Techniques for the mass rearing and culture maintenance of *E. lopezi* (Haug *et al.*, 1987) and automated aerial releases of the parasitoids in fields (Herren *et al.*, 1987) have been developed at IITA.

*Hyperparasitoids.* At least 13 hymenopterous species attack mealybug parasitoids in central and west Africa (*Table 1*). The common hyperparasitoids are: *Prochiloneurus* spp., *Tetrastichus* spp. and *Chartocerus* spp. *Prochiloneurus insolitus* (Alam) is particularly abundant towards the beginning of the dry season. Boussienguet (1986b) reported 47% hyperparasitism by this hyperparasitoid on two African *Anagyrus* species in Gabon, whereas about 10 hyperparasitoids attack *E. lopezi* in Africa (Neuenschwander and Herren, 1988). Thus indigenous parasitoids seem to be attacked by fewer hyperparasitoids than are introduced parasitoids. The hyperparasitoids destroyed up to 50% of *E. lopezi* mummies during the first year of parasitoid establishment but parasitoids were still active on abundant mealybug populations in Nigeria (Neuenschwander and Herren, 1988), the Congo (Ganga, 1984), Ghana (Korang-Amoakoh, Cudjoe and Adjaklde, 1987) and Zaire (Hennessey and Muaka, 1987). Hyperparasitism proved to be positively density dependent (Neuenschwander, Hammond and Hennessey, 1987a). It is expected that a hyperparasitism rate of about 20% will persist in Nigeria for at least a few years, as long as *E. lopezi* populations remain low on stabilized low mealybug populations (Neuenschwander and Herren, 1988).

### *Predators*

*Indigenous predators.* Predators kill the mealybugs by mechanical injury and by preying upon them. About 40 predators are known to attack *P. manihoti* in Africa (Neuenschwander, Hennessey and Herren, 1987). The most common ones are mentioned in *Table 2*. The coccinellid beetles are the major group (>50% of total predators), followed by midges. Among coccinellids, *Exochomus flaviventris* Mader, *E. concavus* Fürsch, *E. flavipes* (Thunberg), *Hyperaspis pumila* Mulsant (*H. marmottani* Fairmaire), *H. senegalensis* Mulsant, *Nephus vetustus* Weise and *N. phenacoccophagus*

**Table 1.** Common hyperparasitoids of indigenous and introduced parasitoids attacking *Phenacoccus manihoti* in Africa

| Order, family, species | Host | Remark | References |
|---|---|---|---|
| Hymenoptera: Aphelinidae | | | |
| *Marietta javensis* (Howard) | *Epidinocarsis lopezi*<br>*Chartocerus* sp.<br>*Anagyrus* sp. | wide distribution but rare | Boussienguet (1986a); Neuenschwander, Hennessey and Herren (1987) |
| Hymenoptera: Encyrtidae | | | |
| *Cheiloneurus cyanonotus* Waterston | *Homalotylus flaminius*<br>*Epidinocarsis lopezi*<br>*Tetrastichus* sp. | wide distribution, 7–10% parasitism | Fabres (1981a); Neuenschwander, Hennessey and Herren (1987) |
| *Ooencyrtus* sp. | *Epidinocarsis lopezi* | | Neuenschwander, Hennessey and Herren (1987) |
| *Prochiloneurus aegyptiacus* (Mercet) | *Epidinocarsis lopezi*<br>*Anagyrus* sp. | | Fabres (1981a); Boussienguet (1986a)<br>Neuenschwander, Hennessey and Herren (1987) |
| *P. bolivari* Mercet | *Epidinocarsis lopezi*<br>*Anagyrus* sp. | | Nenon (1985); Neuenschwander, Hennessey and Herren (1987) |
| *P. insolitus* (Alam) | *Epidinocarsis lopezi*<br>*Anagyrus* sp. | wide distribution, abundant | Boussienguet (1986a); Neuenschwander, Hennessey and Herren (1987); Neuenschwander and Hammond (1988) |
| *P. pulchellus* Silvestri | *Anagyrus* sp. | 25% parasitism | Fabres (1981a) |
| Hymenoptera: Eulophidae | | | |
| *Tetrastichus* sp. | *Epidinocarsis lopezi*<br>*Anagyrus* sp. | abundant, solitary | Neuenschwander, Hennessey and Herren (1987) |

| | | | |
|---|---|---|---|
| Hymenoptera: Pteromalidae | | | |
| *Pachyneuron aphidis* (Bouché) | *Epidinocarsis lopezi* | | Neuenschwander, Hennessey and Herren (1987) |
| *P.* sp. nr *muscarum* (L.) | *Anagyrus* sp. | | Neuenschwander, Hennessey and Herren (1987) |
| *Pachyneuron* sp. | *Epidinocarsis lopezi*<br>*Anagyrus* sp. | rare | Boussienguet (1986a); Neuenschwander, Hennessey and Herren (1987) |
| Hymenoptera: Signiphoridae | | | |
| *Chartocerus subaeneus* (Forster) | *Epidinocarsis lopezi*<br>*Anagyrus* sp. | abundant, gregarious | Neuenschwander, Hennessey and Herren (1987) |
| *Chartocerus* sp. | *Epidinocarsis lopezi*<br>*Anagyrus* sp. | | Neuenschwander, Hennessey and Herren (1987) |

Fürsch, are active and abundant (Fabres, Kiyindou and Epouna-Mouinga, 1981; Kiyindou, 1986). Beetle development at different times of the year makes their action complementary (Fabres, 1981a; Boussienguet, 1986a). Fabres (1981b) monitored the populations of *E. flaviventris* and *H. senegalensis* with the help of sticky traps. Another beetle, *H. pumila*, completed its life cycle in a month (Umeh, 1982) and reduced the mealybug populations to about one-tenth in Nigeria during certain seasons (Umeh, 1983). However, indigenous coccinellids are not effective predators for the following reasons:

1. Whenever mealybug population density drops, there is a general decline in the number of *Exochomus* spp. and *Hyperaspis* spp. because they require numerous mealybugs for food, survival and oviposition and to meet their minimum respiration costs. Predators emigrate or die when mealybug populations are greatly reduced (Fabres and Kiyindou, 1985; Gutierrez *et al.*, 1988b; Neuenschwander and Hammond, 1988).
2. Beetle consumption is less than the ability of mealybugs to produce biomass (Gutierrez *et al.*, 1988b).
3. Beetles are generally active at the end of the dry season, when the crop has already been heavily damaged. Thus, beetles do not provide timely control (Hennessey and Muaka, 1987).
4. Beetles are not efficient at low pest density and react to mealybug odour only (van Alphen *et al.*, 1989).
5. Beetles attack all life stages of the mealybug (Gutierrez *et al.*, 1988b).

A midge, *Coccodiplosis citri* Barnes, is the most abundant single predator species and accounts for up to 52% of the total predator population in some countries (Boussienguet, 1986a). Another midge, *Dicrodiplosis manihoti* Harris, is also common but it remains at low population levels (Harris, 1981; Neuenschwander, Hennessey and Herren, 1987).

Other common predators, though of minor importance, include an anthocorid bug, *Cardiastethus exiguus* Poppius, (Fabres, 1981a; Boussienguet, 1986a) and a lycaenid caterpillar, *Spalgis lemolea* Druce (Fabres, 1981a; Iheagwam, 1981a; Boussienguet, 1986a; Neuenschwander, Hennessey and Herren, 1987; Neuenschwander and Herren, 1988). Chrysopids of frequent occurrence are *Ceratochrysa antica* Wlk. (Barnard and Brooks, 1984), *Anisochrysa boninensis* Okamoto (Boussienguet, 1986a) and *Chrysopa* spp. (Iheagwam, 1981a; Neuenschwander, Hennessey and Herren, 1987).

*Introduced predators.* An exotic coccinellid predator, *Hyperaspis raynevali* Mulsant (*H. jucunda* Mulsant) was introduced from South America into the Congo (Kiyindou and Fabres, 1987) and Nigeria (Nsiama-She, Odebiyi and Herren, 1984) where its biology has been studied in laboratory conditions [26–27°C and 70–75% relative humidity (RH)]. A female lays about 450 eggs. The incubation period lasts for 5 days, larval development through four instars is completed in 13.8 days and beetles emerge from pupae in 7 days. Beetles may live 100 days but each generation takes an average of 32.8 days (Nsiama-She, Odebiyi and Herren, 1984) and the sex ratio is typically 1.5 male : 1 female. The beetle multiplies by about 270 times in 69 days but

**Table 2.** The most common indigenous insect predators of *Phenacoccus manihoti* in Africa

| Order, family, species | Status | References |
|---|---|---|
| Coleoptera: Coccinellidae | | |
| *Exochomus flavipes* (Thunberg) group | major | Boussienguet (1986a); Neuenschwander, Hammond and Hennessey (1987); Neuenschwander, Hennessey and Herren (1987); Neuenschwander and Hammond (1988) |
| *E. promptus* Weise | minor | Neuenschwander, Hennessey and Herren (1987) |
| *Hyperaspis delicatula* Mulsant | minor | Neuenschwander, Hennessey and Herren (1987); Neuenschwander and Hammond (1988) |
| *H. pumila* Mulsant | major | Iheagwam (1981a); Neuenschwander, Hennessey and Herren (1987); Neuenschwander and Hammond (1988) |
| *H. senegalensis* Mulsant | minor | Boussienguet (1986a); Neuenschwander, Hennessey and Herren (1987) |
| *Nephus phenacoccophagus* Fürsch | major | Boussienguet (1986a); Neuenschwander, Hammond and Hennessey (1987) |
| *Scymnus kibonotensis* Weise | minor | Neuenschwander, Hennessey and Herren (1987) |
| Diptera: Cecidomyiidae | | |
| *Coccodiplosis citri* Barnes | major | Boussienguet (1986a) |
| *Dicrodiplosis manihoti* Harris | major | Neuenschwander, Hennessey and Herren (1987); Neuenschwander and Hammond (1988) |
| Hemiptera: Anthocoridae | | |
| *Cardiastethus exiguus* Poppius | minor | Fabres (1981a); Boussienguet (1986a); Neuenschwander, Hennessey and Herren (1987) |
| Lepidoptera: Lycaenidae | | |
| *Spalgis lemolea* Druce | minor | Fabres (1981a); Iheagwam (1981a); Boussienguet (1986a); Neuenschwander, Hennessey and Herren (1987); Neuenschwander and Hammond (1988) |

**Table 3.** Common indigenous insect parasitoids attacking predators of *Phenacoccus manihoti* in Africa

| Order, family, species | Host (stage attacked) | Remarks | References |
|---|---|---|---|
| Hymenoptera: Ceraphronidae | | | |
| *Aphanogmus* sp. | *Coccodiplosis citri* (pupa) | major, 40–60% parasitism during certain seasons | Boussienguet (1986a); Neuenschwander, Hennessey and Herren (1987) |
| Hymenoptera: Encyrtidae | | | |
| *Cheiloneurus cyanonotus* Waterston | *Exochomus flavipes* (larva) | minor, rare | Boussienguet (1986a) |
| *Coccidoctonus* sp. nr *trinidadensis* Crawford | *Coccodiplosis citri* (pupa) | minor, rare | Boussienguet (1986a); Neuenschwander, Hammond and Hennessey (1987); |
| | *Diadiplosis hirticornis* (pupa) | minor | Neuenschwander, Hennessey and Herren (1987) |
| *Homalotylus africanus* Tiberlake | Coccinellids (?) | major, often abundant | Neuenschwander, Hennessey and Herren (1987); Neuenschwander and Hammond (1988) |
| *H. flaminius* (Dalman) | *Exochomus flaviventris* (larva)<br>*Cheiloneurus* sp. (larva) | minor, 7–10% parasitism | Fabres (1981a); Neuenschwander, Hennessey and Herren (1987) |
| *H. quaylei* Timberlake | *Nephus phenacoccophagus* (larva)<br>*N. vetustus* (larva) | minor, 15–20% parasitism | Boussienguet (1986a); Neuenschwander, Hennessey and Herren (1987) |
| *Ooencyrtus* sp. | Coccinellids (?) | minor | Neuenschwander, Hennessey and Herren (1987) |
| *Prochiloneurus aegyptiacus* (Mercet) | Coccinellids (?) | minor | Neuenschwander, Hennessey and Herren (1987) |
| *Xyphigaster pseudococci* Risbec | *Coccodiplosis citri* (pupa)<br>*Dicrodiplosis manihoti* (pupa) | minor<br>minor | Fabres (1981a); Neuenschwander, Hennessey and Herren (1987) |

| | | | |
|---|---|---|---|
| Hymenoptera: Eulophidae | | | |
| *Pedobius* sp. nr *coddinellae* Kurdyumov | *Exochomus* sp. (pupa) | minor | Neuenschwander, Hennessey and Herren (1987) |
| *P.* sp. nr *marjoriae* Kerrich | *Exochomus* sp. (pupa) | minor | Neuenschwander, Hennessey and Herren (1987) |
| *Tetrastichus* spp. | Anisochrysa boninensis (pupa) | major, 20–23 adults/cocoon | Boussienguet (1986a) |
| | *Exochomus flavipes* (pupa) | major, 15–20 adults/cocoon | Neuenschwander, Hennessey and Herren (1987) |
| | *Dicrodiplosis manihoti* (?) | status not known | |
| | *Diadiplosis hirticornis* (?) | status not known | |
| Hymenoptera: Pteromalidae | | | |
| *Metastenus* sp. | *Exochomus* sp. | status not known | Umeh (1984) |
| | *Hyperaspis* spp. | status not known | Neuenschwander, Hennessey and Herren (1987) |

development is slow at temperatures below 25°C (Kiyindou and Fabres, 1987). There are two advantages of this beetle over indigenous predators:

1. its long life span of 100 days may prove a desirable character in predator–prey interaction; and
2. it starts laying eggs quite early (about 45 days after emergence).

However, its successful utilization seems unlikely because:

1. embryonic mortality is high at 20–30°C;
2. its development period increases with lower temperatures (e.g. 50 days at 20°C), which affects its growth index in the off-season (July–August) and slows the rate of population increase when mealybugs start increasing;
3. the monophagy of this beetle can result in male sterility in mass rearing;
4. the influence of food on egg development is not known;
5. it is not yet an established predator under African conditions (Gutierrez *et al.*, 1988b).

Therefore, detailed studies are necessary on its adaptation and effectiveness in mealybug control.

Exotic predatory coccinellids of the genus *Diomus* were released in Nigeria where they rapidly increased in number and reduced numbers of the pest within the first dry season (Herren, Lema and Neuenschwander, 1983) but the beetles did not return after the wet season (Hammond, Neuenschwander and Herren, 1987) because they failed to survive the rainy season (van Alphen *et al.*, 1989). However, in Zaire beetles persist during the wet season on a cassava hybrid, because leaves are continuously harvested and fresh shoots and leaves are always available to mealybugs for food. The leaves are horizontal and so provide protection against rains for the mealybugs (Hennessey and Muaka, 1987). Another factor is that the beetles react not only to the wax of *P. manihoti* as a contact kairomone but also to wax produced by another mealybug, *P. citri* Ckll., which is not a pest (van Alphen *et al.*, 1989).

*Parasitoids of mealybug predators.* The efficiency of the predators, both indigenous and introduced, of *P. manihoti* is hampered by about 14 hymenopterous parasitoids (*Table 3*). The rate of parasitism may be as low as 1.3–1.4% (Neuenschwander and Hammond, 1988) or as high as 40–60% (Boussienguet, 1986b). The reasons for this variation have not been studied. The major parasitoids are: *Homalotylus* spp. on coccinellids (Fabres, 1981a; Boussienguet, 1986a; Neuenschwander, Hennessey and Herren, 1987; Neuenschwander and Hammond, 1988), *Tetrastichus* spp. on chrysopids and coccinellids (Boussienguet, 1986a) and midges (Neuenschwander, Hennessey and Herren, 1987), and *Aphanogmus* sp. on *C. citri* (Boussienguet, 1986a; Neuenschwander, Hennessey and Herren, 1987).

### *Insect diseases*

In the Congo, infection of *P. manihoti* by a fungus, *Neozygites fumosa* (Speare) Remaudiere and Keller (*Entomophthora fumosa* Speare) was observed by LeRu, Silvie and Papierok (1985) and LeRu (1986). Because the rate of pest mortality was low, further studies are necessary to determine the usefulness of this fungus in mealybug control.

## PEST MANAGEMENT

### *Biological control*

Since the predators, insect diseases and indigenous parasitoids are not effective in controlling cassava mealybug populations, only releases of an exotic parasitoid, *E. lopezi*, have been made. In laboratory and field experiments, the parasitism level seldom exceeded 30% (Neuenschwander and Madojemu, 1986; Hammond, Neuenschwander and Herren, 1987; Biassangama, Fabres and Nenon, 1988). Also, according to Odebiyi and Bokonon-Ganta (1986), the age-specific life tables are not sufficient for evaluation and *E. lopezi* is probably not a good control agent. But there have been several advantages with the introduction of this parasitoid over other natural enemies:

1. It is host specific. The feeding of adults extends adult parasitoid life and thereby causes additional mealybug mortality. Its preference for later instars of *P. manihoti* may help to produce more parasitoids and may have the combined effects of not only killing mealybugs but also killing the developing ova in mealybugs that survive.
2. The parasitoid finds its food at a low density and reacts strongly to an increase in the pest population. It adapts its reproductive strategy to highly variable host density and population compositions. The parasitoid's adaptability and success at colonization, good host-finding capacity and its attraction to the odour of infested plants are all factors which are favourable to the parasitoid's survival on low and uneven populations of mealybug during the wet season (Loehr *et al.*, 1988).
3. Predators sometimes indiscriminately destroy the active mealybugs containing parasitoid larvae but *E. lopezi* persists, indicating that it stands up well to competition (Neuenschwander and Herren, 1988).
4. It is a good disperser because, while searching for *P. manihoti* wax, it is not arrested by the wax of other mealybugs. Therefore, it aggregates in patches of high pest density in the fields.
5. The parasitoid's intrinsic rate of increase is higher than that of cassava mealybug and is well synchronized with its host. However, a large variance in development time helps the parasitoid to bridge the host-free periods.
6. Hyperparasitism does not prevent *E. lopezi* being an efficient parasitoid of *P. manihoti*.

The following techniques have been used to evaluate the efficiency of *E. lopezi*.

*Physical exclusion of parasitoids.* The cassava tips were infested artificially with mealybugs and then covered with cages. Two months later, the mealybug densities were up to 7.0 times lower on tips covered with open cages than on tips in completely closed cages which did not allow the parasitoids to enter. On similarly infested but uncovered tips, the mealybug populations were up to 37.5 times lower and parasitism rates were higher (Neuenschwander, Schulthess and Madojemu, 1986).

*Chemical exclusion of parasitoids.* When a field artifically infested by cassava mealybugs was treated weekly with carbaryl sprays (1 kg a.i./ha), the peak mealybug population was 200 individuals/leaf with only 10% parasitism because the adult parasitoids were killed by insecticide. In an untreated field, parasitism was 25% and only 10 mealybugs/tip were noted (Neuenschwander, Schulthess and Madojemu, 1986).

*Monitoring of mealybug populations.* In south-western Nigeria, populations of mealybugs after two releases of *E. lopezi* showed slight differences in dynamics (Lema, Hennessey and Herren, 1984). Later, large-scale surveys during a two-year period following parasitoid release showed that the initial infestation rate of 88% at the end of the dry season decreased to 23% the next year. Ultimately, the parasitoid number per field declined but parasitoid density per infested tip remained the same in both years (Neuenschwander and Hammond, 1988). When the peaks of mealybug population density in some fields were missed during surveys due to sharp fluctuations (Hammond, Neuenschwander and Herren, 1987; Schulthess, 1987), the damage caused on growing tips had lower variances than density estimates. Similar observations on pest reduction have been reported from Malawi, Rwanda, Zambia, Zaire and Ghana (Neuenschwander and Herren, 1988) and other countries in a region covering about 200 000 km$^2$ (Hammond, Neuenschwander and Herren, 1987). Mealybug populations were significantly lower when *E. lopezi* had been present for more than half the growing season as compared to areas where the parasitoid was lacking or only recently introduced (Neuenschwander *et al.*, 1989). The impact of *E. lopezi* appeared to be greater on pest populations in savanna zones than forest areas (Neuenschwander *et al.*, 1989).

*Monitoring the populations of indigenous natural enemies.* Before the establishment of *E. lopezi* in the region, the indigenous predators formed about 98% of the total natural enemies but after *E. lopezi* establishment this proportion declined to 84–88% within a few years (Neuenschwander and Herren, 1988). Neuenschwander and Hammond (1988) observed the following after the release of *E. lopezi* in south-western Nigeria:

1. a decline in population densities of predatory beetles of the genera *Hyperaspis* and *Exochomus* and parasitoids of the genus *Anagyrus;*
2. no difference in abundance of the predatory midge, *D. manihoti;*
3. a decline in the rate of hyperparasitism, e.g. from 35.7% in December

1983 to 17.5% in December 1984, and from 43.8% in March 1983 to 30.7% in March 1984.

*Establishment of a simulation model.* A computer model was established to simulate cassava growth and development and included the effects of abiotic factors (rainfall, solar radiation, temperature, soil nitrogen, water) on population dynamics of cassava mealybug and pest–parasitoid–predator interactions (Gutierrez *et al.*, 1987, 1988a, b). Gutierrez *et al.* (1988b) concluded that, before the establishment of *E. lopezi*, rainfall and plant carbohydrate stress were the major pest population regulatory factors and that indigenous coccinellids (*Hyperaspis* spp. and *Exochomus* spp.) were less important. At present, *E. lopezi* represents the most important factor in the determination of pest mortality in the dry season, and rainfall the most important factor in the rainy season, and the contribution of predatory coccinellids may not be necessary.

*Comparison of benefits.* Cassava tuber yield losses due to cassava mealybug attack in the absence of *E. lopezi* were estimated at 463 g/plant in the savanna region. The parasitoid reduced average damage scores significantly by 1.55 and 0.45 units on a scale of 1–5 in the savanna and forest regions, respectively (Neuenschwander *et al.*, 1989). The increased yield due to parasitoid presence was 228 g/plant or about 2.48 tonnes/ha in the savanna (Neuenschwander *et al.*, 1989). It is speculated that the overall loss of 20% estimated in 1984–85 could fall to zero by 2003–04; thus, between a 50 and 90% reduction in cassava yield loss would have been achieved by the introduction of *E. lopezi* (Norgaard, 1988). A preliminary analysis indicated a benefit : cost ratio of 149 : 1 (Norgaard, 1988), which confirms the potential for global beneficial impact of biological control.

### *Cultural practices*

Using infested cuttings as planting material increases the chance of pest spread. Early plantings are attacked less because the tubers develop before the pests become abundant (James, 1987). Harvesting tubers before the beginning of the dry season for two-year-old plantations, and the removal of infested plants if the dry season is early (Fabres and LeRu, 1988), help to reduce contact between highly infested old fields and new uninfested fields.

### *Resistant cultivars*

Local cultivars in some countries show susceptibility to mealybug attack (James, 1987). Therefore, resistant cultivars (TMS-30572, TMS-91934, etc.) are being bred at IITA and in national programmes. This aspect needs the urgent attention of both entomologists and breeders.

*Chemicals*

As an immediate measure, *P. manihoti* has been controlled by insecticides applied as foliar sprays, viz. phosphamidon, monocrotophos, diazinon, parathion, malathion, dimethoate, methidathion, etc. Even though insecticides, in some instances, gave about a 50% increase in crop yield (Alu and Okeke, 1981a, b), their use is limited in subsistence farming due to their cost, dispersed crop cultivation and the use of tubers/leaves as food all the year round (Nenon, 1985).

*Integrated control*

Strict avoidance of transfer of infested cuttings and delayed plantings, associated with use of natural enemies, should reduce pest damage below economic threshold levels. In Nigeria, a combination of *E. lopezi* and *Diomus* sp. proved quite effective (Herren, Lema and Neuenschwander, 1983). Resistant cultivars may form one of the components in pest management once they become available.

## Other insect pests

### COLEOPTERA

The larger grain borer, *Prostephanus truncatus* Horn, is a recent introduction from Central America into Africa. Pest outbreaks had been noted during 1980–83 in Tanzania and Kenya (Dunstan and Magazine, 1981; Hodges *et al.*, 1983; Hodges, Meik and Denton, 1985; Makundi, 1987) and later in Togo, Benin and Ghana (Walter-Echols, Agounke and Akpalao, 1983; Harnish and Krall, 1984). Generally, the pest spreads through the transfer of infested commodities. The beetles feed on dried cassava which turns into powder. Adults prefer fermented cassava and cause about 74% loss during four months' storage, whereas unfermented tubers are attacked less and the equivalent loss incurred is 52% (Hodges, Meik and Denton, 1985). Shires (1980) and Hodges (1986) studied the biology and control of this insect. In optimum conditions (32°C and 80% RH), a female lays about 50 eggs. The mean development periods for eggs, larvae and pupae are 5, 25 and 5 days, respectively. Male beetles live for 45 days and females for 61 days. Preventive measures include avoidance of transfer of infested products, strict hygiene in stores, improvement in local storage structures and control practices and strict application of quarantine regulations (Makundi, 1987). Rees (1985) recorded a histerid beetle, *Teretriosoma nigrescens* Lewis, as a predator but its efficacy is unknown. If possible, warehouses may be disinfected with fumigants (phosphine, methyl bromide, etc.) as a curative measure.

Other storage pests attacking dried cassava are: *Rhizopertha dominica* Fb., *Lasioderma serricorne* Fb., *Araecerus fasciculatus* De Geer, *Sitophilus* sp. and *Tribolium* spp. (Sylvestre and Arraudeau, 1983).

HETEROPTERA

A coreid bug, *Pseudotheraptus devastans* Dist., is present in Zaire and the Congo. Adults and nymphs feed on leaves which become crumpled and later wither. There is a linear relationship between number of lesions and bug population and duration of bug feeding (Muimba-Kankolongo, Terry and Adeniji, 1987). These punctures/lesions favour the entry of *Colletotrichum gloeosporioides* Fenz., a pathogen responsible for anthracnose disease (Dubois and Mostade, 1973). The disease incidence is directly associated with leaf fall since the fungus infects cortical parenchyma and vascular tissues (Boher *et al.*, 1983). The biology and ecology of this bug on cassava have not been studied and no measures are undertaken to check the damage due to its limited occurrence.

The minor pests, *Mirperus torridus* Westw. and *Homoeocerus* sp. have been recorded occasionally in cassava fields.

HOMOPTERA

The tobacco whitefly, *Bemisia tabaci* Genn., is the most common whitefly species found in cassava fields. Adults and nymphs suck the sap from young leaves which become deformed and may eventually be shed. Also, whiteflies transmit cassava mosaic virus (Leuschner, 1978a; Dengel, 1981; Seif, 1981b; Bock, 1983). In Kenya, 70–80% of cassava mosaic incidence was due to whiteflies (Bock, 1983) because there exists a direct positive relationship between pest density and successful disease transmission (Seif, 1981b).

About 100 eggs (0.2 mm, pear-shaped) are laid by a female on leaf tissues. Hatching takes about a week. Young nymphs wander and settle down at one place on the underside of leaves and start feeding; larger nymphs are greenish-white, oval, scale-like, spiny insects. Total nymphal development takes 2–4 weeks. The adult is a 1 mm long, yellowish insect covered with a waxy white bloom and lives for 2–4 weeks. The pest is present all year round but is abundant in dry months. Pest population distribution is generally determined by the physiological stage of the host plant (Dengel, 1981). Relative humidity affects insect development and rainfall diminishes numbers (Seif, 1981a).

Variation in susceptibility to pest attack among cassava cultivars facilitated the identification of resistant sources in Kenya, viz. 5318.34, 46106 (Seif, 1981b). Among several coccinellids predating on *B. tabaci*, *Clitostethus neuenschwanderi* Fürsch, is a new record reported from Zaire (Fürsch, 1987). Similarly, an aphelinid parasitoid, *Eretmocerus* sp., is present in Zimbabwe (Taylor, 1981). The effectiveness of natural enemies in reducing pest damage has not yet been worked out. *Bemisia tabaci* is a cosmopolitan pest attacking several crops for which insecticides have generally been recommended.

Cassava is attacked by some mealybugs other than *P. manihoti*, viz. *P. madeirensis* Green, *Ferrisia virgata* Ckll. (Neuenschwander, Hennessey and Herren, 1987), and *P. citri* Ckll. (Fabres and Matile-Ferrero, 1980). A diaspid scale, *Aonidomytilus albus* (Ckll.), is sometimes found at a quite low density on cassava plants. The mealybugs have abundant parasitoids and predators. Neuenschwander, Hennessey and Herren (1987) reported a midge, *Kalodiplosis* sp. on *F. virgata* and two coccinellids, *Nephus phenacoccophagus* Fürsch and *N. flavimaculatus* Fürsch on *P. madeirensis*. Among the parasitoids, the encyrtids are the major ones controlling the populations of these mealybugs. For example, *F. virgata* by *Aenasius advena* Compere, *Aenasius* sp. nr *acuminatus* Kerrich, *Anagyrus pseudococci* (Girault), *Anagyrus* sp. nr *bugandaensis* Compere, *Anagyrus* sp., *Blepyrus insularis* (Cameron) and *Gyranusoides* sp.; *P. madeirensis* by *Anagyrus* spp., *Gyranusoides* sp. nr *litura* Prinsloo, *Leptomastix* sp. and *Blepyrus insularis*; *P. citri* by *Leptomastix dactylopii* Howard (Fabres and Matile-Ferrero, 1980). A platygasterid parasitoid, *Allotropa* sp., attacks *P. madeirensis* (Neuenschwander, Hennessey and Herren, 1987). No control measures are undertaken against these pests.

## ISOPTERA

Termites can infest cassava plantations in dry areas and feed on plant bases or tubers. In Nigeria and Ghana, six species have been reported by Malaka (1983), viz. *Macrotermes subhyalinus* Ramb., *M. bellicosus* Smeatn, *Amitermes evuncifer* Silv., *Odontotermes* sp., *Microtermes* sp., *Ancistrotermes* sp. Dust of aldrin, dieldrin, chlordane, heptachlor or other soil insecticides may be used in cases of heavy infestation.

## LEPIDOPTERA

Cassava plants are sometimes attacked by defoliators such as *Euproctis fasciata* Wlk. (Apeji, 1980) and *Spodoptera* spp. The damage is negligible and no attention is given to their control.

## ORTHOPTERA

The variegated grasshopper, *Zonocerus variegatus* Linn., is a polyphagous sporadic pest in west and central Africa (Godfrey-Sam-Aggrey and Bundu, 1978; Page, Harris and Youdeowei, 1980). Cassava leaves are eaten voraciously but plants may compensate by new growth of side shoots. Both nymphs and adults are gregarious in feeding. Its spread in central Africa might be related to the abundance of Laos weed, *Chromolaena odorata* L., which is widely distributed in the region. Page, Harris and Youdeowei (1980) estimated a 36% loss with natural infestation, 63% loss after six weeks of simulated defoliation and 49% loss after simulated debarking, but defoliation during the dry season did not produce any significant loss in tuber yield.

From March to May, each female lays about 300 eggs in the soil in masses of froth which harden to form sponge-like packets. Nymphs are sluggish, black, short-horned grasshoppers with yellow- or white-ringed appendages. After passing five instars, an adult measuring 3.5 cm is formed. It has a dark greenish body patterned with black, yellow and orange bands. The adult grasshopper lives for 3–4 months and a generation is completed in a year.

Tolerance in cassava cultivars to grasshopper attack, particularly first-instar nymphs, depends upon the hydrocyanic acid content of leaves (Chapman, Page and McCaffery, 1986). Poison baits and insecticide applications are the most effective remedies available at present. Modder (1986) suggested an integrated control system which consists of:

1. marking oviposition sites in April;
2. exposing egg pods in August–September;
3. spraying insecticides on roosting nymphs in November–December;
4. trapping nymphs and adults on Laos weed used as a bait.

In Madagascar, *Phymateus puniceus* Bolivar, attacks the crop during certain seasons and is a minor pest.

THYSANOPTERA

Deformation of leaves can be the result of sap sucking by thrips, viz. *Scirtothrips manihoti* Bondar in Uganda, *Nesothrips marshalli* Priesner (*Bolothrips marshalli* Priesner) in Sierra Leone and *Retithrips syriacus* Mayet in the Congo (Teng, 1987). Pest incidence remains low and they are not controlled.

## Cassava green mite

The green mite complex attacking the cassava crop in Africa may consist of several species (Byrne, Bellotti and Guerrero, 1983). At present, *Mononychellus tanajoa* (Bondar) *sensu lato* and *M. progresivus* Doreste (Tetranychidae: Acarina) are key pests, although they are also sometimes found on host plants related to cassava. Confusion on species predominance exists. The presence of only *M. progresivus* has been speculated by Gutierrez (1987) and that of *M. tanajoa* by Yaninek and Herren (1988). Thus, the cassava green mite may be referred to simply as *M. tanajoa* until distinct species can be demonstrated, as most researchers follow this nomenclature. The cassava green mite is an exotic pest of neotropical origin, introduced into Africa in the 1970s and first discovered in Uganda in 1971 (Nyiira, 1972, 1982a, b; Lyon, 1973, 1974). Since then it has spread to at least 27 countries. Yaninek (1988) and Yaninek and Herren (1988) gave information on the chronology of spread of this pest throughout the African cassava belt. Yield losses of 13–80% have been estimated (Nyiira, 1972, 1976, 1982a; Lyon, 1974; Shukla, 1976; Ndayiragije, 1984). This great variation in loss estimates is attributed to the non-standardization of methodology. Recently, Markham, Robertson and Kirkby (1987) conducted regional trials on a comparison of yield from insecticide-treated and untreated plots and reported a loss of 17–30% in Uganda, 10–31% in Burundi, 10–37% in Malawi and 5–40% in Kenya. This method did not, however, appear satisfactory due to the great variation in maturity period and planting time of the several cultivars used, heterogeneous soil conditions and severe attacks of diseases and other pests. Estimates of losses under controlled conditions may therefore be required. Apart from tuber loss, mite feeding reduces leaf chlorophyll content, increases the transpiration rate (Ayanru and Sharma, 1983) and affects the distribution of cyanide in the plant (Ayanru and Sharma, 1984). Mite-infested or defoliated plants have high contents of calcium in mature lamina or stem and tuber parenchymatous tissues, of magnesium and sodium in roots and of nitrogen in roots and stems; they have low contents of phosphorus, iron, potassium, magnesium and zinc (Ayanru and Sharma, 1984).

## BIONOMICS

### *Nature of damage*

Mites suck the sap from the undersurface of leaves, which show the symptoms of leaf chlorosis and become mottled. Young leaves and shoots may die, and branches resemble candlesticks. Flowers and petioles are rarely attacked (Nyiira, 1977). Weak plants produce short internodes and have slow-growing shoots and leaves (Nyiira, 1972), and have an increased incidence of root rot (Hahn *et al.*, 1980). Young leaves and young plants support greater mite populations than do mature leaves and plants (Nyiira, 1976; Yaninek, Herren and Gutierrez, 1987; Yaninek and Herren, 1988).

### *Pest biology and ecology*

During the life span of 3–4 weeks, a female lays 20–90 eggs singly on the underside of leaves. Hatching takes place in 3–4 days. The development of a six-legged larva, protonymph and deutonymph is completed in 1–2, 1–2 and 2–3 days, respectively. Each generation takes about 2 weeks (Nyiira, 1978). The mites disperse upon plants by walking, and within and between fields by aerial drifting. The major cause of mite spread is the widespread transportation and use of infested cuttings as planting material (Nyiira, 1972). Young buds favour mite survival but the pest can persist on cuttings stripped of all leaves and buds, and on leaves and stems used as organic refuse in the fields (Yaninek, 1988).

Yaninek, Gutierrez and Herren (1989) evaluated the effects of temperature, season, plant age and leaf age on development time, fecundity, survival and intrinsic rate of increase ($r_m$) of the mite on cassava using age-specific life-table statistics. Pre-adult development times were similar regardless of treatment. However, adult longevity, fecundity and $r_m$ were enhanced on young leaves of young plants. Patterns of survival and fecundity were similar at 24°, 27° and 31°C on a day-degree basis but lower at 20° and 34°C, while estimates of $r_m$ were similar at all temperatures except 34°C. A simple population model based on empirical $r_m$ values predicted a high population on host plants during the wet season. However, when rainfall-induced mortality was added, the population levels were similar to those observed in the field. Generally, the high $r_m$ is enhanced by high temperature and suitable host plants. In Nigeria, mite populations showed two peaks: the first peak in December (dry season) on plants subjected to severe drought stress and a second peak towards the onset of the wet season (April) when plants regenerate and favour mite growth (Yaninek, Herren and Gutierrez, 1987). This peak lasts until rainfall reduces their population (Yaninek and Animashaun, 1987). Regular observations on population dynamics in each country of the region are to be continued on a long-term basis.

### *Factors affecting mite population abundance and distribution*

*Crop season.* The pest is abundant in the dry season and its population declines in the wet season (Nyiira, 1972; Leuschner, 1980; Akinlosotu, 1982).

*Rainfall.* Rainfall is the major cause of mite mortality during the wet season (Lyon, 1974; Akinlosotu, 1982; Yaninek, Herren and Gutierrez, 1987, 1989; Yaninek, Gutierrez and Herren, 1989). Mortality follows a dose-response curve which increases from 0 to more than 50% at less than 5 mm rainfall per hour, then levels off at about 75% between 10 and 60 mm per hour (Yaninek and Animashaun, 1987). Strong rains may also be unfavourable to mite development through physiological changes in the host plant during the wet season, as demonstrated by Nyiira (1976).

*Temperature.* The temperature threshold for cassava green mite is 14.4°C (Yaninek, Gutierrez and Herren, 1989). The effect of temperature on mites is relatively constant regardless of season, particularly in lowland tropics, because the accumulation of day degrees for a given life stage is similar throughout the year (Yaninek and Animashaun, 1987). However, temperature has a more pronounced effect on mite populations as latitudes and elevations increase (Yaninek, Herren and Gutierrez, 1987).

*Relative humidity.* Atmospheric relative humidity shows a weak positive correlation with pest density (Akinlosotu, 1982).

*Radiation.* Mite population density is little affected by solar radiation (Akinlosotu, 1982).

*Host plant age and leaf stage.* Availability of new foliage is a primary factor limiting mite populations (Yaninek, Herren and Gutierrez, 1989). Pest growth rates are highest on young leaves of young plants, especially during the wet season, and populations decrease when suitable host plants are not available (Yaninek, Herren and Gutierrez, 1987). The mite population also increases on new leaf growth during the early dry season (Yaninek, Herren and Gutierrez, 1989).

*Water stress.* Increasing levels of defoliation due to drought stress bring about abrupt disruptions of mite populations (Gutierrez *et al.*, 1987). In fact, during drought or water stress periods, mites survive for 4–6 days (the period required for leaf buds to generate) and attack new shoots (Yaninek, 1988). Thus, plants affected by drought can harbour greater populations but their suitability for mite growth declines because:

1. the production of new foliage drops as the available soil moisture either evaporates or is used by the plant; and
2. the mite population density on the remaining leaves depends upon the age of leaves and the density-dependent effect of mite feeding activity on the nutrient content of the leaf (Yaninek, Herren and Gutierrez, 1987).

*Plant nutrients.* When the plants were deprived of water and external nutrients for two months, the mites survived and their populations fluctuated according to nitrogen availability, which normally depends on shed foliage cycling between plant and soil (Gutierrez *et al.*, 1988a). Low levels of nitrogen may slow population growth rates at all times of the year (Gutierrez *et al.*, 1987). Thus, pest density may be correlated with leaf quality, which is determined by the amount of nitrogen, phosphorus and, sometimes, potassium that they contain (Nyiira, 1977, 1982a, b).

*Natural enemies.* Mite populations are checked by indigenous predators, but these are rather opportunistic feeders, being efficient only at high mite density. The phytoseiid mites are important predators whose role in population reduction is discussed in the next section.

## PEST MANAGEMENT

### *Biological control*

Natural enemies regulate the mite populations in the neotropics. Therefore, a classical biological control programme was initiated in Africa (Girling, Bennett and Yaseen, 1978). ABCP undertook surveys of natural enemies and basic studies of field ecology and biology of the natural enemies and the pest (Megevand, Yaninek and Friese, 1987; Yaninek, Herren and Gutierrez, 1987). Likewise, techniques of colony maintenance for exotic predators, mass rearing, transportation and releasing and post-release surveys and simulation modelling of cassava ecosystems have been developed recently (Herren, Bird and Nadel, 1987; Haug *et al.*, 1987; Gutierrez *et al.*, 1987, 1988a). Predators are the major natural enemies of cassava green mite in Africa.

Surveys in eastern, central and western African countries indicated that the complex of indigenous predators consists of about 20 species (*Table 4*). The major predators are phytoseiid mites (*Euseius* spp.), coccinellid beetles (*Stethorus* spp.) and staphylinid beetles (*Holobus* spp.). These indigenous predators do not provide adequate mite control under present farming conditions (Markham, Robertson and Kirkby, 1987; Yaninek and Herren, 1988). Therefore, phytoseiid mites were introduced from South America. Releases of *Neoseiulus idaeus* Denmark and Muma, *N. anonymus* (Chant and Baker) and *Galendromus annectens* (Deleon) were made in 1984 in Nigeria, Ghana, Zaire, Rwanda, Zambia and Uganda. Either poor or no recoveries were made and no impact on pest population was detectable (Markham, Robertson and Kirkby, 1987; Megevand, Yaninek and Friese, 1987; Yaninek and Herren, 1988). Similarly, an introduction of staphylinid beetles [*Holobus*

**Table 4.** Indigenous mite and insect predators of *Mononychellus tanajoa* in Africa (based on Markham, Robertson and Kirkby, 1987; Yaninek and Herren, 1988)

| Order, family and species | Remarks |
|---|---|
| Acarina: Phytoseiidae | |
| *Euseius fustis* (Pritchard and Baker) | frequently found |
| *E. lokele* (Pritchard and Baker) | |
| *E. talinga* (Pritchard and Baker) | |
| *E. papayana* (van der Merwe) | |
| *Euseius* sp. | |
| *Neoseiulus teke* (Pritchard and Baker) | |
| *Typhlodromalus saltus* (Denmark and Matthysse) | |
| *Iphiseius degenerans* (Berlese) | abundant |
| Coleoptera: Coccinellidae | |
| *Stethorus aethiops* Weise | widespread |
| *S. jejunus* Casey | |
| *Stethorus* spp. | widespread but rare |
| Coleoptera: Corylophidae | |
| ? *Arthrolips* sp. | |
| Coleoptera: Lathridiidae | |
| *Melanophthalma* sp. | |
| Coleoptera: Staphylinidae | |
| *Holobus fageli* (Williams) | widespread, common |
| *H. pallidicornis* (Cameron) | widespread |
| Diptera: Cecidomyiidae | |
| *Arthrocnodax* sp. | widespread but rare |
| Hemiptera: Anthocoridae | |
| *Orius* sp. | possible predator |
| Thysanoptera: Thripidae | |
| *Frankliniella* sp. | possible predator |
| *Scolothrips* spp. | possible predator |

*(Oligota) minuta* Cameron] made in Kenya first in 1977 and later in 1981, did not result in establishment (Murphy, 1984) probably because they are effective only at high mite densities (Nyiira and Mutinga, 1977; Girling, Bennett and Yaseen, 1978; Murphy, 1984; Ingram, 1986). Also, exotic predators dispersed at a slower rate (maximum rate being 1600 km/year from Zaire to the Congo in 1975) than the cassava green mite (Yaninek and Herren, 1988). Therefore, large-scale and long-range releases would be necessary for better pest control. At present, five phytoseiid mites, viz. *Neoseiulus idaeus, N. anonymus, Euseius concordis* (Chant), *Galendromus annectens* and *Typhlodromalus (Amblyseius) limonicus* (Garman and McGregor), are being mass reared at IITA for future releases (Friese, Megevand and Yaninek, 1987).

### Cultural practices

As a preventive measure, the transportation of infested cuttings to new areas should be stopped by adopting strict legislation. This is virtually impossible under African conditions as the movement of people from one country to another is liberal. The planting period may be adjusted as 2 to 9-month-old plants are the most susceptible to pest attack (Nyiira, 1976, 1982a). Removal of infested leaves can be practised on a limited area (Lyon, 1974), though this practice induces lateral shoots and produces more leaves which are ultimately attacked (Yaninek and Herren, 1988). Gutierrez *et al.* (1988c) predicted a slow build-up of mite damage on nitrogen-poor soils but a combination of fertilizer with irrigation in arid areas could result in high pest densities. Therefore, a judicious use of fertilizers may be necessary.

### *Resistant cultivars*

Resistant sources, such as TMS-91934, TMS-4(2)-1425, TMS-30572, TMS-30001, have been identified at IITA (Yaninek, Herren and Gutierrez, 1989). The improved bitter cultivar TMS-30572 supports lower mite growth rates than the local sweet cultivar Isunikankiyan, mainly due to a lower fecundity of the cassava mite during the dry season (Yaninek, Herren and Gutierrez, 1987). This is a consequence of the antibiosis effect of cyanide in the bitter cassava leaves. Cultivars with hairy leaves show tolerance to mite attack (Hahn *et al.*, 1980). Intensive studies on the mechanisms of resistance are under way at several institutes.

Until appropriate and effective control methods are evolved, the most practical strategy is to plant early in the wet season and harvest as soon as possible.

## Red mites

Red spider mites are found in the colonies of green mites and cause similar crop damage. The red mites recorded on cassava are: *Eutetranychus monodi* André and *Oligonychus gossypii* Zacher in Senegal (Gutierrez and Etienne, 1981), *O. gossypii* in Nigeria (Ezulike and Odebiyi, 1985), *Tetranychus neocaledonicus* André in Madagascar (Chazeau and Gutierrez, 1974) and *T. lombardinii* Baker and Pritchard in Kenya (Van den Berg and Markham, 1987). The populations of these mites rarely reach damaging levels but they may provide food to the predators of green mites.

The red mites are controlled by the phytoseiid mites. Ezulike and Odebiyi (1985) studied the biology of *Amblyseius fustis* Pritchard and Baker, which is an indigenous predator of *O. gossypii* in Nigeria. Development from egg to adult takes a week and an adult mite lives for about 19 days. The sex ratio in progeny of mated females is 1 male : 1.4 females. *Amblyseius fustis* is a potentially useful predator and a good searcher for its prey. When predators were fed on different life stages of the host mite, there were no significant differences in longevity and fecundity of predators. *Amblyseius fustis* has a shorter development period and lives longer than its prey; these characters may offset the higher fecundity of *O. gossypii* and make it the better option for red mite control.

In Kenya, a neotropical phytoseiid mite, *Neoseiulus idaeus*, was introduced as a candidate predator for classical biological control. The predator consumed equal numbers of *M. tanajoa* and *T. lombardinii* in single-species populations, but showed a distinct preference for green mite when both species were presented together at the same density (Van den Berg and Markham, 1987). They developed and reproduced equally well for at least 10 weeks on both species. Another aspect of adaptation of this predator is its possible attraction to prey colonies by volatile kairomones and its ability to move within the prey webs. The presence of red mites is ignored in the field because they are minor pests of the cassava crop. If there were to be a change in the pest status of red mites in the future, use of the exotic predator *N. idaeus* might prove beneficial.

## Limitations to and perspectives of cassava pest management in traditional agriculture

Cassava is a family crop cultivated in a traditional manner without sophisticated crop management practices. It is mixed with other crop species of different maturity cycles according to the family need. It is, therefore, difficult to undertake weeding, inter-row hoeings, application of fertilizers, etc. Farmers use the planting material from neighbours' fields or their own plots without thought of pest attack. Tubers are not inspected for pests or disease until they are harvested for home consumption or sale in the market. Large cassava fields away from houses are not regularly visited by farmers to check for pest infestation unless other crops are standing in the same fields. Chemical control cannot be recommended to the African farmer due to the length of the crop's growing period; the high cost; scarcity of water during off seasons; non-availability of transport, pesticides and equipment in the region; poor knowledge of pesticide application techniques, toxicity and residue problems, etc. Pesticides, when used, are often stored in houses, where hazardous incidents sometimes occur. Difficulty in estimating the exact period of population peaks makes it hard to fix the ideal time for intervention. Knowledge of population monitoring techniques for pests and their natural enemies needs to be exploited and data on economic thresholds and economic injury levels for each pest gathered.

Biocontrol of mealybug and green mite has shown promising results with exotic predators and parasitoids, but long-term data on their recovery/establishment and yield data in different agroecosystems of cassava cultivation are needed. Initial high costs and provision of technical training are some of the problems which African countries may be able to meet through regional collaboration. Cassava-infested cuttings are transported from distant places in a country or even from outside the country since quarantine measures are not adequately respected. Some families choose mosaic-infected leaves for cooking, so it would be unwise to suggest control of whiteflies by application of chemicals. Extension services are poor or non-existent and pest control packages are not available except in cases of severe pest outbreaks when emergency campaigns are arranged.

Cultural practices such as optimum plant density, timely weeding, adjustment of planting and harvesting periods, use of uninfested cuttings, if properly followed, should help to reduce the pest infestation levels. Disinfection of cuttings in insecticide solutions before planting is an effective preventive measure. Resistant cultivars are being developed by international projects based at IITA and other national and regional programmes. Once available, the resistant and improved high-yielding cultivars may be distributed to farmers through government agencies by introducing price incentive schemes, credits, subsidies, etc.

The necessity, potential and perspectives of biocontrol of cassava mealybug and cassava green mite in Africa have been repeatedly stressed (Bennett and Greathead, 1978; Girling, Bennett and Yaseen, 1978; Akinlosotu, 1983; Bennett, 1986; Greathead, 1986; Ingram, 1986; Korang-Amoakoh, Cudjoe and Adjaklde, 1987). According to Norgaard (1988), *E. lopezi* has been quite successful because:

1. it was discovered in a short time at low cost;
2. its mass rearing is possible with simple techniques;
3. it appeared capable of permanent establishment over large areas;
4. it disperses effectively and rapidly;
5. good pest control was achieved under diverse climatic conditions; and
6. no institutional, economical or technological developments are necessary as it disperses itself.

Following this success, ABCP is presently engaged in the exploration of indigenous natural enemies and the enhancement of technical capacity to quarantine, to rear and to promote the safe and timely introduction of effective natural enemies (Herren, 1987). The Inter-African Phytosanitary Council of Organization of African Unity, based at Yaounde (Cameroon) is responsible for the administration of legislation. Demonstrations on farmers' fields by rural development agencies and ministries should form a regular part of the programme of agriculture departments and agencies in the areas of heaviest infestations so that recommended practices are communicated to and easily adopted by the farmers.

## Conclusions

The cassava mealybug and green mite, the key pests of cassava, presently occur in 31 out of 34 countries of the African cassava belt. They cause yield losses of up to 80% and are spreading fast to eastern and southern Africa.

Since mealybugs develop very quickly and high numbers can occur even in the presence of biocontrol agents after 2–3 generations, biological control should be combined with other methods such as varietal resistance, perhaps for more than one pest. Teng (1987) described the guidelines for strengthening the plant protection services in Africa. The Biological Control Programme regularly arranges practical training courses in biological control (Wodageneh, 1985). Integration of other disciplines (plant pathology, crop physiology, agronomy, plant breeding, weed science, economics, etc.) facilitates implementation of management strategies. The joint efforts of national, regional and international organizations (Wodageneh and Herren, 1987) would certainly help to reduce yield losses in cassava and increase food production in Africa.

**Acknowledgements**

I thank Drs P. Neuenschwander and J. S. Yaninek, Entomologists at the International Institute for Tropical Agriculture, Ibadan, Nigeria, and Dr G. Fabres, Entomologist, ORSTOM Centre, Montpellier, France, for their valuable suggestions and comments on the manuscript. Thanks are also due to the Biosystematics Laboratory of CIRAD, Montpellier, France, and CAB International Institute of Entomology, London, UK, for checking the scientific names of some insects.

## References

AKINLOSOTU, T. A. (1982). Seasonal trend of green spider mite, *Mononychellus tanajoa*, population on cassava, *Manihot esculenta*, and its relationship with weather factors at Moor plantation. *Insect Science and its Application* **3**, 251–254.

AKINLOSOTU, T. A. (1983). Studies on the control of cassava mealybug (*Phenacoccus manihoti*) and the green spider mite (*Mononychellus tanajoa*) in south-western Nigeria. *Journal of Root Crops* **9**, 33–43.

AKINLOSOTU, T. A. AND LEUSCHNER, K. (1981). Outbreak of two new cassava pests (*Mononychellus tanajoa* and *Phenacoccus manihoti*) in south-western Nigeria. *Tropical Pest Management* **27**, 247–250.

ALU, U. G. AND OKEKE, J. E. (1981a). Evaluation of insecticides for control of cassava mealybug (*Phenacoccus manihoti*). *Tropical Pest Management* **27**, 251–253.

ALU, U. G. AND OKEKE, J. E. (1981b). Effect of insecticide application on cassava yield and control of cassava mealybug (*Phenacoccus manihoti*). *Tropical Pest Management* **27**, 434–435.

APEJI, S. A. (1980). *Euproctis fasciata* Walker (Lep., Lymantridae) feeding on cassava in northern Nigeria. *Entomologist's Monthly Magazine* **116**, 92.

AYANRU, D. K. G. (1987). Effects of mealybug (*Phenacoccus manihoti*) infestation on cassava yield components and plant tissue quality. *Tropenland Wirt* **68**, 5–10.

AYANRU, D. K. G. AND SHARMA, V. C. (1983). Chlorophyll depletion in leaves of field grown cassava clones infested by cassava green spider mites, *Mononychellus tanajoa* (Bondar) (Acarina: Tetranychidae). *Tropical Agriculture (Trinidad)* **60**, 85–88.

AYANRU, D. K. G. AND SHARMA, V. C. (1984). Changes in total cyanide content of tissues from cassava plants infested by mite (*Mononychellus tanajoa*) and mealybug (*Phenacoccus manihoti*). *Agriculture, Ecosystems and Environment* **12**, 35–46.

AYANRU, D. K. G. AND SHARMA, V. C. (1986). Effects of mite (*Mononychellus tanajoa*) and mealybug (*Phenacoccus manihoti*) infestation on the mineral status of a cassava clone. *Agriculture, Ecosystems and Environment* **18**, 73–81.

BARNARD, P. C. AND BROOKS, S. J. (1984). The African lacewing genus *Ceratochrysa* (Neuroptera: Chrysopidae): a predator on the cassava mealybug, *Phenacoccus manihoti* (Hemiptera: Pseudococcidae). *Systematic Entomology* **9**, 359–371.

BAUMGAERTNER, J. AND SCHULTHESS, F. (1986). Development of sampling plans for *Phenacoccus manihoti* (Homoptera: Pseudococcidae) in cassava fields. In *Proceedings of the Workshop on Biological Control and Host Plant Resistance to Control the Cassava Mealybug and Green Mite in Africa, Ibadan, Nigeria, 6–10*

*December 1982* (H. R. Herren, R. N. Hennessey and R. Bitterli, Eds), pp. 128–134. IITA, Ibadan, Nigeria.

BENNETT, F. D. (1986). Biological control of mealybugs: past successes and safety records. In *Proceedings of the Workshop on Biological Control and Host Plant Resistance to Control the Cassava Mealybug and Green Mite in Africa, Ibadan, Nigeria, 6–10 December 1982* (H. R. Herren, R. N. Hennessey and R. Bitterli, Eds), pp. 15–24. IITA, Ibadan, Nigeria.

BENNETT, F. D. AND GREATHEAD, D. J. (1978). Biological control of mealybug, *Phenacoccus manihoti* Matile-Ferrero: prospects and necessity. In *Proceedings of the Cassava Protection Workshop, Cali, Colombia, 7–12 November 1977* (T. Brekelbaum, A. Bellotti and J. C. Lozano, Eds), pp. 181–194. CIAT, Cali, Colombia.

BIASSANGAMA, A., FABRES, G. AND NENON, J. P. (1988). Parasitisme au laboratoire et au champ d'*Epidinocarsis lopezi* (Hym., Encyrtidae), auxiliaire exotique introduit au Congo pour la régulation de l'abondance de *Phenacoccus manihoti* (Hom., Pseudococcidae). *Entomophaga* **33**, 453–465.

BOCK, K. R. (1983). Epidemiology of cassava mosaic disease in Kenya. In *Plant virus epidemiology: the spread and control of insect borne viruses* (R. T. Plumb and J. M. Thresh, Eds), pp. 337–347. Blackwell Scientific, Oxford.

BOHER, B., DANIEL, J. F., FABRES, G. AND BANI, G. (1983). Action de *Pseudotheraptus devastans* Distant (Het., Coreidae) et de *Colletotrichum gloeosporioides* Penz. dans le développement de chancres et la chute des feuilles chez le manioc (*Manihot esculenta* Cranz). *Agronomie* **3**, 989–993.

BOUSSIENGUET, J. (1986a). Le compléxe entomophage de la cochenille du manioc, *Phenacoccus manihoti* (Hom., Pseudococcidae) au Gabon. I. Inventaire faunistique et les relations trophiques. *Annales de la Société Entomologique de France (Nouvelle Série)* **22**, 35–44.

BOUSSIENGUET, J. (1986b). Etude de populations d'insectes. II. Méthodologie d'études des paramètres biologiques du compléxe entomophage de la cochenille du manioc, *Phenacoccus manihoti*. In *Compte rendu de l'atélier sur la recherche entomologique dans les ecosystémes forestiers africains, Abidjan, Ivory Coast, 25 November–6 December 1985* (G. Couturier, J. Boussienguet and H. Dosso, Eds), pp. 97–99. UNESCO, Paris.

BYRNE, D. H., BELLOTTI, A. C. AND GUERRERO, J. M. (1983). The cassava mites. *Tropical Pest Management* **29**, 378–394.

CHAPMAN, R. F., PAGE, W. W. AND MCCAFFERY, A. R. (1986). Bionomics of the variegated grasshopper, *Zonocerus variegatus*, in West and Central Africa. *Annual Review of Entomology* **31**, 479–505.

CHAZEAU, J. AND GUTIERREZ, J. (1974). Evolution des populations de *Tetranychus neocaledonicus* André (Acariens, Tetranychidae) et de trois de ses prédateurs sur manioc dans le sud-ouest de Madagascar. *Cahiers de l'ORSTOM, Série Biologie* **25**, 3–11.

DENGEL, H. J. (1981). Investigations on the incidence of adults of *Bemisia tabaci* (Genn.) on various cassava varieties. *Zeitschrift für Pflanzenkrankheiten und Pflanzenschutz* **88**, 355–366 (in German).

DUBOIS, J. AND MOSTADE, J. M. (1973). La maladie des cierges du manioc provoquée par *Pseudotheraptus devastans* Dist. *Bulletin d'information INERA (Zaire)* **1**, 2–13.

DUNSTAN, W. R. AND MAGAZINE, I. A. (1981). Tanzania: the larger grain borer on stored products. *FAO Plant Protection Bulletin* **29**, 80–81.

EZULIKE, T. O. AND ODEBIYI, J. A. (1985). Life history of *Amblyseius fustis* (Pritchard and Baker): an indigenous predator of the cassava red mite, *Oligonychus gossypii* (Zacher), in south-western Nigeria. *Insect Science and its Application* **6**, 193–197.

EZUMAH, H. C. AND KNIGHT, A. (1978). Some notes on the mealybug (*Phenacoccus manihoti* Mat.-Ferr.) incidence on manioc (*Manihot esculenta*) in Bas Zaire. In

*Proceedings of the Workshop on the Cassava Mealybug (*Phenacoccus manihoti *Mat.-Ferr.) (Pseudococcidae), Mvuazi, Zaire, 26–29 June 1977* (K. F. Nwanze and K. Leuschner, Eds), pp. 7–14. IITA, Ibadan, Nigeria.

FABRES, G. (1981a). Entomophagous insects associated with the cassava mealybug in Congo. *Tropical Pest Management* **27**, 145–146.

FABRES, G. (1981b). Première quantification du phénomène de gradation des populations de *Phenacoccus manihoti* Matile-Ferrero (Hom., Pseudococcidae) en République Populaire du Congo. *Agronomie* **1**, 483–486.

FABRES, G. (1981c). Bioécologie de la cochenille du manioc (*Phenacoccus manihoti*) (Homoptera: Pseudococcidae) en République Populaire du Congo. 2. Variations d'abondance et facteurs de régulation. *Agronomie Tropicale* **36**, 369–377.

FABRES, G. AND BOUSSIENGUET, J. (1981). Bioécologie de la cochenille du manioc (*Phenacoccus manihoti*) (Homoptera: Pseudococcidae) en République Populaire du Congo. 1. Cycle évolutif et paramètres biologiques. *Agronomie Tropicale* **36**, 82–89.

FABRES, G. AND KIYINDOU, A. (1985). Comparaison du potentiel biotique de deux coccinelles (*Exochomus flaviventris* et *Hyperaspis senegalensis hottentotta*, Col., Coccinellidae) predatrices de *Phenacoccus manihoti* (Hom., Pseudococcidae) au Congo. *Acta Oecologica, Oecologia Applicata* **6**, 339–348.

FABRES, G. AND LERU, B. (1988). Etude des relations plante-insecte pour la mise au point des méthodes de régulation des populations de la cochenille du manioc. In *Proceedings of the 7th Symposium of the International Society for Tropical Food Crops, Gosier, Guadeloupe, 1–6 July 1985*, pp. 563–577. INRA, Paris.

FABRES, G. AND MATILE-FERRERO, D. (1980). Les entomophages inféodés à la cochenille du manioc, *Phenacoccus manihoti* (Hom., Pseudococcidae) en République Populaire du Congo. I. Les composantes de l'entomocoénose et leurs inter-relations. *Annales de la Société Entomologique de France (Nouvelle Série)* **16**, 509–515.

FABRES, G., KIYINDOU, A. AND EPOUNA-MOUINGA, S. (1981). Les entomophages inféodés à la cochenille du manioc, *Phenacoccus manihoti* (Hom., Pseudococcidae), en République Populaire du Congo. 2. Etude morphologique comparative de trois éspèces dominantes de Coccinellidae (Col.). *Cahiers de l'ORSTOM (Série Biologie)* **44**, 3–8.

FAO (1987). *Production Yearbook*, volume 40. Food and Agriculture Organization, Rome.

FRIESE, D. D., MEGEVAND, B. AND YANINEK, J. S. (1987). Culture maintenance and mass production of exotic phytoseiids. *Insect Science and its Application* **8**, 875–878.

FÜRSCH, H. (1987). New African species of *Scymnini* (Coleoptera: Coccinellidae) as predators of cassava pests. *Revue de Zoologie Africaine* **100**, 387–394.

GANGA, T. (1984). Possibilitiés de régulation de la cochenille du manioc, *Phenacoccus manihoti* (Hom., Coccoidea, Pseudococcidae) par un entomophage exotique, *Epidinocarsis lopezi* De Santis (Hym., Encyrtidae), en République Populaire du Congo. Document, ORSTOM, Brazzaville, Congo.

GIRLING, D. J., BENNETT, F. D. AND YASEEN, M. (1978). Biological control of the green mite, *Mononychellus tanajoa* (Bondar) (Acarina: Tetranychidae), in Africa. In *Proceedings of the Cassava Protection Workshop, Cali, Colombia, 7–12 November 1977*, (T. Brekelbaum, A. Bellotti and J. C. Lozano, Eds), pp. 165–170. CIAT, Cali, Colombia.

GODFREY-SAM-AGGREY, W. AND BUNDU, H. S. (1978). Cassava production in Sierra Leone. *World Crops* **31**, 188–189.

GREATHEAD, D. J. (1986). Opportunities for biological control of insect pests in tropical Africa. *Revue de Zoologie Africaine* **100**, 85–96.

GUTIERREZ, J. (1987). The cassava green mite in Africa: one or two species (Acarina: Tetranychidae). *Experimental and Applied Acarology* **3**, 163–168.

GUTIERREZ, J. AND ETIENNE, J. (1981). Quelques données sur les acariens tetranychidae attaquant les plantes cultivées au Sénégal. *Agronomie Tropicale* **36**, 391–394.

GUTIERREZ, A. P., WERMELINGER, B., SCHULTHESS, F., BAUMGAERTNER, J. U., YANINEK, J. S., HERREN, H. R., NEUENSCHWANDER, P., LOEHR, B., HAMMOND, W. N. O. AND ELLIS, C. K. (1987). An overview of a systems model of cassava and cassava pests in Africa. *Insect Science and its Application* **8**, 919–924.

GUTIERREZ, A. P., WERMELINGER, B., SCHULTHESS, F., BAUMGAERTNER, J. U., HERREN, H. R., ELLIS, C. K. AND YANINEK, J. S. (1988a). Analysis of biological control of cassava pests in Africa. I. Simulation of carbon, nitrogen and water dynamics in cassava. *Journal of Applied Ecology* **25**, 901–920.

GUTIERREZ, A. P., NEUENSCHWANDER, P., SCHULTHESS, F., HERREN, H. R., BAUMGAERTNER, J. U., WERMELINGER, B., LOEHR, B. AND ELLIS, C. K. (1988b). Analysis of biological control of cassava pests in Africa. II. Cassava mealybug, *Phenacoccus manihoti. Journal of Applied Ecology* **25**, 921–940.

GUTIERREZ, A. P., YANINEK, J. S., WERMELINGER, B., HERREN, H. R. AND ELLIS, C. K. (1988c). Analysis of biological control of cassava pests in Africa. III. The cassava green mite, *Mononychellus tanajoa. Journal of Applied Ecology* **25**, 941–950.

HAHN, S. K., LEUSCHNER, K., EZEILO, W., CARPENTER, A. J., KHATIBU, A. I. AND CONSTANTIN, C. A. (1980). Resistance of cassava clones to cassava green mite, *Mononychellus tanajoa. Tropical Pest Management* **26**, 265–267.

HAMMOND, W. N. O., NEUENSCHWANDER, P. AND HERREN, H. R. (1987). Impact of the exotic parasitoid *Epidinocarsis lopezi* on cassava mealybug (*Phenacoccus manihoti*) populations. *Insect Science and its Application* **8**, 887–891.

HARNISH, R. AND KRALL, S. (1984). Togo: further distribution of the larger grain borer in Africa. *FAO Plant Protection Bulletin* **32**, 113–114.

HARRIS, K. (1981). *Dicrodiplosis manihoti* sp. n. (Diptera: Cecidomyiidae): a predator of cassava mealybug, *Phenacoccus manihoti* Matile-Ferrero (Homoptera: Coccidoea: Pseudococcidae) in Africa. *Annales de la Société Entomologique de France (Nouvelle Série)* **17**, 337–344.

HAUG, T., HERREN, H. R., NADEL, D. J. AND AKINWUMI, J. B. (1987). Technologies for the mass rearing of cassava mealybugs, cassava green mites and their natural enemies. *Insect Science and its Application* **8**, 879–881.

HENNESSEY, R. D. AND MUAKA, T. (1987). Field biology of the cassava mealybug, *Phenacoccus manihoti*, and its natural enemies in Zaire. *Insect Science and its Application* **8**, 899–903.

HERREN, H. R. (1987). The Africa-wide biological control programme to control cassava mealybugs and cassava green mites: a review of objectives and achievements. *Insect Science and its Application* **8**, 837–840.

HERREN, H. R. AND BENNETT, F. D. (1984). Cassava pests: their spread and control. In *Proceedings of the Conference on Advancing Agricultural Production in Africa, Arusha, Tanzania, 12–18 February 1984* (D. L. Hawksworth, Ed.), pp. 110–114. CAB, Slough.

HERREN, H. R., BIRD, T. J. AND NADEL, D. J. (1987). Technology for automated aerial release of natural enemies of the cassava mealybug and cassava green mite. *Insect Science and its Application* **8**, 883–885.

HERREN, H. R., LEMA, K. M. AND NEUENSCHWANDER, P. (1983). Biological control of the mealybug, *Phenacoccus manihoti*, and the green spider mite complex, *Mononychellus* spp., on cassava, *Manihot esculenta*, in Africa. In *Proceedings of the 10th International Congress of Plant Protection, London, 20–25 November 1983*, p. 782. British Crop Protection Council, London.

HERREN, H. R., NEUENSCHWANDER, P., HENNESSEY, R. D. AND HAMMOND, W. N. O. (1987). Introduction and dispersal of *Epidinocarsis lopezi* (Hym., Encyrtidae), an exotic parasitoid of the cassava mealybug, *Phenacoccus manihoti* (Hom., Pseudococcidae), in Africa. *Agriculture, Ecosystems and Environment* **19**, 131–144.

HODGES, R. J. (1986). The biology and control of *Prostephanus truncatus* (Horn) (Coleoptera: Bostrichidae): a destructive storage pest with an increasing range. *Journal of Stored Products Research* **22**, 1–14.

HODGES, R. J., MEIK, J. AND DENTON, H. (1985). Infestation of dried cassava (*Manihot esculenta* Crantz) by *Prostephanus truncatus* (Coleoptera: Bostrichidae). *Journal of Stored Products Research* **21**, 73–77.

HODGES, R. J., DUNSTAN, W. R., MAGAZINE, I. AND GOLOB, P. (1983). An outbreak of *Prostephanus truncatus* Horn (Coleoptera: Bostrichidae) in East Africa. *Protection Ecology* **5**, 183–194.

IHEAGWAM, E. U. (1981a). Natural enemies and alternate host plants of the cassava mealybug, *Phenacoccus manihoti* (Homoptera: Pseudococcidae), in south-eastern Nigeria. *Revue de Zoologie Africaine* **95**, 433–438.

IHEAGWAM, E. U. (1981b). The influence of temperature on increase rate of the cassava mealybug, *Phenacoccus manihoti* (Hom., Pseudococcidae). *Revue de Zoologie Africaine* **95**, 959–967.

IHEAGWAM, E. U. AND ELUWA, M. C. (1983). The effects of temperature on the development of the immature stages of the cassava mealybug, *Phenacoccus manihoti* Mat.-Ferr. (Homoptera: Pseudococcidae). *Deutsche Entomologische Zeitschrift* **30**, 17–22.

INGRAM, W. R. (1986). Potential for the biocontrol of green cassava mites in Africa. In *Proceedings of the Conference on Biological Control and Host Plant Resistance to Control the Cassava Mealybug and Green Mite in Africa, Ibadan, Nigeria, 6–10 December 1982* (H. R. Herren, R. N. Hennessey and R. Bitterli, Eds), pp. 103–115. IITA, Ibadan, Nigeria.

JAMES, B. D. (1987). The cassava mealybug, *Phenacoccus manihoti* Mat.-Ferr. (Hemiptera: Pseudococcidae), in Sierra Leone: a survey. *Tropical Pest Management* **33**, 61–66.

JONES, W. O. (1959). *Manioc in Africa*. Stanford University Press, Stanford, USA.

KIYINDOU, A. (1986). Utilisation des coccinelles prédatrices locales pour la lutte contre la cochenille au Congo. In *Compte rendu de l'atelier sur la recherche entomologique dans les ecosystemes forestiers africains, Abidjan, Ivory Coast, 25 November–6 December 1985* (G. Couturier, J. Boussienguet and H. Dosso, Eds), pp. 133–135. UNESCO, Paris.

KIYINDOU, A. AND FABRES, G. (1987). Etude de la capacité d'accroissement chez *Hyperaspis raynevali* (Col., Coccinellidae) prédateur introduit au Congo pour la régulation des populations de *Phenacoccus manihoti* (Hom., Pseudococcidae). *Entomophaga* **32**, 181–189.

KORANG-AMOAKOH, S., CUDJOE, R. A. AND ADJAKLDE, R. K. (1987). Biological control of cassava pests in Ghana. *Insect Science and its Application* **6**, 905–907.

KRAAIJEVELD, A. R. AND VAN ALPHEN, J. J. M. (1986). Host stage selection and sex allocation by *Epidinocarsis lopezi* (Hymenoptera: Encyrtidae), a parasitoid of the cassava mealybug, *Phenacoccus manihoti* (Homoptera: Pseudococcidae). *Mededelingen van de Faculteit voor Landbouwwetenschappen van de Rijksuniversiteit Gent* **51**, 1067–1078.

LANGENBACH, G. E. J. AND VAN ALPHEN, J. J. M. (1986). Searching behaviour of *Epidinocarsis lopezi* (Hymenoptera: Encyrtidae) on cassava: effect of leaf topography and a kairomone produced by its host, the cassava mealybug, *Phenacoccus manihoti*. *Mededelingen van de Faculteit voor Landbouwwetenschappen van de Rijksuniversiteit Gent* **51**, 1057–1065.

LEMA, K. M. AND HERREN, H. R. (1985a). Influence of constant temperature on population growth rates of the cassava mealybug, *Phenacoccus manihoti*. *Entomologia Experimentalis et Applicata* **38**, 165–169.

LEMA, K. M. AND HERREN, H. R. (1985b). Release and establishment in Nigeria of *Epidinocarsis lopezi*: a parasitoid of the cassava mealybug, *Phenacoccus manihoti*. *Entomologia Experimentalis et Applicata* **38**, 171–175.

LEMA, K. M., HENNESSEY, R. D. AND HERREN, H. R. (1984). The cassava mealybug front hypothesis: role of indigenous natural enemies. In *Tropical root crops: production and uses in Africa* (E. R. Terry, K. A. Oduru and F. Caveness, Eds), pp. 90–92. IDRC, Ottawa.

LERU, B. (1986). Etude de l'évolution d'une mycose à *Neozygites fumosa* (Zygomycetes, Entomophthorales) dans une population de la cochenille du manioc, *Phenacoccus manihoti* (Hom., Pseudococcidae). *Entomophaga* **31**, 79–89.

LERU, B. AND FABRES, G. (1987). Influence de la temperature et de l'hygrométrie relative sur la capacité d'accroissement et le profil d'abondance des populations de la cochenille du manioc. *Acta Oecologica, Oecologia Applicata* **8**, 165–174.

LERU, B. AND PAPIEROK, B. (1987). Taux intrinséque d'accroissement naturel de la cochenille du manioc, *Phenacoccus manihoti* Matile-Ferrero (Homoptère: Pseudococcidae): intérêt d'une méthode simplifiée d'estimation de $r_m$. *Acta Oecologica, Oecologia Applicata* **8**, 3–14.

LERU, B., SILVIE, P. AND PAPIEROK, B. (1985). L'entomophthorale *Neozygites fumosa*, pathogène de la cochenille du manioc, *Phenacoccus manihoti* (Hom., Pseudococcidae) en République Populaire du Congo. *Entomophaga* **30**, 23–29.

LEUSCHNER, K. (1978a). Whiteflies: biology and transmission of African mosaic disease. In *Proceedings of the Cassava Protection Workshop, Cali, Colombia, 7–12 November 1977* (T. Brekelbaum, A. Bellotti and J.C. Lozano, Eds), pp. 51–58. CIAT, Cali, Colombia.

LEUSCHNER, K. (1978b). Preliminary observations on the mealbybug (Homoptera: Pseudococcidae) in Zaire and a projected outline for subsequent work. In *Proceedings of the Workshop on Cassava Mealybug,* Phenacoccus manihoti *(Pseudococcidae), Mvuazi, Zaire, 26–29 June 1977* (K. F. Nwanze and K. Leuschner, Eds), pp. 15–19. IITA, Ibadan, Nigeria.

LEUSCHNER, K. (1980). Biology, ecology and control of the green spider mite and cassava mealybug. In *Proceedings of the Workshop on Cassava Production and Extension in Central Africa, Mbanza-Ngungu, Zaire, 19–22 May 1980* (H.C. Ezumah, Ed.), pp. 123–128. IITA, Ibadan, Nigeria.

LOEHR, B., NEUENSCHWANDER, P., VARELA, A. M. AND SANTOS, B. (1988). Interactions between the female parasitoid, *Epidinocarsis lopezi* (Hym., Encyrtidae) and its host, the cassava mealybug, *Phenacoccus manihoti* (Hom., Pseudococcidae). *Journal of Applied Entomology* **105**, 403–412.

LYON, W. F. (1973). A plant feeding mite, *Mononychellus tanajoa* (Bondar) (Acarina: Tetranychidae), new to the African continent threatens cassava (*Manihot esculenta* Crantz) in Uganda, East Africa. *Pest Articles and News Summaries* **19**, 36–37.

LYON, W. F. (1974). A green cassava mite recently found in Africa. *FAO Plant Protection Bulletin* **22**, 11–13.

MAKUNDI, R. H. (1987). The spread of *Prostephanus truncatus* (Horn) in Africa and measures for its control. *FAO Plant Protection Bulletin* **35**, 121–126.

MALAKA, S. L. O. (1983). Economic importance of termites: six case studies in Nigeria and Ghana. *Nigerian Field* **47**, 222–230.

MARKHAM, R. H., ROBERTSON, I. A. D. AND KIRKBY, R. A. (1987). Cassava green mite in East Africa: a regional approach to research and control. *Insect Science and its Application* **8**, 900–914.

MATILE-FERRERO, D. (1977). Une cochenille nouvelle nuisible au manioc en Afrique Equatoriale, *Phenacoccus manihoti* n. sp. (Homoptera: Coccoidea: Pseudococcidae). *Annales de la Société Entomologique de France (Nouvelle Série)* **13**, 145–152.

MATILE-FERRERO, D. (1978). Cassava mealybug in the People's Republic of the Congo. In *Proceedings of the Workshop on the Cassava Mealybug,* Phenacoccus manihoti *Mat.-Ferr. (Pseudococcidae), Mvuazi, Zaire, 26–29 June 1977* (K. F. Nwanze and K. Leuschner, Eds), pp. 29–46. IITA, Ibadan, Nigeria.

MEGEVAND, B., YANINEK, J. S. AND FRIESE, D. D. (1987). Classical biological control of the cassava green mite. *Insect Science and its Application* **8**, 871–874.

MODDER, W. W. D. (1986). An integrated pest management strategy for the African grasshopper, *Zonocerus variegatus. Nigerian Field* **51**, 41–52.

MUIMBA-KANKOLONGO, A., TERRY, E. R. AND ADENIJI, M. D. M. (1987). A disease like injury in cassava caused by *Pseudotheraptus devastans* Dist. (Heteroptera: Coreidae). *Tropical Pest Management* **33**, 35–38.

MURPHY, S. T. (1984). Biological control of the cassava green mite, *Mononychellus* spp. in East Africa. In *Proceedings of the Regional Workshop on Integrated Pest Management of Cassava Green Mite in East Africa* (D. J. Greathead, R. H. Markham, R. J. Murphy, S. T. Murphy and I. A. D. Robertson, Eds), pp. 55–61. CIBC, London.

NADEL, H. AND VAN ALPHEN, J. J. M. (1986a). The role of host and host plant odours in the attraction of a parasitoid, *Epidinocarsis lopezi* (Hymenoptera: Encyrtidae), to the habitat of its host, the cassava mealybug, *Phenacoccus manihoti. Entomologia Experimentalis et Applicata* **45**, 181–186.

NADEL, H. AND VAN ALPHEN, J. J. M. (1986b). The role of host and host plant odours in the attraction of a parasitoid, *Epidinocarsis lopezi* (Hymenoptera: Encyrtidae), to its host, the cassava mealybug, *Phenacoccus manihoti* (Homoptera: Pseudococcidae). *Mededelingen van de Faculteit voor Landbouwwetenschapen van de Rijksuniversiteit Gent* **51**, 1079–1086.

NDAYIRAGIJE, P. (1984). Cassava green mite (*Mononychellus tanajoa*) (Bondar) in Burundi. In *Proceedings of the Regional Workshop on Integrated Pest Management of Cassava Green Mite in East Africa* (D. J. Greathead, R. H. Markham, R. J. Murphy, S. T. Murphy and I. A. D. Robertson, Eds), pp. 67–73. CIBC, London.

NENON, J. P. (1985). La cochenille du manioc. *Phytoma* **367**, 54–55.

NENON, J. P., GUYOMARD, O. AND HEMON, G. (1988). Encapsulement des oeufs et des larves de l'hymenoptère encyrtidae, *Epidinocarsis (Anagyrus) lopezi*, par son hôte pseudococcidae, *Phenacoccus manihoti*: effet de la temperature et du superparasitisme. *Comptes Rendus de l'Academie des Sciences, Paris* **306**, 325–331.

NEUENSCHWANDER, P. AND HAMMOND, W. N. O. (1988). Natural enemy activity following the introduction of *Epidinocarsis lopezi* (Hymenoptera: Encyrtidae) against the cassava mealybug, *Phenacoccus manihoti* (Homoptera: Pseudococcidae) in south western Nigeria. *Environmental Entomology* **17**, 894–902.

NEUENSCHWANDER, P. AND HERREN, H. R. (1988). Biological control of the cassava mealybug, *Phenacoccus manihoti*, by the exotic parasitoid, *Epidinocarsis lopezi*, in Africa. *Philosophical Transactions of Royal Society of London (B)* **318**, 319–333.

NEUENSCHWANDER, P. AND MADOJEMU, E. (1986). Mortality of the cassava mealybug, *Phenacoccus manihoti* Mat.-Ferr. (Hom., Pseudococcidae) associated with an attack of *Epidinocarsis lopezi* (Hym., Encyrtidae). *Mitteilungen der Schweizerischen Entomologischen Gesellschaft* **59**, 57–62.

NEUENSCHWANDER, P. AND SULLIVAN, D. (1987). Interactions between the endophagous parasitoid, *Epidinocarsis lopezi* and its host, *Phenacoccus manihoti. Insect Science and its Application* **8**, 857–859.

NEUENSCHWANDER, P., HAMMOND, W. N. O. AND HENNESSEY, R. D. (1987a). Changes in the composition of the fauna associated with the cassava mealybug, *Phenacoccus manihoti*, following the introduction of the parasitoid, *Epidinocarsis lopezi. Insect Science and its Application* **8**, 893–898.

NEUENSCHWANDER, P., HAMMOND, W. N. O. AND HENNESSEY, R. D. (1987b). Biological control of the cassava mealybug (*Phenacoccus manihoti*) by the exotic parasitoid, *Epidinocarsis lopezi*. In *Proceedings of the 3rd Symposium of the International Society for Tropical Root Crops, Owerri, Nigeria, 17–23 August 1986* (E. R. Terry, M. O. Akoroda and D. B. Arene, Eds), pp. 98–104. IDRC, Ottawa.

NEUENSCHWANDER, P., HENNESSEY, R. D. AND HERREN, H. R. (1987). Foodweb of insects associated with the cassava mealybug, *Phenacoccus manihoti* Matile-Ferrero (Hemiptera: Pseudococcidae) and its introduced parasitoid, *Epidinocarsis lopezi* (De Santis) (Hymenoptera: Encyrtidae), in Africa. *Bulletin of Entomological Research* **77**, 177–189.

NEUENSCHWANDER, P., SCHULTHESS, F. AND MADOJEMU, E. (1986). Experimental evaluation of the efficacy of *Epidinocarsis lopezi*, a parasitoid introduced into Africa against the cassava mealybug, *Phenacoccus manihoti. Entomologia Experimentalis et Applicata* **42**, 133–138.

NEUENSCHWANDER, P., HAMMOND, W. N. O., GUTIERREZ, A. P., CUDJOE, A. R., BAUMGAERTNER, J. U., REGEV, U. AND ADJAKLDE, R. (1989). Impact of assessment of the biological control of the cassava mealybug, *Phenacoccus manihoti* Matile-Ferrero (Hemiptera: Pseudococcidae) by the introduced parasitoid, *Epidinocarsis lopezi* (De Santis) (Hymenoptera: Encyrtidae). *Bulletin of Entomological Research* **79**, 579–594.

NORGAARD, R. B. (1988). The biological control of cassava mealybug in Africa. *American Journal of Agricultural Economics* **70**, 366–371.

NSIAMA-SHE, H. D. (1985). *The bioecology of the predator,* Hyperaspis jucunda *Muls. (Col., Coccinellidae), and the temperature responses of its prey, the cassava mealybug,* Phenacoccus manihoti *Mat.-Ferr. (Hom., Pseudococcidae).* Unpublished PhD thesis, University of Ibadan, Nigeria.

NSIAMA-SHE, H. D., ODEBIYI, J. A. AND HERREN, H. R. (1984). The biology of *Hyperaspis jucunda* (Col., Coccinellidae), an exotic predator of the cassava mealybug, *Phenacoccus manihoti* (Hom., Pseudococcidae) in southern Nigeria. *Entomophaga* **29**, 87–93.

NWANZE, K. F. (1978). Biology of the cassava mealybug, *Phenacoccus manihoti* Mat.-Ferr. in Republic of Zaire. In *Proceedings of the Workshop on Cassava Mealybug,* Phenacoccus manihoti *Mat.-Ferr. (Pseudococcidae), Mvuazi, Zaire, 26–29 June 1977* (K. F. Nwanze and K. Leuschner, Eds), pp. 20–28. IITA, Ibadan, Nigeria.

NWANZE, K. F. (1982). Relationships between cassava root yields and crop infestation by the mealybug, *Phenacoccus manihoti. Tropical Pest Management* **28**, 27–32.

NWANZE, K. F., LEUSCHNER, K. AND EZUMAH, H. C. (1979). The cassava mealybug, *Phenacoccus* sp. in Republic of Zaire. *Pest Articles and News Summaries* **25**, 125–130.

NYIIRA, Z. M. (1972). Report of investigation of cassava mite, *Mononychellus tanajoa* (Bondar). Unpublished paper, Kwanda Research Station, Kamphala, Uganda.

NYIIRA, Z. M. (1976). Advances in research on the economic significance of the green cassava mite *(Mononychellus tanajoa)* in Uganda. In *Proceedings of the Interdisciplinary Workshop on the International Exchange and Testing of Cassava Germplasm in Africa, Ibadan, Nigeria, 17–21 November 1975* (E. R. Terry and R. MacIntyre, Eds), pp. 27–29. IDRC, Ottawa.

NYIIRA, Z. M. (1977). Population dynamics of the green mite and its predator, *Oligota* sp. In *Proceedings of the 4th Symposium of the International Society for Tropical Root Crops, Cali, Colombia, 1–7 August, 1976* (J. Cork, R. MacIntyre and M. Graham, Eds), pp. 193–197. IDRC, Ottawa.

NYIIRA, Z. M. (1978). *Mononychellus tanajoa* (Bondar): biology, ecology and economic importance. In *Proceedings of the Cassava Protection Workshop, Cali, Colombia, 7–12 November 1977* (T. Brekelbaum, A. Bellotti and J. C. Lozano, Eds), pp. 155–159. CIAT, Cali, Colombia.

NYIIRA, Z. M. (1982a). Cassava green mite: its distribution and possible control. In *Proceedings of the Workshop on Root Crops in Eastern Africa, Kigali, Rwanda, 23–27 November 1980*, pp. 65–67. IDRC, Ottawa.

NYIIRA, Z. M. (1982b). *Mononychellus tanajoa* (Bondar): implications and approach to its management. In *Proceedings of the International Workshop on Control of Cassava Mealybug and Green Spider Mite, Umudike, Nigeria, 12–16 October 1981*

(O. O. Okoli, T. Enyinnia, M. C. Ogbokwe, S. O. Odurukwe, J. E. Okeke and H. E. Okereke, Eds), pp. 47–64. National Root Crops Research Institute, Nigeria.

NYIIRA, Z. M. AND MUTINGA, M. J. (1977). *Tetranychidae*: Pests of cassava, *Manihot esculenta* Crantz, in Uganda and their natural enemies. *East African Agriculture and Forestry Journal* **43**, 1–4.

ODEBIYI, J. A. AND BOKONON-GANTA, A. H. (1986). Biology of *Epidinocarsis (Apoanagyrus) lopezi* (Hymenoptera: Encyrtidae), an exotic parasite of cassava mealybug, *Phenacoccus manihoti* (Homoptera: Pseudococcidae), in Nigeria. *Entomophaga* **31**, 251–260.

OKIGBO, B. N. (1978). *Cropping systems and related research in Africa*. Association for the Advancement of Agricultural Sciences in Africa, Addis Ababa, Ethiopia.

PAGE, W. W., HARRIS, J. R. W. AND YOUDEOWEI, A. (1980). Defoliation and consequent crop loss in cassava caused by the grasshopper, *Zonocerus variegatus* (L.) (Orthoptera: Pyrgomorphidae), in southern Nigeria. *Bulletin of Entomological Research* **70**, 151–163.

REES, D. P. (1985). Life history of *Teretriosoma nigrescens* Lewis (Coleoptera: Histeridae) and its ability to suppress populations of *Prostephanus truncatus* (Horn) (Coleoptera: Bostrichidae). *Journal of Stored Products Research* **21**, 115–118.

SCHULTHESS, F. (1987). *The interactions between cassava mealybug (*Phenacoccus manihoti *Mat.-Ferr.) populations and cassava (*Manihot esculenta *Crantz) as influenced by weather*. Unpublished PhD thesis, Swiss Federal Institute of Technology, Zurich, Switzerland.

SCHULTHESS, F., BAUMGAERTNER, J. U. AND HERREN, H. R. (1987). Factors influencing the life table statistics of the cassava mealybug, *Phenacoccus manihoti*. *Insect Science and its Application* **8**, 851–856.

SCHULTHESS, F., BAUMGAERTNER, J. U. AND HERREN, H. R. (1989). Sampling *Phenacoccus manihoti* in cassava fields in Nigeria. *Tropical Pest Management* **35**, 193–200.

SEIF, A. A. (1981a). Seasonal fluctuations of adult population of the whitefly, *Bemisia tabaci*, on cassava. *Insect Science and its Application* **1**, 363–364.

SEIF, A. A. (1981b). Transmission of cassava mosaic virus by *Bemisia tabaci*. *Plant Disease* **65**, 606–607.

SHIRES, S. W. (1980). Life history of *Prostephanus truncatus* (Horn) (Coleoptera: Bostrichidae) at optimum conditions of temperature and humidity. *Journal of Stored Products Research* **16**, 147–150.

SHUKLA, P. P. (1976). Preliminary report on the green mite (*Mononychellus tanajoa* Bondar) resistance in Tanzanian local cassava varieties. *East African Agriculture and Forestry Journal* **42**, 55–59.

SULLIVAN, D. J. AND NEUENSCHWANDER, P. (1988). Melanization of eggs and larvae of the parasitoid, *Epidinocarsis lopezi* (De Santis) (Hymenoptera: Encyrtidae), by the cassava mealybug, *Phenacoccus manihoti* Matile-Ferrero (Homoptera: Pseudococcidae). *Canadian Entomologist* **120**, 63–71.

SYLVESTRE, P. AND ARRAUDEAU, M. (1983). *Le manioc. Techniques Agricoles et Productions Tropicales*. Maisonneuve et Larose, Paris.

TAYLOR, D. E. (1981). Whiteflies. *Zimbabwe Agricultural Journal* **78**, 25.

TENG, P. S. (1987). *Action plan for the improvement of plant protection services in central Africa*. FAO, Rome.

UMEH, E. D. N. N. (1982). Biological studies on *Hyperaspis marmottani* Fairm. (Col., Coccinellidae), a predator of the cassava mealybug, *Phenacoccus manihoti* Mat.-Ferr. (Hom., Pseudococcidae). *Zeitschrift für Angewandte Entomologie* **94**, 530–532.

UMEH, E. D. N. N. (1983). Towards a biological control of cassava mealybug, *Phenacoccus manihoti* Mat.-Ferr. (Homoptera: Pseudococcidae). *Revue de Zoologie Africaine* **97**, 60–64.

UMEH, E. D. N. N. (1984). Biological control of the cassava mealybug, *Phenacoccus manihoti* Mat.-Ferr. (Homoptera: Pseudococcidae), by *Hyperaspis marmottani* Fairm. (Coleoptera: Coccinellidae). Paper presented at the annual conference of the Entomological Society of Nigeria, Ibadan, Nigeria.

VAN ALPHEN, J. J. M., NEUENSCHWANDER, P., DIJKEN, M. J., HAMMOND, W. N. O. AND HERREN, H. R. (1989). Insect invasions: the case of the cassava mealybug and its natural enemies evaluated. *The Entomologist* **108**, 38–55.

VAN DEN BERG, H. AND MARKHAM, R. H. (1987). Prey preference and reproductive success of the predatory mite, *Neoseiulus idaeus*, on the prey species *Mononychellus tanajoa* and *Tetranychus lombardinii*. *Insect Science and its Application* **8**, 867–869.

WALKER, P. T., HEYDON, D. L. AND GUTHRIE, E. J. (1985). *Report of a survey of cassava losses caused by mealybug and green mite in Africa, with special reference to Ghana.* Tropical Development and Research Institute, London.

WALTER-ECHOLS, G., AGOUNKE, D. AND AKPALAO, Y. (1983). Importance of cassava, cowpea and maize pests in Togo, West Africa. In *Proceedings of the 10th International Congress on Plant Protection, London, 20–25 November 1983*, p. 103. British Crop Protection Council, London.

WODAGENEH, A. (1985). Cassava and cassava pests in Africa. *FAO Plant Protection Bulletin* **33**, 101–108.

WODAGENEH, A. AND HERREN, H. R. (1987). International cooperation: training and initiation of national biological control programmes. *Insect Science and its Application* **8**, 915–918.

YANINEK, J. S. (1988). Continental dispersal of the cassava green mite, an exotic pest in Africa, and implications for biological control. *Experimental and Applied Acarology* **4**, 211–224.

YANINEK, J. S. AND ANIMASHAUN, A. (1987). Why cassava green mites are dry season pests. Paper presented at WHO/IITA seminar on agrometeorology and crop protection in the lowland humid and subhumid tropics, Cotonou, Benin, 7–11 July 1986.

YANINEK, J. S. AND HERREN, H. R. (1988). Introduction and spread of the cassava green mite, *Mononychellus tanajoa* (Bondar) (Acari: Tetranychidae): an exotic pest in Africa and the research for appropriate control methods: a review. *Bulletin of Entomological Research* **78**, 1–13.

YANINEK, J. S., GUTIERREZ, A. P. AND HERREN, H. R. (1989). Dynamics of the cassava green mite, *Mononychellus tanajoa* (Bondar) (Acari: Tetranychidae), in Africa: experimental evidence of temperature and host plant effects on population growth rates. *Environmental Entomology* **18**, 633–640.

YANINEK, J. S., HERREN, H. R. AND GUTIERREZ, A. P. (1987). The biological basis for the seasonal outbreaks of cassava green mites in Africa. *Insect Science and its Application* **8**, 861–865.

YANINEK, J. S., HERREN, H. R. AND GUTIERREZ, A. P. (1989). Dynamics of the cassava green mite, *Mononychellus tanajoa* (Bondar) (Acari: Tetranychidae), in Africa: seasonal factors affecting phenology and abundance. *Environmental Entomology* **18**, 625–632.

5

# From Behaviour to Control: The Development of Trap and Target Techniques for Tsetse Fly Population Management

RICHARD WALL AND PETER LANGLEY

*ODA/University of Bristol, Tsetse Research Laboratory, Langford, Bristol BS18 7DU, UK*

Introduction
The tsetse life cycle
Early approaches to tsetse control
The lesson of Principe Island
Insecticidal methods of tsetse control
The sterile male technique
Modern trap development
Exploiting tsetse behaviour
    Olfactory responses—The role of vision—The response of tsetse to traps and targets
The development of autosterilizing techniques
Tsetse control with traps and targets
Conclusions
Acknowledgement
References

## Introduction

Tsetse flies (*Glossina* spp.) belong to the family Glossinidae within the order Diptera. They are found in the tropics throughout sub-Saharan Africa, where they feed exclusively on the blood of vertebrates and, in so doing, may transmit parasitic protozoan trypanosomes from host to host. The normal hosts of tsetse flies are wild mammals and reptiles (occasionally birds) which experience few if any ill-effects from the presence of the trypanosomes in their blood, unless subjected to stress. When man or his domestic animals become infected, however, the pathogenic effects of the trypanosomes are debilitating and often ultimately fatal, the disease being known as sleeping sickness in humans and nagana in domestic animals.

The exclusively haematophagous habit of both sexes of adult tsetse and the relatively low energy value of blood as a food source means that flies must

*Agricultural Zoology Reviews* — Vol. 4, February 1991
0269–0543 $20.00 + $0.00 © Intercept Ltd, PO Box 716, Andover, Hampshire SP20 1YG, UK

feed frequently, ingesting more than their own body weight at a single meal every few days (Bursell and Taylor, 1980). This behaviour, coupled with the relatively long life of individuals as adults, enhances the spread of trypanosomiasis among the animals, including man, upon which they feed.

Because of the disease they transmit, tsetse flies have had profound effects on the cultural and economic development of Africa; tsetse flies were responsible for halting the southward and inland advances of Arabic peoples into Africa between AD 750 and 1500 by infecting their horses with trypanosomiasis (Curtin *et al.*, 1978). They were also responsible for halting the northward advance of the Matabele armies of Mzilikazi at the Zambezi in the early nineteenth century. Perhaps more importantly, they have delimited the areas available for human settlement and animal husbandry, by preventing development of settled agricultural systems, by depriving farmers of draught animals and of manure for the land. At present they infest some 11 million square kilometres of sub-Saharan Africa and continue to cause significant human suffering and to retard African rural development (Jordan, 1986).

Chemotherapeutic treatment of trypanosomiasis is used widely, but has changed little in almost 40 years. It can be successful, but the benefits do not always justify the costs and the logistics of cattle treatment are often prohibitively complex. For these reasons, one of the primary means of controlling trypanosomiasis in the long term may be to control the tsetse fly but, at present, only a few African countries employ this strategy.

## The tsetse life cycle

Unlike most other insects that produce large numbers of eggs, the tsetse fly has developed an advanced form of adenotrophic viviparity. After mating, a mature egg is ovulated alternately from right and left ovaries from each of the paired ovarioles in each ovary in turn. An ovulated egg is fertilized by spermatozoa stored (generally from a single mating) in the spermathecae of the female. This egg is then retained in the oviduct instead of being laid. After 4 days, at a temperature of 25°C, embryogenesis is complete and a first-instar larva hatches, which commences to feed on a nutritive secretion produced by the highly modified accessory glands of the female known as 'uterine glands' or 'milk glands'. Over the next 2 to 3 days the larva continues to feed, moults twice and the third-instar larva ingests large quantities of 'milk' before parturition occurs on about the ninth or tenth day after ovulation (Denlinger and Ma, 1974). Although at favourable times of the year larvae may be deposited almost anywhere where the soil is soft and the moisture content high, during prolonged dry seasons discrete larviposition sites can be found, generally in the shade of fallen logs or in dry river banks. The larva does not feed after parturition, but buries itself a few centimetres below the soil surface where it forms a puparium, within which the relatively long period of metamorphosis proceeds. About 30 days later a new adult emerges to begin the life cycle again.

The unusual life cycle of tsetse presents one of the central problems for tsetse control, since the eggs and larvae are protected inside the parent female and the puparia are buried in the ground. Juvenile life stages, therefore, are not available for direct attack. However, the reproductive strategy of tsetse also provides the means of controlling them. The mode of reproduction represents a high degree of maternal investment and, as a result, relatively few offspring are produced. In the laboratory a fertilized, well-fed female can produce up to 10 larvae at approximately 9–10 day intervals during its lifetime. In the field, however, a female's reproductive output is likely to be considerably lower and has been estimated as being about four larvae (Phelps and Vale, 1978). The small number of offspring produced by each female gives the population a relatively low potential for increase. Only a small increase in the burden of mortality acting on the population, thought to be in the region of an additional 2–4% of the population per day, is required to bring about its decline (Rogers, Randolph and Kuzoe, 1984; Hargrove, 1988). For most other species of insect, where the reproductive rate may be up to 200 or 300 times higher than tsetse, a higher proportion of the population would have to be killed per day to bring about effective control.

## Early approaches to tsetse control

Over the past century, considerable efforts have been made to control or eradicate tsetse. Early approaches included habitat destruction by clearing the vegetation in which the various species of tsetse lived, completely eliminating areas of scrub or woodland which, as a result, were replaced by grassland in the savannah areas, or by simply removing the forest vegetation along watercourses in riverine areas. Such practices, coupled with deliberate culling, removed the game animals which act as reservoirs for the pathogenic trypanosomes. Large areas have been freed for human settlement in this way. Today, however, wholesale bush-clearing and game-culling are unacceptable on environmental grounds and have been abandoned as control practices. Selective habitat clearance, culling of favoured host species, or exclusion of game animals from agricultural areas by reinforcing natural barriers, such as escarpments, with fences, have met with limited success but are expensive and inappropriate in many areas.

## The lesson of Principe Island

Between 1911 and 1914, on the island of Principe, sleeping sickness and its vector, the tsetse fly *Glossina palpalis palpalis* (Robineau-Desvoidy), were completely eliminated by the use of dark squares of cloth covered with glue ('bird-lime') worn on the backs of the labourers in the fields, coupled with the almost complete eradication of wild pigs (Azevedo *et al.*, 1961). The flies were attracted to the labourers and large numbers became fatally stuck to the sticky cloths. Swine-breeding on the island was forbidden by government decree and all vagrant swine were hunted to extinction. However, the gradual return of pig-breeding to the island, together with the reintroduction of tsetse from the neighbouring island of Fernando Po (renamed Malabo) saw the eventual re-establishment of *Glossina palpalis palpalis* populations, the fly reappearing there in 1956. The early success, however, indicated the considerable potential for alternative methods of tsetse control to those involving the conventional use of chemical insecticides.

Simple traps, designed to catch tsetse without being attached to a human bait, copied animal shapes (Harris, 1932; Swynnerton, 1933; Morris and Morris, 1949). Almost 2000 Morris traps treated with gamma BHC were deployed on Principe between 1956 and 1958, contributing greatly to the second elimination of *G. p. palpalis* from the island (Azevedo *et al.*, 1961; Azevedo, Costa Maurão and Castro Salazar, 1962). The success of these tsetse eradication campaigns was due in large part to the isolated nature of the tsetse populations and to the thoroughness with which the operation was enforced by the colonial authorities.

With the development of the organochlorines, insecticides were seen as the panacea for tsetse control. The use of fly-trapping techniques for control largely ceased and, when employed at all, they were used generally for survey work.

## Insecticidal methods of tsetse control

Initially, most insecticidal control was achieved by ground-spraying teams with pressurized knapsack sprayers, delivering residual organochlorine insecticides such as dieldrin or DDT, with which they attempted selectively to cover favoured tsetse resting sites on trees and bushes. The advent of aerial spraying techniques permitted the more widespread application of insecticides but with little or no discrimination, although the technique could kill tsetse either at rest or in flight. The development of ultra-low-volume spray techniques and the introduction of improved organochlorine compounds, such as endosulfan, or the synthetic pyrethroids, which require much lower doses of active ingredient to be dispensed per unit area than with the older insecticides, ensured reduced environmental contamination. Generally, organophosphates and carbamates have proved to be unsuitable for tsetse control.

A wide variety of effects of the insecticides used for tsetse control has been noted on non-target insects, birds, mammals, reptiles and fish (Graham, 1964; Wilson, 1972; Koeman *et al.*, 1978; Takken *et al.*, 1978; Baldry *et al.*, 1981). It has been argued that in many cases such non-specific effects of insecticides are transitory and that no long-term damage to the environment results (Smies, 1980). Nevertheless, with organochlorines long-term accumulation in food chains has been shown to be particularly damaging to predators at the top of the food chain. In Zimbabwe, accumulation of DDT and dieldrin used to control tsetse fly has been implicated in the significant reduction in the thickness of the shells of bird and crocodile eggs (Wessels *et al.*, 1980; Tannock, Howells and Phelps, 1983).

Insecticides and the complex logistics associated with their storage, handling, deployment and disposal are costly, involving the expenditure of foreign currency which the countries concerned can ill-afford. Insecticide applications are rarely 100% effective and tsetse populations have been recorded as recovering to their pretreatment levels in as little as 12 months, once the control measure ceases (Turner and Brightwell, 1986). Hence, in many places annual applications of insecticide are required to maintain tsetse populations at acceptably low levels.

As a result, attention has refocused increasingly on the development of alternative methods of tsetse control which are sustainable over long periods of time. It is probably true to say that cost and effectiveness motivated much of the more recent return to favour of traps and targets for tsetse control and that only lately have environmental acceptability and appropriate technology been seen as important considerations.

## The sterile male technique

The release of large numbers of sterilized but otherwise healthy adult male insects to control a natural population was proposed by Knipling (1955) and has been used successfully to control a variety of pest insects since then. The insects are generally subjected to $\gamma$-irradiation before being released. Sterile males compete with normal males for the available females, and a population decline results as each female mated with a sterile male fails to produce viable offspring. Sequential releases of sterile males are required, but the main advantage of the technique is that it is non-polluting.

The large-scale rearing of tsetse in the laboratory using goats or rabbits as hosts (Nash, Jordan and Boyle, 1968), followed by elimination of the need for a living host by feeding blood under a synthetic membrane (Mews *et al.*, 1977), has enabled the sterile male technique to be implemented successfully in several parts of Africa against several species of tsetse (Politzar and Cuisance, 1982; Williamson *et al.*, 1983; Takken *et al.*, 1986; Olandunmade *et al.*, 1990). However, the naturally low rate of reproduction and special feeding requirements of tsetse make the rearing operation to produce large numbers of sterile males a very expensive and complicated business. Additionally, the technique generally requires a drastic reduction to be achieved in the natural population, using a non-residual insecticide or catching devices, before the release of sterile males commences.

## Modern trap development

Recent development of traps for tsetse control began in the early 1970s, initially in West Africa with the production of the 'biconical' trap by Challier and Laveissière (1973). This consists of a white netting cone, apex uppermost, attached over an inverted blue cloth cone. There are usually four slit-like entrances in the lower blue cone, with black cloth baffles inside behind the entrance slits. The trap is about 1–1.5 m in length and is mounted on a pole (*Figure 1*) within the tsetse habitat. It is highly effective against tsetse flies of the *G. palpalis* group and its efficacy can be enhanced by treatment of the cloth with an insecticide to kill flies that alight on its surface. This basic design is still in widespread use, although recent developments have simplified the construction and reduced the amount of material used, to cut costs, resulting in what is known as a 'monoconical' trap (Lancien, 1981).

**Figure 1.** West African, Challier-Laveissière, 'biconical' trap at Nguruman, Kenya.

The biconical and monoconical traps, however, are relatively ineffective against what are generally known as the 'savannah' tsetse species, of the *G. morsitans* group. Detailed quantitative work in Zimbabwe led to the production of a generation of traps which are able to attract and catch these species (Vale, 1974, 1982a), often with predictable efficiencies (Hargrove,

1980). The most commonly used of the more recent designs is known as the 'F3' trap (Flint, 1985; Green and Flint, 1986) and consists of a cube (each side 1 m long), four sides of which are covered by blue cloth (*Figure 2*). The lower half of one side is cut away to form an entrance and the inside back wall opposite the entrance is covered with black cloth. Inside the trap there is a

**Figure 2.** Cubic, blue 'F3' trap in the Zambezi valley, Zimbabwe.

netting cone leading to a collecting chamber at its apex. Further economies in design have been achieved by constructing traps on a triangular rather than a rectangular module; the so-called 'epsilon' trap (designed by G.A. Vale), described by Hargrove and Langley (1990) (*Figure 3*).

A key development in the production of trapping devices for the savannah

**Figure 3.** Triangular 'epsilon' trap in the Zambezi valley, Zimbabwe.

species of tsetse, has been the combination of visual and olfactory stimuli by the addition of synthetic host odours to traps.

Targets or screens were also developed in both Zimbabwe and West Africa for tsetse control. These consist principally of a central insecticide-covered cloth, usually about 1 $m^2$ in area. The version used successfully to suppress an artificially introduced, island population of tsetse in Zimbabwe (Vale *et al.*, 1986), was pivoted so that the face of the cloth was always at right angles to the wind direction, presenting the maximum visual target for flies approaching upwind. A feature of this target was that, in addition to the visual component, an invisible flanking net was incorporated (*Figure 4*). Flies either alighting on the cloth or colliding with the netting as they fly around the device pick up a lethal dose of insecticide. This target was then used to suppress a natural population of *G. pallidipes* Austen in Zimbabwe (Vale *et*

**Figure 4.** R-type target in the Rifa Triangle, Zimbabwe, with protective roof and counterbalanced pivot, to keep the face of the cloth target at right angles to tsetse approaching upwind.

al., 1988). A simpler and more efficient version of this target was developed (Vale, Bursell and Hargrove, 1985) which has two flanking nets (*Figure 5*) and dispenses with the roof, which was incorporated originally to protect the insecticide from the weather. The most recent operational targets have dispensed with flanking nets, which were never used in the simpler West African insecticide-dosed screens. Targets and screens have the advantage over traps of being simpler in design and are thus less expensive to construct.

It can be seen that the traps and targets developed in West Africa and in Zimbabwe, as with their derivatives developed in other parts of Africa such as the 'Ngu' trap used in Kenya (Brightwell *et al.*, 1987), share a large number of features in common. This reflects the fact that the designs are based upon the principle of exploiting natural patterns of fly behaviour.

**Figure 5.** S-type target in the Zambezi valley, Zimbabwe; a simple central cloth target with two flanking nets.

## Exploiting tsetse behaviour

### OLFACTORY RESPONSES

Until the late 1960s most information about tsetse populations was obtained by catching flies in hand nets as they alighted to feed on a man or an animal, or as they approached and landed on a hand-held dark-cloth screen. Work on the *G. morsitans* group of flies suggested that these hand-net samples were not representative of the general tsetse population and were in fact biased, especially for female flies. Hence, the data for males only were used to estimate 'apparent densities'. Not until the development of electrified nets, which consist of a nearly invisible vertical grid of wires that kill tsetse that collide with them, was it possible to sample tsetse populations around baits in the absence of human interference and demonstrate that human catchers are repellent to a number of species of tsetse (Vale, 1974; Hargrove, 1976).

Using electrified nets, the effectiveness of a variety of visual stimuli could be measured and, by hiding host animals in underground pits from which their smell could be extracted and released at a trap, the importance of host odours in attracting flies was demonstrated. Additionally, it was shown that an upwind response to such odours played a central role in host location by tsetse (Bursell, 1984, 1987). Increasing the quantity of ox-odour released at a trap, by increasing the numbers of cattle in an underground pit, increased the numbers of flies caught (Hargrove, 1977; Hargrove and Vale, 1978). However, the precise mechanisms involved in the recruitment of tsetse into odour plumes and in the way they navigate to the source of such odours continues to be debated (Bursell, 1987; Gibson and Brady, 1988; Brady, Gibson and Packer, 1989; Torr, 1989).

It was realized that identification of the various components of host odour that were attractive to flies would allow the development of highly specific and effective baits to control tsetse populations. A synthetic chemical cocktail would be simple to use with a catching device and might be able to present an odour stimulus equivalent to large numbers of host animals. Intensive work, therefore, began to attempt to isolate the attractive chemicals involved. Development of a laboratory bioassay for candidate odours involved studying the electrophysiological responses of antennal receptors and the behavioural responses of intact flies. Promising compounds or mixtures were then assayed in the field in Zimbabwe to evaluate their ability to influence the effectiveness of a standard capture technique.

Initially, it was thought that the odours from host animal breath were of prime importance in the attraction of flies (Vale, 1974). Later, buffalo urine was also found to be a powerful olfactory attractant for *G. pallidipes* in Kenya (Owaga, 1985). Various phenols have subsequently been identified as comprising the most attractive fractions of urine (Hassanali *et al.*, 1986; Bursell *et al.*, 1988; Vale, Hall and Gough, 1988).

A combination of 2 l of carbon dioxide per min, 5 mg of acetone per h and 0.05 mg of 1-octen-3-ol per h was found to be nearly as attractive as natural ox-odour for *G. morsitans morsitans* Westwood, although only half as attractive as ox-odour for *G. pallidipes* (Vale and Hall, 1985a). Octenol, acetone and carbon dioxide approximately double the catch at doses 10 times greater than that produced by an ox (Vale and Hall, 1985b). Increases above this have little additional effect on the catch. Recent work indicates that upwind flight in response to host odour occurs only in the presence of carbon dioxide and that acetone and octenol on their own appear to increase the responsiveness of flies to visual cues (Torr, 1990). Unfortunately, carbon dioxide is too expensive and inconvenient to use on a large scale in control operations. The production of pure phenolic fractions is also expensive and difficult, although those which are available commercially can be mixed in appropriate ratios for slow release from polythene sachets. A standard bait of cow urine with acetone may provide the simplest, most cost-effective odour attractant for most species of savannah tsetse (Brightwell *et al.*, 1987). However, the differing effects of various odours on different tsetse species and the complex behaviour underlying their responses to odour remain to be clarified.

## THE ROLE OF VISION

Tsetse vision is strongly integrated with olfaction in the location of host animals. Many features of the design of traps and targets are thought to exploit components of the responses of tsetse to visual properties of their hosts.

Strong contrast of an object against the background is an important factor in attraction of host-seeking tsetse (Chapman, 1961). However, tsetse appear to employ colour information and not intensity-contrast information alone in their responses to traps and presumably also to hosts. Spectral sensitivity curves of *G. m. morsitans* indicate a peak in the ultraviolet (350–365 nm) and a broad peak at 450–550 nm in the blue-green (Green and Cosens, 1983). However, the behavioural responses of tsetse to colour in the field are complex. In general, trap effectiveness depends mainly on reflectivity in four different wavelength bands, blue-green and red correlating positively with trap performance and ultraviolet and green-yellow-orange negatively. Broadly speaking, therefore, royal blue is the most attractive trap colour, followed by black and white, while yellow and green are consistently unattractive (Green and Flint, 1986).

Challier *et al.* (1977) increased the catch of tsetse flies some 2.5 times by changing the trap covering from a white to a blue material. In Zimbabwe, Green and Flint (1986) compared traps made from 53 different coloured cloths and identified a royal blue as giving the highest trap catch. The tendency for flies to alight is stronger with black than blue cloth (Green, 1986). Hence, blue traps increase the catch by maximizing the attraction while reducing the likelihood that flies will sit on their outside surfaces. On the other hand, black material improves the performance of targets, where the objective is to cause the fly to land on the cloth portion or to collide with the flanking nets to pick up a lethal dose of insecticide. The black material positioned inside the entrances of the biconical, F3, epsilon and Ngu traps is thought to increase efficiency by causing flies that have been attracted and are flying round the trap, to enter (Vale, 1982a; Green, 1986).

Tsetse have a strong photopositive response at temperatures below about 32°C. Inside the trap the strongest light stimulus comes from above, through the mesh cone and, as a result, once inside the trap the flies tend to move upwards towards the light. In doing so they move into the mesh cone and are directed into the collecting device at its apex.

Tsetse are highly responsive to moving objects, the response of *G. m. morsitans* depending on time of day, starvation, angular velocity and size of the stimulus (Brady, 1972). Mobile traps and targets catch greater numbers than do stationary ones, especially of young tsetse and especially of *G. m. morsitans* (Vale, 1974; Owaga and Challier, 1985; Langley, Hargrove and Wall, 1990). However, mobile catching devices would be too impractical and too expensive for large-scale use for tsetse control.

The range of tsetse vision depends on the size of the object and varies between species. Napier Bax (1937) found that *G. swynnertoni* Austen responded to visual stimuli from a moving herd of cattle at a maximum distance of 137–183 m, while Chapman (1961) found that the limit at which *G. medicorum* Austen responded to a large moving target was 8–15 m while for *G. morsitans* it was 46 m. Perhaps unsurprisingly, larger traps tend to attract more flies than smaller traps (*Figure 6*), though larger traps may also be more efficient at capturing attracted tsetse (Hargrove, 1977). However, in using traps for control, size must be offset against increased construction costs and decreased portability.

The proportion of attracted flies that are caught by a biconical trap was estimated at 40–70% for *G. m. morsitans* and 50–80% for *G. pallidipes* when used in the presence of the odour from one ox (Hargrove, 1980). Vale

**Figure 6.** Large-sized 'H-11' trap in the Zambezi valley, Zimbabwe.

(1982b) showed that a substantial proportion of attracted flies circumnavigate a trap or target within half a metre. Mesh flanking nets either side of a target exploit this behaviour since the mesh is almost invisible to the approaching flies, which collide with it and come into contact with insecticide. The efficiency of targets, therefore, is generally higher than that of traps.

## THE RESPONSE OF TSETSE TO TRAPS AND TARGETS

### *Food acquisition*

For the female tsetse the dedication of resources to larval growth is offset by an economy in egg production, but still the nutritional requirements are large. The magnitude of the burden becomes clearer when one considers the relatively low energy value of vertebrate blood as a diet and the metabolic cost of its transformation into utilizable substrates (Bursell *et al.*, 1974). Flight activity must be fuelled for both sexes, but for males the overall burden of acquiring adequate nutrition is considerably less than that of females (Bursell and Taylor, 1980).

Feeding intervals of 3 or 4 days have been estimated for some species of tsetse (Randolph and Rogers, 1978, 1986; Rogers and Randolph, 1986) but recent work suggests that they may feed more frequently (Langley and Stafford, 1990; Langley and Wall, 1990). Hargrove (1976) calculated that, in the absence of man, virtually all *G. pallidipes* adults attracted to a stationary ox in the field alighted on it and that most attempted to feed. Similarly, Vale (1980) observed that between 80 and 100% of flying tsetse responded to the host stimuli they encountered while flying into and out of a woodland clearing. Flies attracted to traps and targets contain a wide range of nutritional reserves in the form of stored lipids and blood meal residues (Randolph and Rogers, 1978; Randolph, Dransfield and Rogers, 1989; Langley and Wall, 1990) but those caught, for the most part, contain lower than average lipid levels (Vale, 1974; Langley, Hargrove and Wall, 1990). Hence, it is very likely that a substantial component of the response to an odour-baited trap or a target represents behaviour associated with host location and feeding.

### *Mating behaviour*

Tsetse flies live at relatively low densities and have no known mating strategy that could enhance the probability of encounters occurring between the sexes other than that which happens when they approach host animals.

Mating behaviour is only initiated upon contact and it seems that the male responds to visual cues, resulting in pursuit of any object of roughly similar size and shape to a female (Langley, Pimley and Carlson, 1975; Wall, 1989a). The role of the female appears to be passive, except that mated females are able to reject males which attempt to mate with them. Mating refractoriness in females is induced by both physical and chemical aspects of mating (Gillott and Langley, 1981).

Following contact with a female, the male is stimulated chemically via receptors on the tarsi and tibiae (Langley, Huyton and Carlson, 1978) by a species-specific contact sex pheromone which is a constituent of the cuticle of females and which is absent from the cuticle of males (Langley, Pimley and Carlson, 1975; Carlson, Langley and Huyton, 1978). The presence of synthetic sex pheromone on small decoys will induce copulatory behaviour in males that make contact with them.

Contact pheromone baited decoys placed on targets or screens, whether baited with host odour or not, are encountered by males that land on the screens, presumably because they provide a suitable visual stimulus. Once contact has been made, males will indulge in copulatory activity with such decoys for several minutes (Langley *et al.*, 1982; Hall, 1988; Wall, 1988a, b; Hall and Langley, 1989). It seems, therefore, that there is also a sexual component in the response of male tsetse to host animals and to traps and targets (Jaenson, 1980; Wall, 1989b).

### *Other behaviours*

The extent to which aspects of behaviour other than those associated with food acquisition and mating are involved in the responses of tsetse towards hosts, traps and targets is unclear. However, a 'refuge response', where flies collect in shaded cool hollows, may also be of significance during periods when the ambient temperature exceeds 32°C (Hargrove and Vale, 1979).

## The development of autosterilizing techniques

Most traps and targets appear to manipulate tsetse behaviour associated with host location and feeding. Unfortunately, there seems to be no way in which sexual behaviour can be exploited directly to increase the efficiency of odour-baited traps or targets. Although the addition of decoys will increase the numbers of males alighting on a target (Wall, 1989b), killing increased numbers of males is likely to have little or no effect on the insemination rate of females and, hence, on the dynamics of the population.

Sterilization of males, using pheromone- and chemosterilant-dosed decoys on targets has been considered (Coates and Langley, 1982; Hall, 1988; Hall and Langley, 1989) but sterilization of males will not be as effective as killing both sexes at the same rate, such as with an insecticide-treated target which kills all flies that make contact with it.

However, it has been estimated that the sterilization of both sexes of tsetse in a population at a certain rate would be more effective at suppressing that population than simply killing them at the same rate (Langley and Weidhaas, 1986; Barclay, 1987). The simple explanation is that killing females only would be as effective as killing both sexes, since it is only females that reproduce. It follows, therefore, that to sterilize females only, would be as effective as killing both sexes. Hence, the sterilization of males as well as females would be a bonus.

The action of density-dependent population regulation would be expected to enhance the superiority of sterilization of both sexes over killing (Barclay, 1987). In addition, if both males and females are sterilized, the area of influence of the sterilizing device should be greater than that of a target that simply kills flies. This could have great practical value in difficult terrain where the deployment of targets in a regular array is not possible. Finally, selection for behavioural resistance to traps and targets would be expected to be reduced if they were designed to sterilize rather than to kill, since attracted sterilized flies would mate with unattracted individuals.

These theoretical advantages of sterilization over killing have prompted the search for chemical sterilants that are environmentally acceptable and that will be a competitive substitute for insecticides on traps or targets used for tsetse control. The juvenile hormones and their mimics have long been known to disrupt metamorphosis in insects by their untimely presence in either the last larval instar or the pupa. They are also able to disrupt embryogenesis and therefore to prevent egg hatch. They have the advantage over other compounds that usually they can be formulated to penetrate the insect cuticle readily.

Since embryogenesis in tsetse occurs within the adult female and metamorphosis occurs underground, there may seem little purpose in pursuing juvenile hormones as an alternative to conventional insecticides for their control. However, the unusual mode of reproduction in tsetse actually favours the use of the juvenile hormone mimic pyriproxyfen (Langley, Felton and Oouchi, 1988; Langley *et al.*, 1990). Following suitable formulation in an oil, a trap or target can be treated so that the compound is absorbed through the adult female cuticle on contact, whereupon it is transferred to the larva 'in utero'. The third-instar larva appears to be normal but after pupation metamorphosis is disrupted. Only very small amounts of pyriproxyfen are required, and a single treatment will ensure that the female is effectively sterilized for life. Laboratory studies have shown that males making contact with a treated surface can transfer sterilizing doses to females when they mate (Langley *et al.*, 1990).

A field trial conducted in Zimbabwe has shown that pyriproxyfen can be used as an alternative to conventional insecticides in traps or on targets for tsetse control (Hargrove and Langley, 1990).

## Tsetse control with traps and targets

A number of experimental operations have shown that tsetse control can be achieved by the deployment of traps and targets. For example, isolated populations of *G. m. morsitans* and *G. pallidipes* were established on a 4.5 km$^2$ island in Lake Kariba in Zimbabwe, where there was no invasion pressure from outside (Vale *et al.*, 1986). Six traps baited with carbon dioxide and acetone were used to capture 0.1–0.3% per day of the *G. m. morsitans* population and 1–4% per day of the *G. pallidipes* population. Captured flies were either killed or sterilized. After 2 years the two populations had declined by 90 and 99%, respectively. The traps were then replaced by 20 insecticide-treated targets baited with acetone and octenol. The targets killed about 2%

per day of the *G. m. morsitans* and 5% per day of the *G. pallidipes*. The populations declined rapidly and disappeared in 9 months and 11 weeks, respectively (Vale, Bursell and Hargrove, 1985). Following this, insecticide-treated targets were used to eliminate natural populations of these two species in the Rifa Triangle, an area of 600 $km^2$ in the Zambezi valley, where reinvasion pressure from the neighbouring wildlife area was immense (Vale *et al.*, 1988).

Odour-baited traps and targets are now in widespread use for tsetse control in other parts of Africa (Filledier and Politzar, 1985; Dagnogo *et al.*, 1986). In West Africa traps or insecticide-treated screens without host odour have been used to control local foci of sleeping sickness for more than 10 years (Laveissière, Gouteux and Couret, 1980; Merot and Filledier, 1985). Attempts are now being made to identify olfactory stimulants for the tsetse species in these areas in order to increase the cost effectiveness of the technique (Cheke and Garms, 1988; Politzar and Merot, 1984; Merot, Filledier and Mulato, 1988).

Tsetse eradication from an area requires that tsetse-proof barriers are maintained around the cleared zone. The maintenance of such barriers can be difficult and expensive. As an alternative to eradication, traps and targets offer the possibility of continuously suppressing tsetse populations to the point where transmission of trypanosomiasis is minimized. Under such circumstances, occasional cases of trypanosomiasis in the human and animal populations can be treated with trypanocidal drugs.

A serious disadvantage of this approach is the continuous cost of trap and target deployment and maintenance. However, such costs can be reduced by involving the local people in disease and vector suppression in their own areas. In five villages in a focus of sleeping sickness in the Republic of Congo, Gouteux *et al.* (1987) showed that fly-trapping could be carried out effectively by the rural communities. In three of the villages with a high tsetse density, a 97–98% reduction in the tsetse populations was obtained within a few months using monoconical traps without insecticide. A similar approach is being applied at Nguruman in Kenya (Brightwell *et al.*, 1987) where Masai herdsmen are being encouraged to take responsibility for the manufacture and deployment of tsetse traps.

The extent to which further improvements in dispensing systems for odours, or the discovery of new olfactory stimulants (Vale and Hall, 1985a, b; Owaga, Hassanali and McDowell, 1988), will lead to more cost-effective ways of killing tsetse remains to be seen. There seems, however, considerable room for improvement in target design in terms of size, shape (Laveissière, Couret and Grebaut, 1987), construction material (Laveissière, Couret and Manno, 1987) and colour (Green, 1986, 1989). It should be noted that the cost of manufacturing traps and targets at their present state of development is relatively low compared to the cost of deploying and maintaining them in the field (A. Shaw, personal communication). It is for this reason that any development which would prolong the effectiveness of a target, thus reducing servicing intervals, would be well worthwhile.

## Conclusions

There seems every reason for optimism that the trypanosomiasis problem could be ameliorated in Africa using trap and target technologies, particularly when such techniques are integrated with the use of trypanocidal drugs and the husbandry of trypanotolerant forms of cattle. However, because of the scale of the African problem, the success or otherwise of the battle against trypanosomiasis also depends critically on the extent to which political, logistical and educational problems can be solved. Furthermore, unplanned human encroachment into tsetse habitats can be environmentally damaging, leading to land degradation and to the disappearance of game animals which are a valuable resource for the countries concerned. Carefully planned utilization of areas freed from tsetse flies is therefore essential.

## Acknowledgement

We would like to thank Dr A.M. Jordan for helpful comments on this manuscript.

## References

AZEVEDO, J.F. DE, TENDEIRO, J., ALMEIDA FRANCO, L.T. DE, COSTA MOURÃO, M. DA AND CASTRO SALAZAR, J.M. DE (1961). O reaparecimento da *Glossina palpalis palpalis* na ilha do Principe: *Estudos, Ensaios e Documentos No. 89*, Lisbon 1961.

AZEVEDO, J.F. DE, COSTA MOURÃO, M. DA AND CASTRO SALAZAR, J.M. DE (1962). A erradicação da *Glossina palpalis palpalis* da ilha do Principe (1956–1958): *Estudos, Ensaios e Documentos No. 91*, Lisbon, 1962.

BALDRY, D.A.T., EVERTS, J., ROMAN, B., VON OCHSSEE, G.A.B. AND LAVEISSIÈRE, C. (1981). The experimental application of insecticides from a helicopter for the control of riverine populations of *Glossina tachinoides* in West Africa. VIII. The effects of two spray applications of OMS-570 (endosulphan) and of OMS-1998 (decamethrin) on *G. tachinoides* and non-target organisms in Upper Volta. *Tropical Pest Management* **27**, 83–110.

BARCLAY, H.J. (1987). Models for pest control using sex pheromones and chemosterilants. *Insect Science and its Application* **8**, 187–196.

BRADY, J. (1972). The visual responsiveness of the tsetse fly *Glossina morsitans* Westw. (Glossinidae) to moving objects: the effects of hunger, sex, host odour and stimulus characteristics. *Bulletin of Entomological Research* **62**, 257–279.

BRADY, J., GIBSON, G. AND PACKER, M.J. (1989). Odour movement, wind direction and the problem of host finding by tsetse flies. *Physiological Entomology* **14**, 369–380.

BRIGHTWELL, R., DRANSFIELD, R.D., KYORKU, C., GOLDER, T.K., TARIMO, S.A. AND MUNGAI, D. (1987). A new trap for *Glossina pallidipes*. *Tropical Pest Management* **33**, 151–159.

BURSELL, E. (1984). Observations on the orientation of tsetse flies (*Glossina pallidipes*) to wind-borne odours. *Physiological Entomology* **9**, 133–137.

BURSELL, E. (1987). The effect of wind-borne odours on the direction of flight in tsetse flies, *Glossina* spp. *Physiological Entomology* **12**, 149–156.

BURSELL, E. AND TAYLOR, P. (1980). An energy budget for *Glossina* (Diptera: Glossinidae). *Bulletin of Entomological Research* **70**, 187–196.

BURSELL, E., BILLING, K.C., HARGROVE, J.W., MCCABE, C.T. AND SLACK, E.

(1974). Metabolism of the bloodmeal in tsetse flies (a review). *Acta Tropica* **31**, 297–320.

BURSELL, E., GOUGH, A.E.J., BEEVOR, P.S., CORK, A., HALL, D.R. AND VALE, G.A. (1988). Identification of components of cattle urine attractive to tsetse flies *Glossina* spp. (Diptera: Glossinidae). *Bulletin of Entomological Research* **78**, 281–291.

CARLSON, D.A., LANGLEY, P.A. AND HUYTON, P.M. (1978). Sex pheromones of the tsetse fly: isolation, identification and synthesis of contact aphrodisiacs. *Science* **201**, 750–753.

CHALLIER, A. AND LAVEISSIÈRE, C. (1973). Un nouveau piége pour la capture des glossines (*Glossina*: Diptera, Muscidae): déscription et essais sur le terrain. *Cahiers ORSTOM. Serie Entomologie Médicale et Parasitologie* **11**, 251–262.

CHALLIER, A., EYRAUD, M., LAFAYE, A. AND LAVEISSIÈRE, C. (1977). Amélioration du rendement du piége biconique pour glossines (Diptera, Glossinidae) par l'emploi d'un cone inferier bleu. *Cahiers ORSTOM. Serie Entomologie Médicale et Parasitologie* **15**, 283–286.

CHAPMAN, R.F. (1961). Some experiments to determine the methods used in host finding by tsetse flies, *Glossina medicorum. Bulletin of Entomological Research* **52**, 83–97.

CHEKE, R.A. AND GARMS, R. (1988). Trials of compounds to enhance trap catches of *Glossina palpalis palpalis* in Liberia. *Medical and Veterinary Entomology* **2**, 199–200.

COATES, T.W. AND LANGLEY, P.A. (1982). Laboratory evaluation of contact sex pheromone and bisazir for autosterilisation of *Glossina morsitans. Entomologia Experimentalis et Applicata* **31**, 276–284.

CURTIN, P., FEIERMAN, S., THOMPSON, L. AND VANSINA, J. (1978). *African History*. Longman, London.

DAGNOGO, M., NEKPENI, E., EOUZAN, J. AND DIAMONDÉ, T. (1986). Utilisation du piégages dans la lutte anti-tsétsé: effets des supports traités et non traités. *Tropical Medical Parasitology* **37**, 295–297.

DENLINGER, D.A. AND MA, W.C. (1974). Dynamics of the pregnancy cycle in the tsetse *Glossina morsitans. Journal of Insect Physiology* **20**, 1015–1026.

FILLEDIER, J. AND POLITZAR, H. (1985). Efficacité relative de différentes formes de leurres sur trois espèces de glossines présentes au Burkina Faso (*G. morsitans submorsitaus, G. tachinoides, G. polpolis gambiensis*). *Revue d'Elevage et de Médecine Véterinaire des Pays Tropicaux* **38**, 358–363.

FLINT, S. (1985). A comparison of various traps for *Glossina* spp. and other Diptera. *Bulletin of Entomological Research* **75**, 529–534.

GIBSON, G. AND BRADY, J. (1988). Flight behaviour of tsetse flies in host odour plumes: the initial response to entering or leaving odour. *Physiological Entomology* **13**, 29–42.

GILLOTT, C. AND LANGLEY, P.A. (1981). The control of receptivity and ovulation in the tsetse fly, *Glossina morsitans. Physiological Entomology* **6**, 269–281.

GOUTEUX, J.P., BANSIMBA, P., BISSADIDI, N. AND NOIREAU, F. (1987). La prise en charge de la lutte contre les tsétsé par les communautés rurales: premiers essais dans cinq villages congolais. *Annales de la Société Belge de Médicine Tropicale* **67**, 37–49.

GRAHAM, P. (1964). Destruction of birds and other wildlife, by dieldrix spraying against tsetse fly in Bechuanaland. *Arnoldia* **1**, 1–14.

GREEN, C.H. (1986). Effects of colours and synthetic odours on the attraction of *Glossina pallidipes* and *G. morsitans morsitans* to traps and screens. *Physiological Entomology* **11**, 411–421.

GREEN, C.H. (1989). The use of two-coloured screens for catching *Glossina palpalis palpalis* (Robineau-Desvoidy) (Diptera: Glossinidae). *Bulletin of Entomological Research* **79**, 81–93.

GREEN, C.H. AND COSENS, D. (1983). Spectral responses of the tsetse fly, *Glossina morsitans morsitans*. *Journal of Insect Physiology* **29**, 795–800.

GREEN, C.H. AND FLINT, S. (1986). An analysis of colour effects in the performance of the F2 trap against *Glossina pallidipes* Austen and *G. morsitans morsitans* Westwood (Diptera: Glossinidae). *Bulletin of Entomological Research* **76**, 409–418.

HALL, M.J.R. (1988). Characterisation of the sexual responses of male tsetse flies, *Glossina morsitans morsitans* to pheromone-baited decoy 'females' in the field. *Physiological Entomology* **13**, 49–58.

HALL, M.J.R. AND LANGLEY, P.A. (1989). The responses of individual males in an isolated population of *Glossina morsitans morsitans* Westwood (Diptera: Glossinidae) to pheromone-baited decoy 'females'. *Bulletin of Entomological Research* **79**, 319–334.

HARGROVE, J.W. (1976). The effect of human presence on the behaviour of tsetse (*Glossina* spp.) near a stationary ox. *Bulletin of Entomological Research* **66**, 173–178.

HARGROVE, J.W. (1977). Some advances in the trapping of tsetse (*Glossina* spp.) and other flies. *Ecological Entomology* **2**, 123–137.

HARGROVE, J.W. (1980). Improved estimates of the efficiency of traps for *Glossina morsitans morsitans* Westwood and *G. pallidipes* Austen (Diptera: Glossinidae) with a note on the effect of the concentration of accompanying host odour on efficiency. *Bulletin of Entomological Research* **70**, 579–587.

HARGROVE, J.W. (1988). Tsetse: the limits to population growth. *Medical and Veterinary Entomology* **2**, 203–217.

HARGROVE, J.W. AND LANGLEY, P.A. (1990). Sterilizing tsetse (Diptera: Glossinidae) in the field: a successful trial. *Bulletin of Entomological Research* **80**, in press.

HARGROVE, J.W. AND VALE, G.A. (1978). The effects of host odour concentration on catches of tsetse flies (Glossinidae) and other Diptera in the field. *Bulletin of Entomological Research* **68**, 607–612.

HARGROVE, J.W. AND VALE, G.A. (1979). Aspects of the feasibility of employing odour-baited traps for controlling tsetse flies (Diptera: Glossinidae). *Bulletin of Entomological Research* **69**, 283–290.

HARRIS, R.H.T.P. (1932). Some facts and figures regarding the attempted control of *Glossina pallidipes* in Zululand. *South African Journal of Science* **29**, 495–507.

HASSANALI, A., MCDOWELL, P.G., OWAGA, M.L.A. AND SAINI, R.K. (1986). Identification of tsetse attractants from excretory products of a wild host animal, *Syncerus caffer*. *Insect Science and its Application* **7**, 5–9.

JAENSON, T.G.T. (1980). Mating behaviour of females of *Glossina pallidipes* Austen (Diptera: Glossinidae). *Bulletin of Entomological Research* **70**, 49–60.

JORDAN, A.M. (1986). *Trypanosomiasis Control and African Rural Development*. Longman, London.

KNIPLING, E.F. (1955). Possibilities of insect population control or eradication through the use of sexually sterile males. *Journal of Economic Entomology* **48**, 459–462.

KOEMAN, J.H., RIJKSEN, H.D., SMIES, M., NA'ISA, B.K. AND MACLENNAN, K.J.R. (1978). Faunal changes in a swamp habitat in Nigeria sprayed with insecticide to exterminate *Glossina*. *Netherlands Journal of Zoology* **21**, 443–463.

LANCIEN, J. (1981). Description du piége monoconique utilisé pour l'élimination des glossines en Republique Populair du Congo. *Cahiers ORSTOM Serie Entomologie Médicale et Parasitologie* **19**, 235–238.

LANGLEY, P.A. AND STAFFORD, K. (1990). Feeding frequency in relation to reproduction in *Glossina morsitans morsitans* and *G. pallidipes*. *Physiological Entomology* **15**, 415–421.

LANGLEY, P.A. AND WALL, R. (1990). The implications of hunger in the Tsetse fly, *Glossina pallidipes*, in relation to its availability to trapping techniques. *Journal of Insect Physiology*, in press.

LANGLEY, P.A. AND WEIDHAAS, D. (1986). Trapping as a means of controlling tsetse, *Glossina* spp. (Diptera: Glossinidae): the relative merits of killing and sterilization. *Bulletin of Entomological Research* **76**, 89–95.

LANGLEY, P.A., FELTON, T. AND OOUCHI, H. (1988). Juvenile hormone mimics as effective sterilants for the tsetse fly *Glossina morsitans morsitans. Medical and Veterinary Entomology* **2**, 29–35.

LANGLEY, P.A., HARGROVE, J.W. AND WALL, R. (1990). Maturation of the tsetse fly *Glossina pallidipes* Austen (Diptera: Glossinidae) in relation to trap-orientated behaviour. *Physiological Entomology* **14**, 179–186.

LANGLEY, P.A., HUYTON, P.M. AND CARLSON, D.A. (1978). Sex pheromone perception by males of the tsetse fly *Glossina morsitans morsitans. Physiological Entomology* **12**, 425–433.

LANGLEY, P.A., PIMLEY, R.W. AND CARLSON, D.A. (1975). Sex recognition pheromone in tsetse fly. *Nature* **254**, 51–53.

LANGLEY, P.A., COATES, T.W., CARLSON, D.A., VALE, G.A. AND MARSHALL, J. (1982). Prospects for autosterilisation of tsetse flies, *Glossina* spp. (Diptera: Glossinidae), using sex pheromone and bisazir in the field. *Bulletin of Entomological Research* **72**, 319–327.

LANGLEY, P.A., FELTON, T., STAFFORD, K. AND OOUCHI, H. (1990). Formulation of pyriproxyfen, a juvenile hormone mimic, for tsetse control. *Medical and Veterinary Entomology* **4**, 127–133.

LAVEISSIÈRE, C., COURET, D. AND GREBAUT, P. (1987). Recherches sur les écrans pour la lutte contre les glossines en region forestiere de Côte d'Ivoire. Mise au point d'un nouvel écran. *Cahiers ORSTOM Serie Entomologie Médicale et Parasitologie* **25**, 145–164.

LAVEISSIÈRE, C., COURET, D. AND MANNO, A. (1987). Importance de la nature des tissus dans la lutte par piégage contre les glossines. *Cahiers ORSTOM Serie Entomologie Médicale et Parasitologie* **25**, 133–144.

LAVEISSIÈRE, C., GOUTEUX, J.P. AND COURET, D. (1980). Essais de methodes de lutte contre les glossines en zone preforestiere de Côte d'Ivoire. 2. Resultats quantitatifs obtenus sur les populations de *Glossina palpalis* s.l. *Cahiers ORSTOM Serie Entomologie Médicale et Parasitologie* **18**, 245–259.

MEROT, P. AND FILLEDIER, J. (1985). Efficacité contre *Glossina morsitans submorsitans* d'écrans de differentes couleurs, avec ou sans adjonction de panneaux en moustiquaire noire. *Revue d'Élevage et de Médicine Veterinaire des Pays Tropicaux* **38**, 64–71.

MEROT, P., FILLEDIER, J. AND MULATO, C. (1988). Pouvoir attractif, pour *Glossina tachinoides*, des produits chimiques isolés des odeurs animales. *Revue d'Élevage et de Médicine Veterinaire des Pays Tropicaux* **41**, 79–85.

MEWS, A.R., LANGLEY, P.A., PIMLEY, R.W. AND FLOOD, M.E.T. (1977). Large scale rearing of tsetse flies (*Glossina* spp.) in the absence of a living host. *Bulletin of Entomological Research* **67**, 119–128.

MORRIS, K.R.S. AND MORRIS, M.G. (1949). The use of traps against tsetse in West Africa. *Bulletin of Entomological Research* **39**, 491–528.

NAPIER BAX, S. (1937). The senses of smell and sight in *Glossina swynnertoni. Bulletin of Entomological Research* **28**, 539–582.

NASH, T.A.M., JORDAN, A.M. AND BOYLE, J.A. (1968). The large scale rearing of *Glossina austeni* Newst. in the laboratory. IV. – The final technique. *Annals of Tropical Medicine and Parasitology* **62**, 336–341.

OLANDUNMADE, M.A., FELDMANN, U., TAKKEN, W., TENABE, S.O., HAMANN, H.J., ONAH, J., DENGWAT, L., VAN DER VLOEDT, A.M.V. AND GINGRICH, R.E. (1990). Eradication of *Glossina palpalis palpalis* (Robineau Desvoidy) (Diptera: Glossinidae) from agropastoral land in central Nigeria by means of the sterile insect technique. In *Sterile Insect Technique for Tsetse Control and Eradication. Proceedings of the final research coordination meeting organized by the joint*

*FAO/IAEA Division of Nuclear Techniques in Food and Agriculture, Vom, Plateau State, Nigeria, June 1988*, pp. 5–23. IAEA, Vienna.

OWAGA, M.L.A. (1985). Observations on the efficacy of buffalo urine as a potent olfactory attractant for *Glossina pallidipes* Austen. *Insect Science and its Application* **6**, 561–566.

OWAGA, M.L.A. AND CHALLIER, A. (1985). Catch composition of the tsetse *Glossina pallidipes* Austen in revolving and stationary traps with respect to age, sex ratio and hunger stage. *Insect Science and its Application* **6**, 711–718.

OWAGA, M.L.A., HASSANALI, A. AND MCDOWELL, P.G. (1988). The role of 4-cresol and 3-*n*-propylphenol in the attraction of tsetse flies to buffalo urine. *Insect Science and its Application* **9**, 95–100.

PHELPS, R.J. AND VALE, G.A. (1978). Studies on populations of *Glossina morsitans morsitans* and *G. pallidipes* (Diptera: Glossinidae) in Rhodesia. *Journal of Applied Ecology* **15**, 743–760.

POLITZAR, H. AND CUISANCE, D. (1982). SIT in the control and eradication of *Glossina palpalis gambiensis*. In *Proceedings of the Symposium on Sterile Insect Technique and Radiation in Insect Control*, pp. 101–109. International Atomic Energy Agency, Vienna.

POLITZAR, H. AND MEROT, P. (1984). Attraction of the tsetse fly *Glossina morsitans submorsitans* to acetone, 1-octen-3-ol and the combination of these compounds in West Africa. *Revue d'Élevage et de Médicine Veterinaire des Pays Tropicaux* **37**, 468–473.

RANDOLPH, S.E. AND ROGERS, D.J. (1978). Feeding cycles and flight activity in field populations of tsetse (Diptera: Glossinidae). *Bulletin of Entomological Research* **68**, 655–671.

RANDOLPH, S.E. AND ROGERS, D.J. (1986). The use of discriminant analysis for the estimation of sampling biases for female tsetse. *Ecological Entomology* **11**, 205–220.

RANDOLPH, S.E., DRANSFIELD, R.D. AND ROGERS, D.J. (1989). Effect of host odours on trap catch composition of *Glossina pallidipes* in Kenya. *Medical and Veterinary Entomology* **3**, 297–306.

ROGERS, D.J. AND RANDOLPH, S.E. (1986). Detection of activity cycles of tsetse from capture recapture data. *Ecological Entomology* **11**, 95–109.

ROGERS, D.J., RANDOLPH, S.E. AND KUZOE, F.A.S. (1984). Local variation in the population dynamics of *Glossina palpalis palpalis* (Robineau-Desvoidy). I. Natural population regulation. *Bulletin of Entomological Research* **74**, 403–423.

SMIES, M. (1980). Environmental impact of tsetse control/eradication operations. *Tropical Pest Management* **26**, 237–240.

SWYNNERTON, C.F.M. (1933). Some traps for tsetse flies. *Bulletin of Entomological Research* **24**, 69–192.

TAKKEN, W., BALK, F., JANSEN, R.C. AND KOEMAN, J.H. (1978). The experimental application of insecticides from a helicopter for the control of riverine populations of *Glossina tachinoides* in West Africa. VI. Observations on side effects. *PANS* **24**, 455–466.

TAKKEN, W., OLANDUNMADE, M.A., DENGWAT, L., FELDMANN, H.H., ONAH, J.A., TENABE, S.O. AND HAMANN, H.J. (1986). The eradication of *Glossina palpalis palpalis* (Robineau-Desvoidy) (Diptera: Glossinidae) using traps, insecticide-impregnated targets and the sterile insect technique in central Nigeria. *Bulletin of Entomological Research* **76**, 275–286.

TANNOCK, J., HOWELLS, W.W. AND PHELPS, R.J. (1983). Chlorinated hydrocarbon pesticide residues in eggs of some birds in Zimbabwe. *Environmental Pollution (B)* **5**, 147–155.

TORR, S.J. (1989). The host-orientated behaviour of tsetse flies (*Glossina*): the interaction of visual and olfactory stimuli. *Physiological Entomology* **14**, 325–340.

TORR, S.J. (1990). Dose responses of tsetse flies (*Glossina*) to carbon dioxide, acetone and octenol in the field. *Physiological Entomology* **15**, 93–103.

TURNER, D.A. AND BRIGHTWELL, R. (1986). An evaluation of a sequential aerial spraying operation against *Glossina pallidipes* Austen (Diptera: Glossinidae) in the Lambwe Valley of Kenya: aspects of post-spray recovery and evidence of natural population regulation. *Bulletin of Entomological Research* **76**, 331–349.

VALE, G.A. (1974). New field methods for studying the responses of tsetse flies (Diptera: Glossinidae) to hosts. *Bulletin of Entomological Research* **64**, 199–208.

VALE, G.A. (1980). Flight as a factor in the host finding behaviour of tsetse flies (Diptera: Glossinidae). *Bulletin of Entomological Research* **70**, 299–307.

VALE, G.A. (1982a). The improvement of traps for tsetse flies (Diptera: Glossinidae). *Bulletin of Entomological Research* **72**, 95–106.

VALE, G.A. (1982b). The trap orientated behaviour of tsetse flies (Glossinidae) and other Diptera. *Bulletin of Entomological Research* **72**, 71–93.

VALE, G.A. AND HALL, D.R. (1985a). The role of 1-octen-3-ol, acetone and carbon dioxide in the attraction of tsetse flies, *Glossina* spp. (Diptera: Glossinidae), to ox odour. *Bulletin of Entomological Research* **75**, 209–217.

VALE, G.A. AND HALL, D.R. (1985b). The use of 1-octen-3-ol, acetone and carbon dioxide to improve baits for tsetse flies, *Glossina* spp. *Bulletin of Entomological Research* **74**, 143–152.

VALE, G.A., BURSELL, E. AND HARGROVE, J.W. (1985). Catching-out the tsetse fly. *Parasitology Today* **1**, 106–110.

VALE, G.A., HALL, D.R. AND GOUGH, A.E.J. (1988). The olfactory responses of tsetse flies, *Glossina* spp. (Diptera: Glossinidae) to phenols in the field. *Bulletin of Entomological Research* **78**, 293–300.

VALE, G.A., HARGROVE, J.W., COCKBILL, G.F. AND PHELPS, R.J. (1986). Field trials of baits to control populations of *Glossina morsitans morsitans* Westwood and *G. pallidipes* Austen (Diptera: Glossinidae). *Bulletin of Entomological Research* **76**, 179–193.

VALE, G.A., LOVEMORE, D.F., FLINT, S. AND COCKBILL, G.F. (1988). Odour baited targets to control tsetse flies, *Glossina* spp. (Diptera: Glossinidae) in Zimbabwe. *Bulletin of Entomological Research* **78**, 31–49.

WALL, R. (1988a). Analysis of the mating activity of male tsetse flies *Glossina m. morsitans* and *G. pallidipes* in the laboratory. *Physiological Entomology* **13**, 103–110.

WALL, R. (1988b). Tsetse mating behaviour: effects of age and hunger in *Glossina morsitans morsitans* and *G. pallidipes*. *Physiological Entomology* **13**, 479–486.

WALL, R. (1989a). The roles of vision and olfaction in mate location by the tsetse fly *Glossina morsitans morsitans*. *Medical and Veterinary Entomology* **3**, 147–152.

WALL, R. (1989b). Sexual responses of males of *Glossina morsitans morsitans* and *G. pallidipes* to traps and targets in the field. *Bulletin of Entomological Research* **79**, 335–343.

WESSELS, C.L., TANNOCK, J., BLAKE, D. AND PHELPS, R.J. (1980). Chlorinated hydrocarbon insecticide residues in *Crocodilus niloticus* Laurentius from Lake Kariba. *Transactions of the Zimbabwe Scientific Association* **60**, 11–17.

WILLIAMSON, D.L., DAME, D.A., GATES, D.B., COBB, P.E., BAKULI, B. AND WARNER, P.V. (1983). Integration of insect sterility and insecticides for control of *Glossina morsitans morsitans* Westwood (Diptera: Glossinidae) in Tanzania. V. The impact of sequential releases of sterilised flies. *Bulletin of Entomological Research* **73**, 391–404.

WILSON, V.J. (1972). Observations on the effect of dieldrin on wildlife during tsetse fly *Glossina morsitans* control operations in eastern Zambia. *Arnoldia* **5**, 1–12.

# 6
# Anhydrobiotic Nematodes

JOHN BARRETT

*Department of Biological Sciences, University College of Wales, Aberystwyth, Dyfed SY24 3DA, UK*

Introduction
Characteristics of anhydrobiotic nematodes
Structural adaptations
Metabolic changes during desiccation
Metabolic changes during rehydration
Membrane function in anhydrobiotes
Commercial aspects
Conclusions
References

## Introduction

Water comprises 50–95% of the fresh weight of most animals and plants and is normally considered essential for life. The majority of biochemical reactions take place in an aqueous environment and water acts both as a solvent for, and a participant in, many enzyme reactions. In addition, water plays an intimate role in the structure of biological membranes and macromolecules, particularly DNA. All organisms can withstand a certain degree of desiccation, survival being dependent on the amount of body water which can be lost without causing irreversible damage. Most animals die if they lose more than 15–20% of their body water, although some annelids can lose up to 70% and still survive. There are, however, organisms that can lose 95–98% of their body water and survive in a desiccated state for long periods of time, 20 or 30 years in some instances. When rehydrated they become active again and resume normal development. The sorts of organisms that can enter this state of anhydrobiosis include bacterial spores, rotifers, tardigrades, arthropod larvae and nematodes. Mosses, algae, lichens, fungi and a few higher plants, such as the resurrection plant (*Craterostigma plantagineum* Hochst) are also able to enter a state of anhydrobiosis. The water content of many plant seeds drops to 5% or less, and they, too, must face the same problems as anhydrobiotes.

Characteristically anhydrobiotic organisms do not secrete a protective cyst around themselves, but stay in direct contact with their environment. When

*Agricultural Zoology Reviews* — Vol. 4, February 1991
0269–0543 $20.00 + $0.00 © Intercept Ltd, PO Box 716, Andover, Hampshire SP10 1YG, UK

desiccated, they have no detectable metabolism and are resistant to extremes of temperature and to metabolic poisons which would normally be lethal to the active organism.

Some of the most successful anhydrobiotic organisms are nematodes, which is perhaps not surprising when one considers the vast range of habitats they can occupy. Amongst those species of nematodes that are able to become anhydrobiotic there is a great deal of variability. Some species must be dried slowly and carefully if they are to survive, others can be dried rapidly *in vacuo* and still be rehydrated successfully. This variability may be related to the different degrees of environmental stress the nematodes normally encounter during their life cycle (Womersley, 1987; Womersley and Ching, 1989). Different stages in the life cycle of the same species of nematode can also differ considerably in their ability to survive desiccation, there usually being one stage which is particularly adapted for anhydrobiosis. In *Ditylenchus dipsaci* (Kuhn, 1857) Filipjev, 1936, for example, it is the fourth-stage juveniles which have the ability to survive extreme desiccation, although other stages, including the adult, are also resistant to drying (Perry, 1977a).

## Characteristics of anhydrobiotic nematodes

Anhydrobiotic nematodes (and anhydrobiotes in general) show a number of common characteristics. Firstly, they survive desiccation much better if they are allowed to dry slowly (Goodey, 1958; Crowe and Madin, 1974; Perry, 1977a; Babatola, 1980; Glazer and Orion, 1983; Saeed and Rossner, 1984; Townshend, 1984). Rates of water loss in excess of 3%/h usually result in reduced survival (Demeure and Freckman, 1981). Adaptations to reduce the rate of water loss are mostly behavioural and include coiling, which reduces the exposed surface area of the nematode (Womersley, 1978; Demeure, Freckman and van Gundy, 1979; Wharton, 1982a), and the formation of clumps of 'eelworm wool'. Clumping slows down the rate of water loss, and worms from the centre of such clumps survive better than worms on the surface (Ellenby, 1969; Evans and Perry, 1976; Perry, 1977a). One of the factors that triggers the coiling response in nematodes is the restriction of lateral movement, brought about by the contracting water film as their habitat dries out; a rise in temperature and increase in osmotic pressure may also be involved (Wharton, 1981a,b, 1982b; Wharton, Perry and Beane, 1983). In *Rotylenchus robustus* Thorne, 1949, desiccation is accompanied by contraction of the cuticular annuli and the animal shortens rather like a concertina, again reducing the exposed surface area (Rossner, 1973). Plant parasitic nematodes can restrict their rate of water loss by drying out inside their plant host, often within specialized galls. In some animal parasitic nematodes the outer cuticle of ensheathed infective stages may also restrict water movement (Ellenby, 1968), although, when hydrated, the outer cuticle of the ensheathed juveniles of *Trichostrongylus colubriformis* Giles, 1892 seems to be freely permeable to water (Wharton, Perry and Beane, 1983). Adults of *Rotylenchulus reniformis* Linford and Oliveira, 1940 retain the moulted cuticles of the juvenile stages until they become parasitic. These moulted cuticles appear to be essential in controlling water loss for desiccation survival (Gaur and Perry, 1991).

There are probably two reasons why slow dehydration is best for survival. Rapid drying can result in mechanical distortion and physical damage. Rapid drying may also not allow sufficient time for any biochemical changes which may be required before entering anhydrobiosis.

There is usually a better percentage recovery if the dried organisms are rehydrated slowly (Crowe and Madin, 1975; Evans and Perry, 1976). That is, if the anhydrobiotic nematodes are allowed to equilibrate at a high relative humidity before being put into water. Again, rapid rehydration may lead to mechanical damage and physical distortion.

When anhydrobiotic nematodes are rehydrated, there is a delay between immersion in water and the commencement of activity (Barrett, 1982). This 'lag phase' increases with the length of time the animals have been kept desiccated. Similarly, the longer the anhydrobiotic nematodes have been stored dried, the lower the final percentage recovery after rehydration. In the anhydrobiotic state there is no metabolism and so there is no turnover of tissue components and no repair. If an anhydrobiotic organism is damaged, this has no effect until it is rehydrated when, if the damage is severe, the organism dies. While the nematodes are dry they accumulate random chemical and physical damage which is not being repaired. This accumulation of random damage with time is probably responsible for the decrease in viability as the dried organisms age. Similarly, as the dried organisms age, the time needed for repair before they can fully recover increases, thus lengthening the lag phase. However, the fact that the anhydrobiotic nematodes are dehydrated does help to reduce adventitious chemical reactions. These reactions can be further reduced by storing the dried material in the dark and at low temperature. Both of these treatments improve survival of dried worms, as does storing under an inert gas to exclude oxygen (Barrett, unpublished).

Anhydrobiotic nematodes are not able to survive repeated cycles of drying and rehydration (Schmidt, Todd and Levine, 1974; Perry, 1977b). This may be due in part to there being insufficient time during the hydrated phase to repair damage caused by desiccation, but it may also be due to the loss of ions and metabolites. During rehydration anhydrobiotic nematodes can lose up to 20% of their metabolite and ion content before membrane function is fully restored (Crowe, O'Dell and Armstrong, 1979; Womersley, 1981a, b; Barrett, 1982).

In the dried state, anhydrobiotic nematodes have an increased resistance to environmental extremes and ionizing radiation, and are unaffected by metabolic poisons, including nematicides (Demeure and Freckman, 1981; Barrett, 1982; Crowe, Jackson and Crowe, 1983; Antoniou, 1989). In particular, anhydrobiotic nematodes show an increased resistance to thermal stress. In the active hydrated state *Anguina tritici* (Steinbuch, 1799) Chitwood, 1939 and *D. dipsaci* are killed by freezing but when anhydrobiotic they can survive exposure to liquid nitrogen. Similarly, *Scutellonema cavenessi* Sher, 1964 dies at 50°C but when anhydrobiotic can survive limited exposure to 80°C

(Demeure and Freckman, 1981) and the anhydrobiotic juveniles of *A. tritici* can survive exposure to 105°C for at least 2 min (Bird and Buttrose, 1974).

**Structural adaptations**

When desiccated, anhydrobiotic nematodes show surprisingly little morphological change, cell membranes remain intact and cellular organelles become packed together in an almost crystalline array (Bird and Buttrose, 1974; Crowe, Lambert and Crowe, 1978). Many anhydrobiotic nematodes have a high lipid content (up to 40% of the dry weight in *D. dipsaci*) and after desiccation the lipid may form a matrix which protects the cell organelles. However, Krusberg (1967) has shown that there is no correlation between the lipid content of different species of nematode and their ability to survive desiccation. In species that can survive desiccation there does seem to be a relationship between the quantity of stored food reserves (lipids and carbohydrate) and desiccation survival (Huang and Huang, 1974; Demeure *et al.*, 1978). However, this may merely show that healthy nematodes can survive stress better than starved ones.

In the dehydrated nematode the main changes are in the cuticle, which decreases in thickness, the epicuticle, which becomes more prominent, and in the muscle cells where the sarcoplasm condenses (Bird and Buttrose, 1974; Wharton and Barrett, 1985; Preston and Bird, 1987; Allan and Wharton, 1990). During rehydration there is a slow and ordered series of changes, the muscle sarcoplasm swells, lipid droplets coalesce and the mitochondria swell, before regaining their normal size and shape.

Anhydrobiotic nematodes shrink during desiccation, although different nematodes may show different patterns of shrinkage in terms of relative changes in length and diameter (Wharton, Barrett and Perry, 1985). In the dry state, anhydrobiotic nematodes are freely permeable to water (at least in the inward direction) and rehydrate rapidly on immersion (*Figure 1*). In the fourth-stage juveniles of *D. dipsaci* this is a two-stage process, an initial rapid uptake over the first 10 min to reach approximately 70% of the final water content, followed by a slower rate to give full hydration after about 5 h (Perry, 1977c; Wharton, Barrett and Perry, 1985). There also seems to be a direct correlation between rate of water uptake and the time taken for the individual worm to become active, those which take up water fastest becoming active first. An interesting feature during rehydration in both *D. dipsaci* and *T. colubriformis* is a marked transient shrinkage after about 60 min hydration, at a time when the worms would be expected to be still swelling as they take up water (Wharton *et al.*, 1988; Allan and Wharton, 1990). This 'anomalous shrinkage' occurs just before the worms start to move and may be related to changes taking place in the muscles as the contractile mechanism becomes active.

As the dehydrated nematodes take up water there is initially a transient increase in cuticular permeability, followed by a rapid fall as the worms hydrate (*Figure 1*) (Preston and Bird, 1987; Wharton *et al.*, 1988). Studies suggest that the re-establishment of the cuticular permeability barrier during rehydration may have two components, a physical effect due to hydration of the cuticle and a metabolic component. Although it has not been definitely established, the site of the cuticular permeability barrier is probably the epicuticle and the metabolic component may involve the synthesis of new lipid constituents. Metabolic energy is also required to maintain cuticular impermeability in the fully hydrated nematode and cuticular permeability increases again if the worms die or are treated with metabolic inhibitors (Rossner and Perry, 1975; Wharton *et al.*, 1988). Interestingly, as *D. dipsaci* juveniles are dehydrated there appears to be a decrease in cuticular permeability, which would have the effect of reducing water loss (Perry, 1977c). One possibility is that the nematode cuticle shows asymmetric permeability, allowing water movement in more rapidly than movement out. Similar cuticular asymmetries have been proposed for insects, but insects, unlike nematodes, are also able to take up water directly from the atmosphere (Beament, 1964).

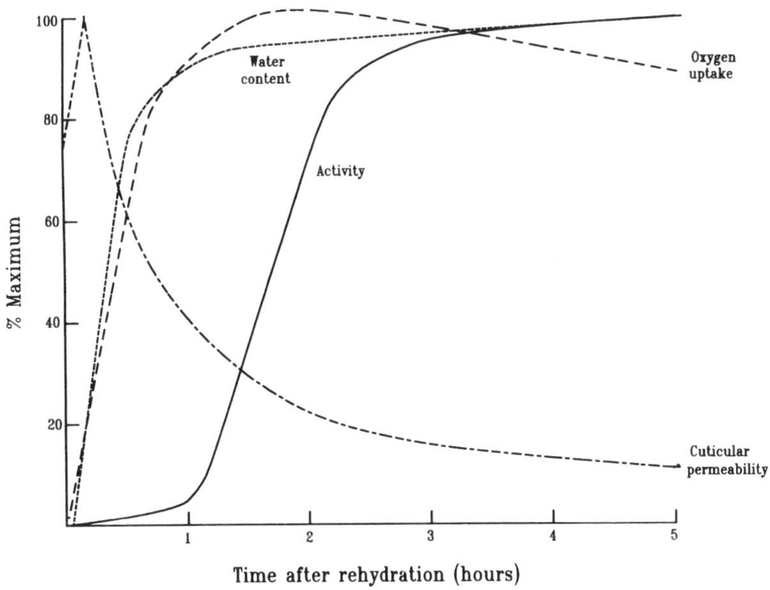


**Figure 1.** Changes in fourth-stage juveniles of *Ditylenchus dipsaci* during rehydration.

## Metabolic changes during desiccation

There have been no systematic studies on the way metabolism changes during dehydration in anhydrobiotic nematodes. As a result of work on the brine shrimp *Artemia*, Clegg (1979) has proposed a general model for anhydrobiotic tissue in which there are three distinct water-dependent metabolic domains, each separated from the other by a critical level of hydration (*Figure 2*). Within the range 100–40% water content, metabolism is essentially normal and there is sufficient water to form a continuous bulk phase. In the range 20–40% water content, there is insufficient water in the cell cytoplasm to form a continuous aqueous phase and water is confined to interfacial locations ('vicinal water'). At these water levels metabolism is restricted to isolated enzyme reactions, since with no bulk phase there can be no transport of metabolites and sequential reactions become impossible. Below about 20% hydration, there is no free water in the cell and consequently no metabolism. This 20% is often referred to as 'bound water'. However, the true amount of structural water in the cell (i.e. water intimately involved in macromolecular structure) is thought to be considerably less than this, in the range 0.15–0.4% (Andronikashvili, Mrevlishvili and Privalov, 1969). The range 0–20% water content can be further divided by NMR into localized, mobile association and gel regions (*Figure 2*). The physical state of bound water in the cell is unclear. It may represent an organized water layer up to several hundred molecules thick over the surface of cell membranes and proteins, and it may or may not be freely exchangeable with the bulk phase. The water content of dry, anhydrobiotic nematodes is of the order of 1–5%, so most of the bound water has been lost. However, there is no evidence that anhydrobiotic organisms can survive the total loss of structural water.

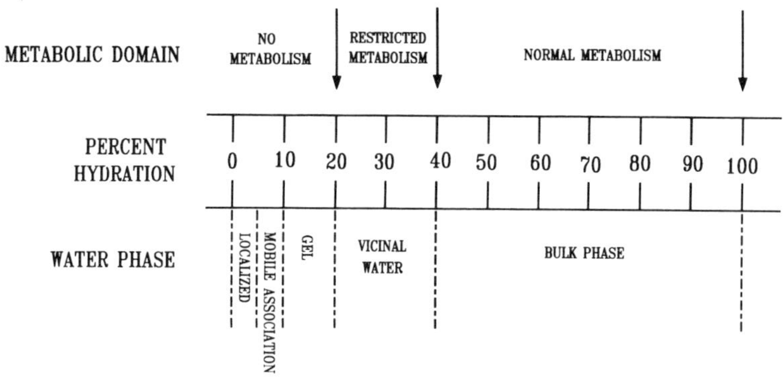


**Figure 2.** Hydration-dependent metabolic events in anhydrobiotic tissue.

One biochemical adaptation that has been noted during water loss in anhydrobiotic nematodes is the accumulation of polyols. In *Aphelenchus avenae* Bastian, 1865, glycogen and lipid reserves are converted during desiccation to glycerol and trehalose (Madin and Crowe, 1975), whilst *Ditylenchus myceliophagus* Goodey, 1958 accumulates glycerol and ribitol (Womersley, 1987). The pathway of trehalose synthesis in *A. avenae* has been studied and seems similar to that in other organisms (Loomis, Madin and Crowe, 1980; Loomis, O'Dell and Crowe, 1980). Lipid depletion during desiccation has been noted in several other anhydrobiotic nematodes (Demeure and Freckman, 1981; Storey, Glazer and Orion, 1982). However, no depletion of lipid reserves during desiccation takes place in *A. tritici* or *D. dipsaci*, which accumulate varying amounts of trehalose. Both of these nematodes also contain glycerol and inositol, but there is no evidence that these compounds are preferentially synthesized prior to anhydrobiosis (Womersley and Smith, 1981; Womersley, Thompson and Smith, 1982). Interestingly, in studies with bacteria, inositol was found to be the most efficient polyol at replacing bound water (Webb, 1964). Pretreatment of *D. dipsaci, D. myceliophagus* and *A. tritici* with inositol improves their ability to survive desiccation (Womersley, 1981c). Finally, some anhydrobiotic nematodes, such as *Plectus* sp. Yeates, 1970 and *Orrina phyllobia* (Thorne, 1934) Brzeski, 1981, which like *D. dipsaci* and *A. tritici*, can survive extremely rapid drying, seem to accumulate little or no extra polyols during desiccation (Hendriksen, 1983; Robinson, Orr and Heintz, 1984). It is possible that nematodes that can survive rapid drying normally contain high levels of tissue polyols, even when unstressed.

Polyols are thought to be able to protect cell membranes and intracellular proteins against denaturation, although the mechanisms involved are unclear. By virtue of their hydroxyl groups, polyols may be able to replace bound water associated with polar side-groups on proteins and phospholipids. This would maintain the balance between hydrophilic and hydrophobic forces acting on the molecules and prevent them from 'collapsing' in on themselves. In replacing bound water, the polyols may also prevent 'proximity effects', i.e. the cross-linking of molecules and membranes that can occur when they are brought into close contact as the bulk aqueous phase contracts. Proximity effects involve the formation of both covalent bonds (sulphydryl bonds and ester links) and of electrostatic and salt bridges. As cell water is lost, the ionic strength in the cytoplasm increases and this in turn can lead to protein denaturation with potassium ions displacing calcium ions from membranes.

In tissues where glycerol is accumulated it may also, to a certain extent, replace bulk water, since glycerol is liquid at ambient temperatures and is a fairly good solvent for hydrophilic molecules. Finally, polyols such as glycerol, ribitol and trehalose can all act as free-radical scavenging agents, and thus reduce random chemical damage.

Trehalose may have an added advantage in anhydrobiotic organisms in that since it is a non-reducing sugar, it is a way of storing glucose in a soluble but inert form. If tissues containing glucose are dried, the free aldehyde group of glucose can react with the amino-acid side-chains of proteins to form brown pigments called melanoidins. This is the basis of the so-called 'browning reaction' of dried blood. Not only does trehalose not participate in the browning reaction, but it also seems to suppress the reaction of other sugars with proteins (Loomis, O'Dell and Crowe, 1979).

Dehydration does not lead to any significant denaturation of the enzyme systems in anhydrobiotic nematodes (Barrett, 1982). Initially this seems rather surprising, since one would have thought that desiccation would have led to considerable loss of enzyme activity. There is no evidence that enzymes from anhydrobiotic nematodes are intrinsically more stable than enzymes from other sources, nor is there any evidence for increased levels of protein glycosylation. Although polyols undoubtedly help, the conditions in anhydrobiotic tissues, namely low water activity and high protein concentration, strongly favour protein stability. There is also no evidence that desiccation leads to any increase in the frequency of DNA breaks in anhydrobiotic nematodes (Barrett and Butterworth, 1985). In contrast, desiccation of bacteria does cause breaks in the DNA and an increased mutation rate is found in bacteria rehydrated after freeze-drying (Ashwood-Smith and Grant, 1976).

Since in the dried state there is no metabolism and, therefore, no turnover of cellular constituents, the survival of the dried organism becomes dependent on the stability of its macromolecules. Biochemical macromolecules are, in fact, considerably more stable than generally realized and it is unlikely that the survival time of an anhydrobiotic organism will be greatly limited by the inherent instability of its constituents. There are data to suggest that isolated proteins can retain their biological activity *in vitro* for at least 50 years (Keilin, 1959). Other biological macromolecules may be even more stable: lipids have been recovered from frozen mammoth tissue ($10^5$ years old), DNA has been isolated from fossil leaves ($10^7$ years old) and amino acids have been extracted from fossil brachiopods ($10^8$ years old). The short half-life of most biological molecules *in vivo* is due not to their instability but to a well-regulated mechanism for their intracellular turnover.

Dried anhydrobiotic nematodes contain measurable amounts of metabolic intermediates, including ATP (Spurr, 1976; Willett, Freckman and van Gundy, 1978; Barrett, 1982). Compared with active hydrated juveniles, the dried anhydrobiotic fourth-stage juveniles of *D. dipsaci* show a much reduced energy charge and lower levels of glycolytic intermediates, but there is little change in the levels of the tricarboxylic acid cycle intermediates (Barrett, 1982). The occurrence of high-energy intermediates in anhydrobiotic tissue is important in enabling metabolism to start again after dehydration, since all of the major catabolic pathways require priming with a high energy compound before they can function; glycolysis requires ATP, the tricarboxylic acid cycle uses acetyl-CoA and beta-oxidation needs fatty acyl-CoA. Similarly, although the levels of glycolytic intermediates are reduced in anhydrobiotic tissue, the substrates and products of the different enzymes do not come into equilibrium. If they did, the pathways would not be able to start up again because there would be no change in free-energy to drive them.

It has been suggested that, during the dehydration of anhydrobiotic tissues, metabolism may be downregulated by change in pH or through the synthesis of specific inhibitory compounds (De Herdt *et al.*, 1981; Busa and Crowe, 1983). However, in anhydrobiotic nematodes the switch-off of metabolism during desiccation does not seem to be a regulated process and there is no evidence for the synthesis of regulatory compounds (Barrett, 1982). The metabolite profile of dry *D. dipsaci* juveniles is not a defined control state. Instead, it is similar to that of any non-anhydrobiotic tissue where metabolism has been disrupted by desiccation. It would, however, be interesting to know what happens to proteases during dehydration since, if uncontrolled, they could cause considerable tissue damage. One way of inhibiting protease activity during drying is, of course, by altering the pH.

## Metabolic changes during rehydration

In the dried state, there is no measurable metabolism and although the worms take up water rapidly, there is a lag phase of 2–3 h before they become spontaneously active (*Figure 1*). However, metabolism (as judged by oxygen uptake, heat output or carbon dioxide release from labelled substrate) begins immediately after hydration and peaks during the lag phase (Bhatt and Rhode, 1970; Crowe, Madin and Loomis, 1977; Madin, Crowe and Loomis, 1978; Barrett, 1982). The metabolite levels also recover quickly and are back to within 30% of normal within 10 min. The keto acids, 2-oxoglutarate and pyruvate, may be reconstituted from glutamate and alanine by transamination. In contrast to the other intermediary metabolites, ATP levels do not recover fully for several hours.

There is no evidence for an orderly switching on of metabolic pathways during rehydration. Glycolysis, the tricarboxylic acid cycle, beta-oxidation and the respiratory chain all seem to start at once and this correlates with the presence, in the dried tissue, of all of the enzyme systems and metabolic intermediates. Immediately after rehydration the mitochondria appear to be essentially uncoupled and so ATP production would be impaired. Evidence for this comes from *D. dipsaci*, where mitochondria isolated from anhydrobiotic tissue show no respiratory control and no stimulation of oxygen uptake by 2,4-dinitrophenol (Barrett, unpublished). During rehydration the mitochondria first swell then shrink, again suggesting changes in respiratory state. The recovery of mitochondrial function correlates with the restoration of ATP levels and the onset of spontaneous activity.

Anhydrobiotic juveniles of *D. dipsaci* can be successfully rehydrated in the presence of the protein synthesis inhibitors cycloheximide and actinomycin D (Barrett, 1982). This suggests that the synthesis of new proteins is not essential for recovery and there is, as yet, no evidence for the production of specific heat-shock proteins. It is also not known if anhydrobiosis is accompanied by a change in the proportion of soluble enzymes which are bound to cell membranes. In the anhydrobiotic moss *Tortula* it has been shown that desiccation results in the dissociation of the polysomes. However, all the components necessary for protein synthesis are present in the dried moss and, following rehydration, synthesis is able to start again immediately (Bewley, 1979). In *Tortula* there is an increase in catalase and superoxide dismutase levels following rehydration and this may be related to membrane protection. A similar increase in superoxide dismutase activity after rehydration has been observed in *A. avenae* (Womersley, 1987).

## Membrane function in anhydrobiotes

A key factor in the ability of anhydrobiotic nematodes to survive cycles of desiccation and rehydration is their capacity to limit membrane disruption during dehydration and to repair damage quickly on recovery. The dried worms are thus able to restore ionic and metabolic gradients sufficiently rapidly during rehydration to prevent lethal effects.

At the ultrastructural level, the cell membranes of anhydrobiotic tissues remain intact during desiccation, although the leakage of metabolites and ions does suggest that some functional damage has occurred. It is not known if the lipid composition of the membranes from anhydrobiotes is in any way different from those of other organisms. Total lipid analyses of anhydrobiotic nematodes shows no unusual features, but subtle differences in membrane composition could easily be masked (Womersley, Thompson and Smith, 1982).

Studies with model membrane systems have shown that when the water content is reduced below about 20%, the membrane can undergo a number of transitions (Crowe and Crowe, 1982a, b). These changes include a phase shift from hydrated liquid crystal to gel (causing a drop in membrane fluidity), a lateral separation of proteins and lipids into separate domains and a transition from laminar (bilayer) to hexagonal structure. These changes result in a total loss of functional integrity on rehydration. However, the inclusion of trehalose, at physiological concentrations, allows the structural and functional integrity of the model membrane systems to be retained during desiccation and rehydration (Crowe, Crowe and Jackson, 1983; Crowe, Crowe and Chapman, 1984a, b). To exert a protective effect, trehalose must be present on both sides of the membrane. Of a range of naturally occurring carbohydrates tested, trehalose was the most effective, whereas glycerol was the least effective and actually promoted the fusion of dried membranes (Crowe *et al.*, 1984; Womersley *et al.*, 1984).

There is some evidence that nematode membranes may be more fluid and hence have lower transition temperatures than mammalian systems (Lyons, Keith and Thomason, 1975). However, this may be a general feature of nematodes and not necessarily specific to anhydrobiotic species. Freeze-fracture studies on *A. avenae* membranes show that during desiccation there is a change in the packing density of intermembrane particles, but there is no good evidence for any phase change in membrane lipids (Crowe and Crowe, 1982b).

## Commercial aspects

The occurrence of anhydrobiotic stages enables nematodes to cope with periods of drought. Anhydrobiosis can also aid dispersal in contaminated seed and plant debris or in dry soil adhering to plants or agricultural machinery. Dissemination in wind-blown dust is also a possibility.

Seed contamination is a major problem with *Aphelenchoides besseyi* Christie, 1942, *D. dipsaci* and *A. tritici*. Ironically, seed storage conditions probably prolong nematode survival. Control measures include seed cleaning (the galls of *A. tritici* can be removed by flotation or winnowing), hot-water treatment, fumigation and quarantine restrictions. Eggs and juveniles of *Meloidogyne* have been shown to be able to survive anhydrobiotically in plant debris, as can the infective stages of *Ditylenchus angustus* (Butler, 1913) Filipjev, 1936 and *D. dipsaci*. Finally, survival in dry soil may be a significant feature in the biology of *S. cavenessi* and *Rotylenchulus reniformis*. One practical consequence of this is that when extracting nematodes from dry soil sufficient time must be allowed for re-activation. Anhydrobiotic stages are probably much more common in tropical and subtropical plant parasitic nematodes than is generally realized and this may help them to survive some non-chemical control measures such as dry fallow or solarization (Luc, Sikora and Bridge, 1990). The artificial conditions created by the early lifting and drying of bulbs favours the survival of the anhydrobiotic stages of *D. dipsaci*, and the bulb industry in East Anglia and Holland was only saved from disaster earlier this century by the development of an effective hot-water control treatment. Amongst animal parasitic nematodes there are reports of outbreaks of trichostrongyle disease in cattle caused by anhydrobiotic stages surviving on hay.

One area of current interest is whether the knowledge gained from the study of anhydrobiotic nematodes can be used to induce anhydrobiosis in normally non-anhydrobiotic tissue or organisms. Trehalose (and to a lesser extent other polyols) has been found to protect purified proteins against denaturation caused by desiccation (Crowe *et al.*, 1987). However, there has been no success in protecting isolated tissues or whole organisms. Considerable effort has been put into trying to induce anhydrobiosis in the infective juveniles of insect parasitic nematodes of the genus *Steinernema* Travossos, 1927 (= *Neoaplectana*). These nematodes have considerable potential as biological control agents, and storage and distribution would be greatly simplified if they could be made anhydrobiotic. However, all attempts have again proved unsuccessful so far.

An unusual use of anhydrobiotic nematodes is as part of an experiment designed to investigate the effects of the space environment on biological materials. Of particular interest is the interaction of tissues with high-energy particles. The latter are essentially atomic nuclei, stripped of their electrons and accelerated to very high voltages ($10^{14}$ MeV). High-energy particles are relatively rare events ($1/cm^2/day$) and they do not penetrate to the Earth's surface, so have to be studied in space. Anhydrobiotic organisms offer a unique marker tissue with which to study these particles. The anhydrobiotic stages are able to survive a six-month space flight without the need for complex environmental control and, since there is no tissue repair, damage will accumulate and can be studied when the organisms return to Earth. So when the European Retrievable Carrier is launched some time in the 1990s it will be carrying with it 10 quartz discs coated with the anhydrobiotic stages of *D. dipsaci*.

## Conclusions

Anhydrobiotic stages have been found in free-living nematodes as well as plant- and animal-parasitic species. Although anhydrobiosis can aid survival and dispersal, there is no evidence that it is ever an obligate stage in the life cycle. There is no single adaptation which characterizes anhydrobiotic nematodes or, indeed, anhydrobiotes in general. Rather, there seems to be a combination of factors involved and it is this variety of factors which is responsible for the wide range of anhydrobiotic abilities found in different species. When dehydrated, anhydrobiotic nematodes show increased resistance to environmental stress and, although anhydrobiosis is usually correlated with drought survival, desiccation is also a way of avoiding freeze damage.

It seems likely that the free water in a cell has to be totally removed before metabolism can come to a complete halt. In a hydrated tissue, if metabolism were to cease completely, biochemical organization would be lost rapidly through diffusion. A key feature of anhydrobiotic organisms seems to be their ability to limit membrane damage during dehydration and to repair any damage and restore ionic and metabolic gradients sufficiently rapidly during rehydration, to prevent lethal effects.

Since there is no metabolism in the desiccated state, energy stores are conserved. Anhydrobiosis, therefore, provides a way of greatly extending the life span of small organisms such as nematodes. Short periods of activity can be interspersed with long periods of anhydrobiosis, although the total 'living time' remains unaltered. Anhydrobiosis thus delays senescence, but it cannot, of course, reverse it.

The cellular components of an active hydrated organism are in a state of dynamic equilibrium, constantly turning over. In the anhydrobiotic state there is no turnover or repair of tissue components but, in the absence of water, most biological molecules are extremely stable and the anhydrobiotic organism retains the structural organization necessary to return to an active state on rehydration. However, if this structure is damaged or destroyed, the tissue dies.

## References

ALLAN, G.S. AND WHARTON, D.A. (1990). Anhydrobiosis in the infective juveniles of *Trichostrongylus colubriformis* (Nematoda: Trichostrongylidae). *International Journal for Parasitology* **20**, 183–192.

ANDRONIKASHVILI, E.L., MREVLISHVILI, G.M. AND PRIVALOV, P.L. (1969). Calorimetric investigation of tissue water. In *Water in Biological Systems* (L.P. Kayushin, Ed), pp. 664–667. Plenum, New York.

ANTONIOU, M. (1989). Arrested development in plant parasitic nematodes. *Helminthological Abstracts B* **58**, 1–19.

ASHWOOD-SMITH, M.J. AND GRANT, E. (1976). Mutation induced in bacteria by freeze-drying. *Cryobiology* **13**, 206–213.

BABATOLA, J.O. (1980). Effect of relative humidity on the rate of desiccation and survival of *Hirschmanniella* spp. *Nematologica* **26**, 289–294.

BARRETT, J. (1982). Metabolic responses to anabiosis in the fourth stage juveniles of *Ditylenchus dipsaci* (Nematoda). *Proceedings of the Royal Society of London Series B* **216**, 159–177.

BARRETT, J. AND BUTTERWORTH, P.E. (1985). DNA stability in the anabiotic fourth-stage juveniles of *Ditylenchus dipsaci* (Nematoda). *Annals of Applied Biology* **106**, 121–124.

BEAMENT, J.W.L. (1964). The active transport and passive movement of water in insects. In *Advances in Insect Physiology* (J.W.L. Beament, J.E. Treherne and V.B. Wigglesworth, Eds), volume 2, pp. 67–129. Academic Press, New York.

BEWLEY, J.D. (1979). Physiological aspects of desiccation tolerance. *Annual Review of Plant Physiology* **30**, 195–238.

BHATT, B.D. AND ROHDE, R.A. (1970). The influence of environmental factors on the respiration of plant-parasitic nematodes. *Journal of Nematology* **2**, 277–285.

BIRD, A.F. AND BUTTROSE, M.S. (1974). Ultrastructural changes in the nematode *Anguina tritici* associated with anhydrobiosis. *Journal of Ultrastructural Research* **48**, 177–189.

BUSA, W.B. AND CROWE, J.H. (1983). Intracellular pH regulates transitions between dormancy and development of brine shrimp (*Artemia salina*) embryos. *Science* **221**, 366–368.

CLEGG, J.S. (1979). Metabolism and the intracellular environment: the vicinal-water network model. In *Cell Associated Water* (W. Drost-Hansen and J.S. Clegg, Eds), pp. 363–413. Academic Press, New York.

CROWE, L.M. AND CROWE, J.H. (1982a). Hydration-dependent hexagonal phase lipid in a biological membrane. *Archives of Biochemistry and Biophysics* **217**, 582–587.

CROWE, J.H. AND CROWE, L.M. (1982b). Induction of anhydrobiosis: membrane changes during drying. *Cryobiology* **19**, 317–328.

CROWE, J.H. AND MADIN, K.A. (1974). Anhydrobiosis in tardigrades and nematodes. *Transactions of the American Microscopy Society* **93**, 513–524.

CROWE, J.H. AND MADIN, K.A.C. (1975). Anhydrobiosis in nematodes: evaporative water loss and survival. *Journal of Experimental Zoology* **193**, 323–334.

CROWE, J.H., CROWE, L.M. AND CHAPMAN, D. (1984a). Preservation of membranes in anhydrobiotic organisms: the role of trehalose. *Science* **223**, 701–703.

CROWE, J.H., CROWE, L.M. AND CHAPMAN, D. (1984b). Infrared spectroscopic studies on interactions of water and carbohydrates with a biological membrane. *Archives of Biochemistry and Biophysics* **222**, 400–407.

CROWE, J.H., CROWE, L.M. AND JACKSON, S.A. (1983). Preservation of structural and functional activity in lyophilized sarcoplasmic reticulum. *Archives of Biochemistry and Biophysics* **220**, 477–484.

CROWE, J.H., JACKSON, S.A. AND CROWE, L.M. (1983). Nonfreezable water in anhydrobiotic nematodes. *Molecular Physiology* **3**, 99–105.

CROWE, J.H., LAMBERT, D.T. AND CROWE, L.M. (1978). Ultrastructural and freeze fracture studies on anhydrobiotic nematodes. In *Dry Biological Systems* (J.H. Crowe and J.S. Clegg, Eds), pp. 23–52. Academic Press, New York.

CROWE, J.H., MADIN, K.A.C. AND LOOMIS, S.H. (1977). Anhydrobiosis in nematodes: metabolism during resumption of activity. *Journal of Experimental Zoology* **201**, 57–64.

CROWE, J.H., O'DELL, S.J. AND ARMSTRONG, D.A. (1979). Anhydrobiosis in nematodes: permeability during rehydration. *Journal of Experimental Zoology* **207**, 431–438.

CROWE, L.M., MOURADIAN, R., CROWE, J.H., JACKSON, S.A. AND WOMERSLEY, C. (1984). Effects of carbohydrates on membrane stability at low water activities. *Biochimica et Biophysica Acta* **769**, 141–150.

CROWE, J.H., CROWE, L.M., CARPENTER, J.F. AND AURELL WISTROM, C. (1987). Stabilization of dry phospholipid bilayers and proteins by sugars. *Biochemical Journal* **242**, 1–10.

DE HERDT, E., DE VOEGHT, F., CLAUWAERT, J., KONDO, M. AND SLEGERS, H. (1981). A cryptobiosis-specific 19S protein complex of *Artemia salina* gastrulae. *Biochemical Journal* **194**, 9–17.

DEMEURE, Y. AND FRECKMAN, D.W. (1981). Recent advances in the study of anhydrobiotic nematodes. In *Plant Parasitic Nematodes* (B.M. Zuckerman and R.A. Rhode, Eds), volume 3, pp. 205–226. Academic Press, New York.

DEMEURE Y., FRECKMAN, D.W. AND VAN GUNDY, S.D. (1979). Anhydrobiotic coiling of nematodes in soil. *Journal of Nematology* **11**, 189–195.

DEMEURE, Y., REVERSAT, G., VAN GUNDY, S.D. AND FRECKMAN, D.W. (1978). The relationship between nematode reserves and their survival to desiccation. *Nematropica* **8**, 7–8.

ELLENBY, C. (1968). Desiccation survival of the infective larva of *Haemonchus contortus*. *Journal of Experimental Biology* **49**, 469–475.

ELLENBY, C. (1969). Dormancy and survival in nematodes. *Symposia of the Society for Experimental Biology* **23**, 83–97.

EVANS, A.A.F. AND PERRY, R.N. (1976). Survival strategies in nematodes. In *The Organization of Nematodes* (N.A. Croll, Ed.), pp. 383–424. Academic Press, New York.

GAUR, H.S. AND PERRY, R.N. (1991). The role of the moulted cuticles in the desiccation survival of adults of *Rotylenchulus reniformis*. *Revue de Nematologie*, in press.

GLAZER, I. AND ORION, D. (1983). Studies on anhydrobiosis of *Pratylenchus thornei*. *Journal of Nematology* **15**, 333–338.

GOODEY, J.B. (1958). *Ditylenchus myceliophagus* n.sp (Nematoda: Tylenchidae). *Nematologica* **3**, 91–96.

HENDRIKSEN, N.B. (1983). Anhydrobiosis in nematodes: studies on *Plectus* sp. In *New trends in Soil Biology. Proceedings of the 8th International Colloquium of Soil Zoology, Louvain-la-Neuve, Belgium, 1982*, pp. 387–394.

HUANG, C.S. AND HUANG, S.P. (1974). Dehydration and the survival of rice white tip nematode *Aphelenchoides besseyi*. *Nematologica* **20**, 9–18.

KEILIN, D. (1959). The problem of anabiosis or latent life: history and current concept. *Proceedings of the Royal Society of London, Series B* **150**, 149–191.

KRUSBERG, L.R. (1967). Analyses of total lipids and fatty acids of plant-parasitic nematodes and host tissues. *Comparative Biochemistry and Physiology* **21**, 83–90.

LOOMIS, S.H., MADIN, K.A.C. AND CROWE, J.H. (1980). Anhydrobiosis in nematodes: biosynthesis of trehalose. *Journal of Experimental Zoology* **211**, 311–320.

LOOMIS, S.H., O'DELL, S.J. AND CROWE, J.H. (1979). Anhydrobiosis in nematodes: inhibition of the browning reaction of reducing sugars with dry protein. *Journal of Experimental Zoology* **208**, 355–360.

LOOMIS, S.H., O'DELL, S.J. AND CROWE, J.H. (1980). Anhydrobiosis in nematodes: control of the synthesis of trehalose during induction. *Journal of Experimental Zoology* **211**, 321–330.

LUC, M., SIKORA, R.A. AND BRIDGE, J. (1990). *Plant parasitic nematodes in subtropical and tropical agriculture*. CAB International, Oxford.

LYONS, J.M., KEITH, A.D. AND THOMASON, I.J. (1975). Temperature-induced phase transitions in nematode lipids and their influence on respiration. *Journal of Nematology* **7**, 98–104.

MADIN, K.A.C. AND CROWE, J.H. (1975). Anhydrobiosis in nematodes: carbohydrate and lipid metabolism during rehydration. *Journal of Experimental Zoology* **193**, 335–342.

MADIN, K.A.C., CROWE, J.H. AND LOOMIS, S.H. (1978). Metabolic transitions in a nematode during induction of and recovery from anhydrobiosis. In *Dry Biological Systems* (J.H. Crowe and J.S. Clegg, Eds), pp. 155–174. Academic Press, New York.

PERRY, R.N. (1977a). Desiccation survival of larval and adult stages of the plant parasitic nematodes *Ditylenchus dipsaci* and *D. myceliophagus*. *Parasitology* **74**, 139–148.

PERRY, R.N. (1977b). The effect of previous desiccation on the ability of 4th-stage larvae of *Ditylenchus dipsaci* to control rate of water loss and to survive drying. *Parasitology* **75**, 215–231.

PERRY, R.N. (1977c). The water dynamics of stages of *Ditylenchus dipsaci* and *D. myceliophagus* during desiccation and rehydration. *Parasitology* **75**, 45–70.

PRESTON, C.M. AND BIRD, A.F. (1987). Physiological and morphological changes associated with recovery from anabiosis in the dauer larva of the nematode *Anguina agrostis*. *Parasitology* **95**, 125–133.

ROBINSON, A.F., ORR, C.C. AND HEINTZ, C.E. (1984). Some factors affecting survival of desiccation by infective juveniles of *Orrina phyllobia*. *Journal of Nematology* **16**, 86–91.

ROSSNER, J. (1973). Anpassung wandern der Wurzelnematoden an Trockenheit im Boden. *Nematologica* **19**, 366–378.

ROSSNER, J. AND PERRY, R.N. (1975). Water loss and associated surface changes after desiccation in *Rotylenchus robustus*. *Nematologica* **21**, 438–442.

SAEED, M. AND ROSSNER, J. (1984). Anhydrobiosis in five species of plant associated nematodes. *Journal of Nematology* **16**, 119–124.

SCHMIDT, J.M., TODD, K.S. AND LEVINE, N.D. (1974). Moisture stress effects on survival of infective *Trichostrongylus colubriformis* larvae. *Journal of Nematology* **6**, 27–29.

SPURR, H.W. (1976). Adenosine triphosphate quantification as related to cryptobiosis, nematode eggs and larvae. *Journal of Nematology* **8**, 152–158.

STOREY, R.M.J., GLAZER, I. AND ORION, D. (1982). Lipid utilization by starved and anhydrobiotic individuals of *Pratylenchus thornei*. *Nematologica* **28**, 373–378.

TOWNSHEND, J.L. (1984). Anhydrobiosis in *Pratylenchus penetrans*. *Journal of Nematology* **16**, 282–289.

WEBB, S.J. (1964). Bound water, inositol, and the effect of X-rays on *Escherichia* cells. *Canadian Journal of Microbiology* **10**, 877–885.

WHARTON, D.A. (1981a). The initiation of coiling behaviour prior to desiccation in the infective larvae of *Trichostrongylus colubriformis*. *International Journal for Parasitology* **11**, 353–357.

WHARTON, D.A. (1981b). The effect of temperature on the behaviour of the infective larvae of *Trichostrongylus colubriformis*. *Parasitology* **82**, 269–279.

WHARTON, D.A. (1982a). The survival of desiccation by the free-living stages of *Trichostrongylus colubriformis* (Nematoda: Trichostrongylidae). *Parasitology* **84**, 455–462.

WHARTON, D.A. (1982b). Observations on the coiled posture of trichostrongyle infective larvae using a freeze-substitution method and scanning electron microscopy. *International Journal for Parasitology* **12**, 335–343.

WHARTON, D.A. AND BARRETT, J. (1985). Ultrastructural changes during recovery from anabiosis in the plant parasitic nematode, *Ditylenchus*. *Tissue and Cell* **17**, 79–96.

WHARTON, D.A., BARRETT, J. AND PERRY, R.N. (1985). Water uptake and morphological changes during recovery from anabiosis in the plant-parasitic nematode, *Ditylenchus dipsaci*. *Journal of Zoology* **206**, 391–402.

WHARTON, D.A., PERRY, R.N. AND BEANE, J. (1983). The effect of osmotic stress on behaviour and water content of infective larvae of *Trichostrongylus colubriformis*. *International Journal for Parasitology* **13**, 185–190.

WHARTON, D.A., PRESTON, C.M., BARRETT, J. AND PERRY, R.N. (1988). Changes in cuticular permeability associated with recovery from anhydrobiosis in the plant parasitic nematode, *Ditylenchus dipsaci*. *Parasitology* **97**, 317–330.

WILLETT, J.D., FRECKMAN, D.W. AND VAN GUNDY, S.D. (1978). Determination of ATP levels in *Aphelenchus avenae* in the free-living and anhydrobiotic states. *Journal of Nematology* **10**, 301–302.

WOMERSLEY, C. (1978). A comparison of the rate of drying of four nematode species using a liquid paraffin technique. *Annals of Applied Biology* **90**, 401–405.

WOMERSLEY, C. (1981a). The effect of dehydration and rehydration on salt loss in the second-stage larvae of *Anguina tritici*. *Parasitology* **82**, 411–419.

WOMERSLEY, C. (1981b). Biochemical and physiological aspects of anhydrobiosis. *Comparative Biochemistry and Physiology* **70B**, 669–678.

WOMERSLEY, C. (1981c). The effect of myo-inositol on the ability of *Ditylenchus dipsaci, D. myceliophagus* and *Anguina tritici* to survive desiccation. *Comparative Biochemistry and Physiology* **68A**, 249–252.

WOMERSLEY, C. (1987). A reevaluation of strategies employed by nematode anhydrobiotes in relation to their natural environment. In *Vistas on Nematology* (J.A.

Veech and D.W. Dickson, Eds), pp. 165–173. Society of Nematologists, Hyattsville, MD.

WOMERSLEY, C. AND CHING, C. (1989). Natural dehydration regimes as a prerequisite for the successful induction of anhydrobiosis in the nematode *Rotylenchulus reniformis*. *Journal of Experimental Biology* **143**, 359–372.

WOMERSLEY, C. AND SMITH, L. (1981). Anhydrobiosis in nematodes. 1. The role of glycerol, myo-inositol and trehalose during desiccation. *Comparative Biochemistry and Physiology* **70B**, 579–586.

WOMERSLEY, C., THOMPSON, S.N. AND SMITH, L. (1982). Anhydrobiosis in nematodes. 2. Carbohydrate and lipid analysis in undesiccated and desiccate nematodes. *Journal of Nematology* **14**, 145–153.

WOMERSLEY, C., USTER, P., RUDOLPH, A. AND CROWE, J. (1984). Inhibition of and dehydration-induced fusion between liposomes by a carbohydrate. *Biophysical Journal* **45**, 191a.

# 7
# The Biology and Control of the Plant Parasitic Nematode *Rotylenchulus reniformis*

H. S. GAUR* AND R. N. PERRY†

**Division of Nematology, Indian Agricultural Research Institute, New Delhi 110012, India and †Entomology and Nematology Department, AFRC IACR, Rothamsted Experimental Station, Harpenden, Hertfordshire AL5 2JQ, UK*

Introduction
Gross morphology and systematics
Biological races
Life history
  Reproduction—Embryogenesis—Hatching and moulting
Infectivity and histopathology
Duration of the life cycle
Pathogenicity and crop losses
Ecology
  Temperature—Moisture—Spatial distribution—Dispersal—Population dynamics
Interaction with other soil micro-organisms
  Other nematodes—Phytopathogenic fungi, bacteria and viruses
Control
  Chemical methods—Cultural methods—Host resistance—Biological control—Non-conventional control methods—Integrated control
References

## Introduction

Nematodes were found parasitizing the roots of cowpea [*Vigna unguiculata* (L.) Walp. aggreg.] grown in pineapple fields in Oahu, Hawaii, USA, in 1931 and were thought to be doing serious damage to pineapple [*Ananas comosus* L. (Merr.)] by the year 1936. They were later described as *Rotylenchulus reniformis*, the reniform nematode, by Linford and Oliveira (1940). *Rotylenchulus reniformis* is a sedentary, semi-endoparasitic pest of a wide range of food, fibre, oilseed, fruit and plantation crops in most subtropical and tropical parts of the world. The distribution extends over more than 38 countries in South America, the southern Gulf-coast of the United States, Africa, the Middle East, South-East Asia, China, Japan and Australia (Siddiqi, 1972;

*Agricultural Zoology Reviews* — Vol. 4, February 1991
0269–0543 $20.00 + $0.00 © Intercept Ltd, PO Box 716, Andover, Hampshire SP10 1YG, UK

Varaprasad, 1986; Luc, Sikora and Bridge, 1990). The host range is very wide, including herbs, shrubs and trees belonging to over 30 plant families (Varaprasad, 1986). These nematodes usually prefer dicotyledons but several monocotyledons are also hosts of *R. reniformis*.

Surveys of agricultural areas in several countries have shown that the reniform nematodes are widespread and persistent pests (Luc, Sikora and Bridge, 1990). The symptoms of damage to host plants are not specific on the above-ground or even on the below-ground parts, necessitating a close examination to confirm their presence and the damage due to these nematodes. Thus, the reniform nematodes have not become as well known as the root-knot nematodes (*Meloidogyne* spp.), but comparative pathogenicity experiments on several crops have shown similar damage due to both groups. Reniform nematodes have also been implicated in disease complexes with bacteria and fungi. In addition to their agricultural importance, the reniform nematodes, due to their sexual dimorphism, moulting behaviour, survival and feeding mechanisms, have attained importance in systematics as links between migratory and sedentary endoparasitic nematodes, and in the study of the biology and ecology of nematodes.

The attention paid to the reniform nematodes during the 50 years since they were first described has produced a considerable amount of information (*see* Varaprasad, 1986). The objective of the present article is to review the applied aspects of the biology, ecology and control of *R. reniformis* in order to assess the present state of knowledge and to consider future priorities for research. During the past two decades over 900 publications referring to *R. reniformis* have been made but only selected research has been cited here.

## Gross morphology and systematics

The reniform nematodes have gross morphology typical of tylenchid nematodes and have been described in detail by Linford and Oliveira (1940), Dasgupta, Raski and Sher (1968), Siddiqi (1972) and Ying (1985). The dimensions and deMan ratios are as follows:

*Migratory young adult females*: length = 0.3–0.45 mm, a = 21–27, b = 3.6–4.3, b′ = 2.4–3.5, c = 13–17, c′ = 2.6–3.4, stylet = 16–18 μm, o = 81–106, V = 68–73%.

*Males*: length = 0.38–0.43 mm, a = 24–29, b′ = 2.8–4.8, c = 12–17, stylet = 12–15 μm, T = 35–45, spicules = 19–23 μm, gubernaculum = 7–9 μm.

The cuticle is finely annulated and the lateral field is marked by four fine incisures; the lip region is moderately sclerotized, high, conoid to rounded and continuous with the body contour, with 4–6 fine labial annules. When killed by heat the juveniles, males and young females assume a curved to open C shape. The stylet has rounded or slightly sloping knobs. The dorsal oesophageal gland opens more than a half, usually about one, stylet length behind the stylet knobs. The median oesophageal bulb is elongate or oval, with strongly developed musculature and valve. The posterior oesophageal glands form a large irregular flattened lobe extending latero-ventrally over the intestine. The excretory pore is located opposite the oesophageal isthmus region and a hemizonid of 2–3 annules in length just precedes it. The tail is short and conoid and more bluntly rounded in second-stage juveniles than in young females. Pore-like phasmids are located slightly anterior to the middle of the tail region. The females are didelphic and both ovaries have double flexures in young females but are very convoluted in fully mature reniform females. After entering host tissue, the young female swells and finally attains an irregular anterior and a kidney-shaped posterior, with a transverse slit-like vulva with raised lips located in a depression in the middle of the kidney-shaped region; a 5–8 μm spike-like tail projects posteriorly. The males remain vermiform and have degenerate cephalic sclerotization, stylet and oesophagus. The tail is short, about two cloacal body widths, with a narrow bursa which does not cover the tail terminus. The paired slightly curved spicules lack distal flanges. The gubernaculum is fixed and lacks titillae or telamon. The cloacal lips are pointed but do not form a tube; hypoptygmae are absent.

There have been controversies regarding the phylogenetic relationships of the genus *Rotylenchulus*, resulting in its placement in different higher taxa, such as subfamilies Pratylenchinae (Thorne, 1949), Nacobbinae (Chitwood and Chitwood, 1950) and Rotylenchulinae (Husain and Khan, 1967). At the family rank these nematodes have been variously accommodated under Tylenchidae, Hopololaimidae, Nacobbidae or Pratylenchidae until the family Rotylenchulidae was proposed by Husain and Khan (1967); this was recognized by Fortuner (1984) and Siddiqi (1986) but Fortuner (1987) and Maggenti *et al.* (1988) include *Rotylenchulus* in the Hoplolaimidae. At present, the taxonomic position of reniform nematodes is as given by Maggenti *et al.* (1988):

Phylum: Nemata (Nematoda)
Class: Secernentea
Order: Tylenchida
Suborder: Tylenchina
Superfamily: Tylenchoidea
Family: Hoplolaimidae
Subfamily: Rotylenchulinae
Genus: *Rotylenchulus*
Type species: *R. reniformis*

The 10 species described under the genus *Rotylenchulus*, along with their type hosts, type localities and authors, are shown in *Table 1*. *Rotylenchulus reniformis* is the most widespread of these species and has a very wide host range (Siddiqi, 1972). *Rotylenchulus borealis, R. parvus, R. macrosoma* and *R. macrodoratus* are important in limited areas, while very little is known about the other five species. Taxonomically this genus has received little attention since the publications of Dasgupta, Raski and Sher (1968) and Germani (1978b), as only one species, *R. sacchari* van den Berg and Spaull (1981), has been described during the past 22 years. Despite wide variations in reproductive and pathogenic behaviour, most reniform nematode populations have been identified as *R. reniformis*.

**Table 1.** Known species of the genus *Rotylenchulus*

| Species | Type locality | Type host | Authors |
|---|---|---|---|
| *R. anamictus* | Merca, Somaliland | *Acacia* sp. | Dasgupta, Raski and Sher (1968) |
| *R. borealis* | Arnhem, The Netherlands | Grass | Loof and Oostenbrink (1962) |
| *R. clavicaudatus* | Transkei, S. Africa | *Strelitzia* sp. | Dasgupta, Raski and Sher (1968) |
| *R. leptus* | Gwelo, S. Rhodesia | Bamboo | Dasgupta, Raski and Sher (1968) |
| *R. macrodoratus* | Bari, Italy | Grapevines | Dasgupta, Raski and Sher (1968) |
| *R. macrosoma* | Hulda, Israel | Olive | Dasgupta, Raski and Sher (1968) |
| *R. parvus* | Mauritius | Sugarcane | Williams (1960); Sher (1961) |
| *R. reniformis* | Idaho, Hawaii | Cowpea | Linford and Oliveira (1940) |
| *R. sacchari* | Natal, S. Africa | Sugarcane | Van den Berg and Spaull (1981) |
| *R. variabilis* | Inyanga, S. Rhodesia | *Rumex* sp. | Dasgupta, Raski and Sher (1968) |

Several features, such as their semi-endoparasitic habit, induction of host hypertrophy and some hyperplasia, marked sexual dimorphism in fully grown males and females, reduced subterminal bursa, gelatinous egg-matrix and parthenogenetic reproduction in some populations, have resulted in the reniform nematodes being considered as connecting links between the migratory endoparasitic pratylenchid and hoplolaimid nematodes and the sedentary endoparasitic root-knot and cyst nematodes (Siddiqi, 1986). However, the degenerate stylet and oesophagus in males indicate some deviations in phylogeny or some missing links. Hence Varaprasad (1986) expected the superfamilial rank, Rotylenchuloidea, to be an evolutionary bridge between Hopololaimoidea and Heteroderoidea. None the less Fortuner (1987) included *Rotylenchulus* within Hoplolaimidae as confirmed by Maggenti *et al.* (1988).

## Biological races

Birchfield and Brister (1962) noted some differences in the parasitism of *R. reniformis* populations on sugarcane (*Saccharum* spp.) in the Dominican Republic, Puerto Rico and Venezuela, and suggested the occurrence of different races, a suggestion supported by Ayala and Román (1963). Dasgupta and Seshadri (1971) found that amongst 10 Indian populations, nine (Race A) multiplied on cowpea, cotton (*Gossypium* spp.) and castor (*Ricinus communis* L.) but one (Race B) only multiplied on cowpea; effective control of Race B may be achieved with crop rotation. They also noted some parthenogenetic reproduction in one population. Germani (1978a) differentiated the *R. reniformis* populations of the West Coast of Africa from those of Senegal on the basis of their development on tomato (*Lycopersicon esculentum* Mill.) cvs Rossol and Roma, and groundnut (*Arachis hypogaea* L.) cv. 28–206. Nakasono (1977) used the type of reproduction as the basis of differentiating three biotypes in Japan, namely, Male Numerous Type (amphimictic), Male Rare Type (mostly parthenogenetic) and Male Absent Type(parthenogenetic).

It is obvious that ecological adaptations have resulted in significant variatons among the *R. reniformis* populations in different parts of the globe. No international comparisons, using standard differential hosts or biochemical approaches (Burrows, 1990), have been carried out. Such studies may demonstrate the existence of several biotypes or even species.

## Life history

### REPRODUCTION

In most populations reproduction is amphimictic with numerous (40–60%) males, but a few populations are parthenogenetic with very few or no males (Triantaphyllou and Hirschmann, 1964; Sivakumar and Seshadri, 1971; Nakasono, 1977, 1983). The young, infective adult females and males are present in the soil but there is no evidence of copulation before the young female establishes a suitable feeding site and matures, within 6–15 days, to a swollen kidney-shaped stage with convoluted ovaries. The males are attracted to the female (Nakasono, 1977). Eggs are laid into a gelatinous matrix, which is secreted by six specialized cells around the vagina and can be seen 7–20 days after invasion, depending on the host and environmental conditions (Linford and Oliveira, 1940; Peacock, 1956; Birchfield, 1962; Sivakumar and Seshadri, 1971). Egg-laying commences soon after the appearance of the gelatinous matrix and may continue for 1–3 weeks. The number of eggs in the gelatinous matrix varies with environmental conditions and host suitability (Gaur, 1986), but is usually between 30 and 200 eggs (Sivakumar and Seshadri, 1971; Johnson and Fassuliotis, 1984). In amphimictic populations, 5–10 males can often be seen entangled within the egg matrix. Occasionally a few eggs may be retained within the female body where they may complete embryonic development (Nakasono, 1966; Sivakumar and Seshadri, 1971). This feature has been used to compare the reniform nematodes with cyst nematodes, but it is likely that, towards the end of the egg-laying period, the female *R. reniformis* simply becomes too old and weak to expel the last few eggs. There is no evidence that the unlaid eggs or juveniles survive for very long periods within the desiccated female body.

### EMBRYOGENESIS

Embryonic development inside the egg was studied in detail by Sivakumar and Seshadri (1971). The single-celled egg laid by the female measures about 86–99 μm in length and 39–45 μm in width. The first cell division occurs transversely within 1–2 h (sometimes even before laying), followed by second and third transverse divisions within 3–4 h, giving rise to four blastomeres within 6–7 h. Further longitudinal divisions and movement of blastomeres produce the 'tadpole' stage in 3 days. The first-stage vermiform juvenile is formed within 3 days.

### HATCHING AND MOULTING

The first moult occurs within the egg on the 5th or 6th day after egg-laying

(Sivakumar and Seshadri, 1971). The second-stage juvenile (J2) is well developed and has a stylet that shows movement. The process of hatching commences within 1–3 days of the formation of the J2. Although juveniles hatch freely in water, a slight enhancement of hatch has been noted in root diffusates of good hosts such as tomato, okra [*Abelmoschus esculentus* (L.) Moench.], castor, egg-plant (*Solanum melongena* L.) and cowpea; root diffusates of poor or non-hosts such as onion (*Allium cepa* L.) and chilli (*Capsicum annuum* L.) suppressed hatch (Khan, 1985).

The post-hatch moulting from J2 to adult is completed in the soil without feeding (Peacock, 1956; Nakasono, 1966, 1983; Sivakumar and Seshadri, 1971; Bird, 1984; Gaur and Perry, 1991). The actual time required for completion of the 2nd, 3rd and 4th moults varies with temperature, moisture and other environmental factors. Generally, the J2 undergoes a pre-moult period of 3–5 days and shows distinct morphological changes during this period. The stylet becomes less prominent and the telorhabdions disappear. The body shows a slight swelling near the excretory duct and the nematode attains an arcuate shape, becoming virtually inactive but showing slight movement when light is focused on it. The stylet, except for the prorhabdions, eventually disappears and the oesophagus, especially the metacorpus, is reduced. The excretory system may be involved in the moulting process (Bird, 1984) but no experimental evidence is yet available.

The second moult occurs 6–7 days after egg-laying. The J3 is also crescent-shaped, inactive, lacks a stylet and a valve in the metacorpus. The sexes can be differentiated at this stage by the presence of a hyaline genital primordial area in the posterior region of the female and a slightly swollen tail in the male. The cuticle of the J2 is separated from the body but is retained as a sheath enclosing the J3. The third moult occurs within the next 1–2 days in males and 2–3 days in females (Sivakumar and Seshadri, 1971; Bird, 1984; Bishnoi and Yadav, 1989). The J4 is also crescent-shaped, inactive and ensheathed in the moulted J2 and J3 cuticles. The stylet, metacorpus and valve are still not present. Differentiation of the sexes becomes more distinct. The J4 frequently shows slight movements in the head and tail regions, indicating the redevelopment of functional somatic musculature. The cephalic framework, new stylet and thick-walled excretory duct are developed and the valve appears in the metacorpus. The ovaries and vulva in the female and testis and spicules in the male can be seen developing. The fourth and final moult occurs within the next 2–3 days.

The young adults are still enclosed in the three cuticular sheaths from the previous stages and lie inactive for 4–5 days to complete their development; after this period, they move actively with the sheaths on. In water, exsheathment does not take place for several weeks but in soil or sand the sheaths are lost within 24–48 h, indicating that frictional forces aid successful shedding of the juvenile cuticles (Gaur and Perry, 1991). Bird (1983, 1984) has illustrated the detailed morphological changes during the moulting process. Bird (1983) also noted that the computed body volume drops from about 95.6 pl for the J2 to 72.9 and 74.2 pl for the young female and male, respectively. As the nematodes do not feed during the inter- or intramoult periods, this reduction may be partly due to the consumption of some of the glycogen and lipid reserves present in the J2 stage.

The rapid completion of all the moults and the retention of all the moulted cuticles of the previous three stages are special adaptations that improve the chances of survival of the reniform nematodes in the rapidly changing moisture and temperature conditions found in the warmer regions where they occur (Birchfield, 1962; Sivakumar and Seshadri, 1971; Gaur and Haque, 1986; Sehgal and Gaur, 1989a; Womersley and Ching, 1989). These adaptations are discussed further in the section on ecology.

## Infectivity and histopathology

Adult males have not been observed feeding on plants. After exsheathment the young adult females are attracted towards the germinating host seedlings and to certain ions that may occur in the root diffusates (Riddle and Bird, 1985). Penetration of the host roots can take place at any point, but the zone of elongation behind the growing tip is preferred. The vermiform young females enter the root epidermis and cortex intracellularly, usually leaving the posterior half or two-thirds of the body outside the root. Some feeding may occur on the cortical cells, but the permanent feeding site is established on the endodermal and pericycle cells (Razak and Evans, 1976; Taha and Sultan, 1979; Khan and Khan, 1985; Vovlas and Ekanayake, 1985; Patel, 1986; Patel, Vyas and Patel, 1988). Phloem parenchyma cells are also included after some days (Sivakumar and Seshadri, 1974; Khan, Khan and Khan, 1985a). Comparisons with root-knot nematodes (*Meloidogyne* spp.) and the related species *Rotylenchulus macrodoratus* have shown that, while these nematodes feed from deeper zones of the stele, the feeding of *R. reniformis* is restricted to the endodermis, pericycle and the periphery of the stele (Cohn, 1976; Vovlas and Ekanayake, 1985; Patel, Vyas and Patel, 1988).

*Rotylenchulus reniformis* induces marked cellular changes, including formation of dense granular cytoplasm, dissolution of cell walls and coalescing of the cytoplasm of adjacent cells, thickening of cell walls by deposition of polysaccharides, enlargement (hypertrophy) of cells and enlargement of nuclei and nucleoli, resulting in the formation of multinucleate syncytia (giant cells) (Rebois, Madden and Eldrige, 1975). The salivary secretions react with the cytoplasm to produce plant polysaccharides, which form the feeding peg near the stylet and a coiled feeding tube, about 120 μm long and 0.5 μm wide and surrounded by the endoplasmic reticulum (Taha and Sultan, 1979; Rebois, 1980).

The cell wall ingrowths observed in syncytia induced by cyst and root-knot nematodes (Jones and Dropkin, 1975; Jones and Payne, 1978) have not been seen in syncytia induced by *R. reniformis*. The histological changes differ slightly in various hosts, depending upon their reaction to the nematode. In grapevines (*Vitis vinifera* L.) the syncytium formed girdles around the vascular tissue (Taha and Sultan, 1979) while in tomato, sweet potato [*Ipomoea batatas* (L.) Lam.] and cauliflower (*Brassica oleracea* v. *botrytis* L.)

the syncytium and adjacent cells may form a curved sheet (Yik and Birchfield, 1979; Khan and Khan, 1985). Generally, the reniform nematodes do not form galls but hyperplasia and slight swellings may be seen on the fine roots of some hosts.

### Duration of the life cycle

After establishing the feeding site, the female develops further to attain the typical reniform kidney-shaped posterior outside the host tissue, with a relatively narrower anterior part. The duration of the life cycle has been determined from egg to egg by some workers but from infection/inoculation to egg-laying by others, making comparisons difficult. Linford and Oliveira (1940) found that the life cycle from egg to egg was completed in about 25 days; periods of 24–29 days on okra (Sivakumar and Seshadri, 1971) and 17–23 days on soybean [*Glycine max* (L.) Merr.] (Rebois, 1973) have also been reported. The duration may vary with host suitability, temperature and other physicochemical properties of the rhizosphere. After invading the host root, the females may begin to lay eggs by 7–9 days at 25–28°C on a good host (Sivakumar and Seshadri, 1971; Johnson and Fassuliotis, 1984) but they may require longer under suboptimal conditions or on poor hosts. Nakasono (1983) noted egg-laying on the fourteenth day on tomato in Japan but in India the eggs are laid within 7–10 days on the same host. Bishnoi and Yadav (1989) found that egg production on castor, egg-plant, okra and clusterbean (*Cyamopsis tetragonaloba* Tayb.) could be seen after 8, 11, 15 and 20 days, respectively, and that the life cycle on these hosts was completed in 19, 22, 26 and 31 days, respectively, under similar conditions. The males complete their life cycle about 1 week before the females, as they mature 1–2 days before the young females are formed and the latter require a further 6–8 days for pre-oviposition development on the host (Sivakumar and Seshadri, 1971).

### Pathogenicity and crop losses

At high population densities, the reniform nematodes can reduce or delay seedling germination and emergence of some vegetables, legumes and tobacco (*Nicotiana tabacum* L.). Seedling growth is suppressed and this results in a reduction of the number of transplantable seedlings, or in poor plant density in directly sown crops (Nanjappa, Panda and Seshadri, 1978; Patel, Desai and Patel, 1986a). The uptake of water is inhibited (Ismail and Saxena, 1980) and excessive damage to roots near the growing tips may cause root death (Ayala, 1962). Some weeks after infection, the roots show poor development, with fewer and shortened secondary roots, and cortical necrosis with brown discolouration (Mahapatra and Padhi, 1986) associated with increasing concentrations of phenolic compounds (Mahmood and Saxena, 1986). The appearance on roots of protruding reniform females covered with a gelatinous egg matrix and adhering soil particles is unmistakable evidence of infection by the reniform nematode.

The growth of above-ground plant parts is stunted, the area of individual leaves is reduced in broad-leaved crops (Nath *et al.*, 1979) and the foliage becomes chlorotic; branches may show progressive dieback with heavy *R. reniformis* infestation (Hutton and Hammerton, 1975). *Rotylenchulus reniformis* reduced the yield of sweet potato and, although nematodes were not found within enlarged fleshy roots, infestation may cause cracking of sweet potatoes (Clark and Wright, 1983). Variation in leaf reflectance in infested cotton has been reported (Gausman, Heald and Escobar, 1975) which may aid remote sensing of hot spots of nematode infestation. These symptoms resemble those of moisture and nutrient deficiency. Crop growth, yield, quality of the produce and life of perennial plants may be reduced.

Clearly, it is not possible to give an exact figure for the tolerance level of each crop against these nematodes, as it varies greatly with the plant growth parameter(s) used, soil types, host cultivars, nutrients and growing conditions. Pot experiments with graded series of nematode inoculation levels have shown that, in most crops, initial population densities above 1 infective female per $cm^3$ of soil can cause a significant reduction in plant growth. Some hosts, such as tomato, egg-plant, greengram (*Phaseolus aureus* Roxb.) and chickpea (*Cicer arietinum* L.), have been reported to be more susceptible, showing damage at 0.1–0.5 infective females per $cm^3$ of soil; others, such as cowpea, pigeonpea [*Cajanus cajan* (L.) Mill.], mothbean (*Phaseolus aconitifolius* Jacq.), French bean (*P. vulgaris* L.), some curcurbits, grapevines and betelvine (*Piper betle* L.), are intermediate, suffering damage at or above 1 female per $cm^3$ of soil. A few hosts, like sweet potato, are more tolerant, with their growth being decreased only at or above 5 females per $cm^3$ of soil. The tolerance level increases with increase in plant age at the time of inoculation (Mahmood and Saxena, 1986).

Usually a near-linear relationship can be seen between the logarithms of the initial population density and plant growth or yield (Verma and Prasad, 1969; Upadhyaya, Khan and Saxena, 1974; Acosta and Ayala, 1975; Rebois, Eldridge and Webb, 1978; Panda and Seshadri, 1979; Mishra and Gaur, 1981; Acharya and Padhi, 1987). In some cases the relationship can be better explained with quadratic or sigmoid curves. An attempt to fit Seinhorst's (1965) model to determine the value of the tolerance level resulted in values as low as 0.0016–0.02 nematodes per $cm^3$ of soil (Sud *et al.*, 1985; Varaprasad, Sud and Swarup, 1987). However, no damage has been reported at such low inoculation levels.

Few estimates are available for crop losses in the field due to *R. reniformis*. Application of nematode control methods has usually given crop improvements in fields infested with these nematodes. For example, Lordello and Brancalion (1986) applied carbofuran (350 FMC) at 0.7% a.i. as seed treatment of cotton, and this reduced *R. reniformis* population density by 82–87%; soil solarization in Egypt increased the yields of broad beans (*Vicia faba* L.), onions, tomato and clover by 25–32%, part of which could be attributed to the control of *R. reniformis*. However, in many cases the yield increase obtained may be due to a variety of factors. This is a general limitation with most groups of plant pests and pathogens in general, and the plant-parasitic nematodes in particular. The damage to host plants inoculated with *R. reniformis* under controlled conditions, and the improvements in crop performance following application of nematicides or other nematode control methods in fields heavily infested with *R. reniformis*, are sufficient to conclude that the yield increases obtained are partly, if not wholly, due to control of these nematodes.

## Ecology

### TEMPERATURE

The geographical distribution, local spatial pattern, seasonal population changes, life cycle and survival are greatly influenced by temperature. The reniform nematodes are distributed mainly in the subtropical and tropical regions, except in subtemperate parts of the USA, China and Japan, the thermal limits for reproduction being 15 and 36°C, with the optimum temperature lying between 25 and 30°C (Rebois, 1973; Khan and Khan, 1974; Rodriguez-Fuentes, 1976; Singh and Khera, 1978; Heald, Inserra and Vovlas, 1988). In Japan the northern limit of occurrence is the 14°C isotherm of annual average temperature (Nakasono, 1983).

At low moisture levels (>15 bars) the nematodes survived at –5 and –1°C for 6 months, whereas plentiful moisture (<1 bar) aided survival at 25°C (Heald, Inserra and Vovlas, 1988). A temperature of 47°C for 1 h or 41°C for 25 h was lethal (Heald and Robinson, 1987). Infectivity is reduced and the duration of the life cycle extended with deviation of the soil temperature from the optimum. Nakasono (1978, 1980) found that the threshold temperature for development was 11.1 and 13.0°C for the amphimictic and parthenogenetic populations, respectively, and that the thermal requirement for completion of the post-infection life cycle varied from 313.8 to 395.6 day degrees. He did not find significant changes in the percentage of males due to temperature. The life cycle after root invasion could be completed in 25–26 days at 21.5°C, but in only 14 days at 27.6°C (Rodriguez-Fuentes, 1976). Heald (1975) found a similar degree of root invasion on cantaloupe (*Cucumis melo* L.) roots within 24 h at 21 and 27°C but development was faster at 27°C. The period from inoculation with young adult females to egg deposition was found to be 41, 13, 7 and 7 days at 15, 20, 25 and 34°C, respectively (Heald, Inserra and Vovlas, 1988).

Like many other nematodes, *R. reniformis* also exhibits thermotaxis but the preferred temperature tends to be close to the previous ambient or storage temperature; storage for 24 h at a different temperature could alter the thermal preference (Robinson, 1989). This reflects the highly adaptive capabilities of the reniform nematodes. Robinson and Heald (1989) found that covering Baermann extraction funnels, which reverses the thermal gradient, improved the extraction of the reniform, root-knot and citrus nematodes.

### MOISTURE

The reniform nematodes have been known to survive in air-dried soil for long periods. Survival in air-dried soil without a host for more than 1 year was recorded by Radewald and Takeshita (1964), and Birchfield and Martin (1967) noted that *R. reniformis* survived for over 7 months in air-dried and moist soil, although survival was better in the latter. Similar observations were made by Apt (1976). Sivakumar and Seshadri (1976) found that the fourth-stage juveniles with moulted cuticular sheaths of the previous stages could survive for 6 and 7 months in air-dried and moist soil, respectively. Survival for 25 months (Rodriguez-Fuentes, 1980) and over 27 months (Gaur and Haque, 1986) in air-dried soil without a host has been recorded, with the nematodes still retaining infectivity and reproductive abilities. Individual young females, males and fourth-stage juveniles survived in a coiled anhydrobiotic state, with most ensheathed in the moulted cuticles of the previous juvenile stages (Gaur and Haque, 1986; Gaur and Sehgal, 1988). During the initial period of 3 months, the population density declined rapidly in moist soil and more rapidly in dry (up to 30 bars) and wet soils (0.001 bar); after this period the nematodes surviving in dry soil became quiescent and persisted for over 2 years. Some individuals also survived in a quiescent state in the soil with continuously high moisture levels (Gaur and Sehgal, 1988).

Tsai and Apt (1979) did not consider coiling as an adaptation towards survival but, although *R. reniformis* cannot survive direct exposure to low relative humidities (RH), Womersley and Ching (1989) found that individuals could survive at 40–80% RH after induction of coiling by slow drying on a model substrate (0.5% agar or 1% agarose) that closely simulates the moisture stress conditions in drying soil (Womersley and Ching, 1989). In addition to coiling, the retained moulted cuticles are important adaptations that enhance survival. Gaur and Perry (1991) studied the survival and rate of water loss of ensheathed and exsheathed adults of *R. reniformis* after exposure to varying RH levels. Exsheathed forms survived poorly compared to ensheathed adults, and the retained moulted cuticles aided desiccation survival by slowing the rate of drying of the enclosed adult. However, the reduced rate of water loss only assisted individuals to survive for the period over which water loss was controlled; they showed no ability for prolonged survival once their water content had been reduced to less than 10%. Thus, while control of water loss enables species such as *Ditylenchus dipsaci* (Kühn) Filipjev to enter anhydrobiosis (Perry, 1977a,b), *R. reniformis* shows little intrinsic ability for anhydrobiotic survival; control of water loss merely prolongs the time taken for the nematode's water content to reach lethal low levels. However, the survival abilities are related to the rates of drying experienced in the soil (Womersley and Ching, 1989), and the combination of adaptations such as coiling and retention of the moulted cuticles, together with an environmentally induced slow rate of drying, considerably enhance the chances of survival in dry soils.

Survival of *R. reniformis* in drying soils is inversely related to the rate of soil moisture loss (Sehgal and Gaur, 1989b). Alternate wetting and drying of the soil caused a greater decline in the population density than continuous dry or wet conditions (Gaur and Sehgal, 1988), perhaps because rehydrating soils caused activation and exsheathment of adults, thus making them vulnerable to subsequent drying. Gaur and Haque (1987) found that moisture stress at some stages of crop growth favoured multiplication of *R. reniformis* but not *Tylenchorhynchus vulgaris* Upadhyay, Swarup and Sethi. The hatching of *R. reniformis* was inhibited at 3.8–9.0 bars, compared to <1 bar for *T. vulgaris*.

Clearly, the reniform nematodes are well adapted to meet the survival requirements of their natural habitat.

## SPATIAL DISTRIBUTION

As with most plant-parasitic nematodes, the horizontal distribution of the reniform nematodes in a field is aggregated, although this distribution appears to be more regular than for root-knot nematodes (Jimenez and Lopez, 1987; McSorley, 1987). The distribution of population densities generally fits the negative binomial or Poisson distribution models. The application of these models, or of Taylor's Power Law (Taylor, 1961; McSorley *et al.*, 1985), can help to determine the optimum sampling frequency for estimation of mean population densities with a given precision. Vertically, the highest population densities during a crop season coincide with the greatest density of young feeder roots, generally in the 0–15 cm zone (Jimenez and Lopez, 1987; Gaur and Sehgal, 1988). Occurrence of some individuals as deep as 2 m indicates that the reniform nematodes are tolerant of low oxygen levels. There is no evidence to show large-scale vertical migration in the soil layers, although some active movement in a temperature or moisture gradient may occur. Vertical movements may occur passively with the movement of water, soil and plant roots.

In a particular region, the distribution may be greatly influenced by soil texture. Robinson *et al.* (1987) found more frequent occurrence of *R. reniformis* in the lower Rio Grande Valley in clay silts and clays, although it also occurred in clay loams and sandy clay loams; by contrast, *M. incognita* (Kofoid and White) Chitwood preferred elevated sandy loams and moderately well-drained soils. However, *R. reniformis* also occurs frequently in sandy loams and laterites as well as relatively heavier soils (Mukhopadhyaya and Haque, 1980; Heald, Inserra and Vovlas, 1988).

## DISPERSAL

As with other nematode species, active dispersal is limited. Movement of infested soil with water, air, farm machinery, cattle, planting material or other similar means is the major mode of dispersal. Gaur (1989) found that, in anhydrobiotic forms, these nematodes can be carried over long distances with dust blowing in wind-storms. Rooted cuttings and seedlings can also carry these nematodes from infested nurseries to previously uninfested areas.

## POPULATION DYNAMICS

Very few studies have been carried out on the population dynamics of the reniform nematodes under varying conditions and cropping systems. The major factors determining population behaviour appear to be temperature, moisture and the suitability and growing periods of the host. In a one-year rotation of maize (*Zea mays* L.)–cowpea–wheat (*Triticum aestivum* L.) in Delhi, India, no major fluctuations in population densities were recorded (Srivastava and Sethi, 1986), although the population densities were generally low. Gaur (1975 and unpublished data) studied the population changes in a number of cropping sequences in Delhi. The population densities peaked in March–April and August–October, which corresponded to soil temperatures of 25–30°C during the preceding month and with maturity of the host crop, especially dicotyledons such as vegetables and legumes. Inclusion of mustard (*Brassica campestris* L.), wheat and some other graminaceous crops in a cropping sequence resulted in low population densities (Gaur and Haque, 1986), although the presence of weeds may maintain high population densities over longer periods (Gaur and Haque, 1987). In southern parts of India peak population densities have been recorded in July and December (Sundaram, Velayutham and Vadivelu, 1987). Jimenez and Lopez (1987) also found temperature- and moisture-dependent seasonal variation in population densities under papaya (*Carica papaya* L.). They found highest densities in December–January, decreasing markedly in March–April when the moisture levels are low and temperature is high; there was a further decline in August–December.

Although the densities of active stages in soil increased suddenly after rainfall following dry periods, probably due to hatching and the activation of anhydrobiotic forms, heavy rains or flooded conditions decreased the numbers (Castillo, Arceo and Litsinger, 1978). Usually, seasonal fluctuations of *R. reniformis* are less than with *M. incognita* (Jimenez and Lopez, 1987).

The seasonal change in population densities varies at different depths in the soil profile, due to differences in the distribution of host roots, temperature, moisture, gaseous composition and related factors, which may cause either differential survival or active and passive movements.

## Interaction with other soil micro-organisms

### OTHER NEMATODES

A large number of studies on the interaction of species such as *M. incognita, M. javanica* (Treub) Chitwood, *Tylenchorhynchus brassicae* Siddiqi, *Radopholus similis* Cobb, *Tylenchulus semipenetrans* Cobb and *Pratylenchus sefaensis* Fortuner with *R. reniformis* have yielded variable results (*see* Varaprasad, 1986). In some studies, host damage and the multiplication of *R. reniformis* were reduced in the presence of other species, especially root-knot nematodes (Taha and Kassab, 1980; Thomas and Clark, 1983; Khan, Khan and Khan, 1985b; Ahmad, Tiyagi and Khan, 1987). In other studies, the reniform nematode suppressed other nematode species (Rao and Prasad, 1971; Pathak, Nath and Haider, 1986). The type of relationship and the extent of competition between *R. reniformis* and other species depends upon many factors, including host crop, the environmental suitability, initial population levels, the nature of the infection and feeding process and the rate of reproduction. In some cases the presence of another nematode species altered the host physiology; for example, the presence of *P. sefaensis* on maize and cowpea made these hosts more vulnerable to invasion by *R. reniformis* (Egunjobi, Akonde and Caveness, 1986) and a cowpea cultivar resistant to *R. reniformis* became susceptible to it in the presence of *M. incognita* (Khan and Husain, 1989a,b).

Gaur and Haque (1986) studied the interspecific relations of *Tylenchorhynchus* spp., *R. reniformis* and *Hoplolaimus indicus* Sher in different crops under field conditions. Regression analyses revealed only very weak interspecific competition among these nematodes, probably because they feed on different root tissues. Interspecific competition may not reduce the total damage to the host, but over a long period it may influence which species predominates. In a field infested with the reniform and root-knot nematodes, larger, more regularly distributed and persistent populations of the former were found after a few crop seasons, unless the crops grown supported root-knot nematodes particularly well (Gaur, unpublished data).

#### PHYTOPATHOGENIC FUNGI, BACTERIA AND VIRUSES

The possibility of interaction between *R. reniformis* and phytopathogenic root fungi was explored as soon as this species was described. The effects of root-rot and wilt diseases of cotton caused by *Fusarium* sp. (Neal, 1954), *F. oxysporum* Schlecht (Elgindi, Oteifa and Khadr, 1974), *Rhizoctonia solani* Kühn (Brodie and Cooper, 1964) and *Verticillium dahliae* Kleb. (Tchatchoua and Sikora, 1978; Prasad and Padagnur, 1980) are all increased in the presence of *R. reniformis*. Carter (1980) noted similar effects with *Macrophomina phaseolina* (Tassi) Goid on cantaloupe. Root injury by nematodes and, perhaps, associated changes in root physiology appear to be responsible for the increase in severity of fungal diseases. Resistance of cowpea cvs RC-8 and EC 4213-A to *R. solani* was broken in the presence of the reniform or root-knot nematodes (Khan and Husain, 1989a).

Increased reproduction of the reniform nematodes has been reported on plants infected with *Verticillium* sp. (Prasad and Padagnur, 1980) and *R. solani* (Brodie and Cooper, 1964), and has also been recorded on okra infected with yellow mosaic virus (Sivakumar and Meerzainuddin, 1973) and chilli with marigold mottle virus (Naqvi and Haseeb, 1978).

High population densities of *R. reniformis* were found on grapevines suffering from the crown gall bacterium, *Agrobacterium tumefaciens* (E. F. Smith and Townsend) Conn., and Lele *et al.* (1978) proposed that these nematodes predisposed the roots to the bacterium, which can infect only injured roots.

### Control

The economic importance of reniform nematodes in a wide variety of commercial crops has resulted in studies on several chemical and non-chemical methods of control, which are briefly summarized here.

CHEMICAL METHODS

Application of a wide range of nematicidal chemicals can reduce nematode populations, including those of *R. reniformis*, and crop performance can often be improved (*see* Varaprasad, 1986). Fumigants containing halogenated C-3 hydrocarbons or methyl iso-thiocyanates have been found to be effective (McSorley, McMillan and Parrado, 1986). A number of non-fumigant organophosphates and carbamates have also been reported to control these nematodes.

Field application of nematicides to control reniform nematodes has been tried in vegetables (Reddy and Seshadri, 1974; Singh, 1975; Birchfield and Martin, 1976; Heald, 1980; Clark, Strand and Watson, 1983; McSorley and Parrado, 1983; Acosta, Cruz and Abreu, 1984), pineapple (Apt, 1979), soybean (Birchfield and Williams, 1974), cotton (Bost, 1984; Lordello and Brancalion, 1986), peanuts (Mahapatra and Padhi, 1988) and grapevines (Rajendran and Naganathan, 1979).

The dosage of non-fumigant nematicides used has varied from 1 to 16 kg a.i./ha, but generally 2–5 kg a.i./ha gives good control. Recently aldicarb, carbofuran, nemacur, oxamyl, phorate and fenamiphos have been tried more frequently. However, very few attempts have been made to assess the economics of nematicide treatment and, in many cases, the effects on crop yield have been due to the control of more than one nematode species. Field application is often not economical in many low-value field crops, due to the high dose required.

Since the young females are the infective stage, the reniform nematodes can mature and deposit eggs within 7–10 days of planting, before systemic nematicides applied to the soil become fully functional in the host tissue. The eggs are more resistant to the nematicides than the free-living stages, and by the time they hatch, the nematicide has started to degrade. Some ovistatic and ovicidal effect of fenamiphos, isofenphos and terbufos at 250–750 p.p.m. has been noted (Thakar and Patel, 1988), although very few eggsacs were used in this study.

Attempts to reduce the amount of nematicide required and to target the nematode at an early stage have been made through nursery-bed, seed-dressing, root-dip or foliar-spray treatments. Application of carbofuran to tobacco nursery-beds provided healthier seedlings, and root-dip treatment of egg-plant seedlings in 500–1000 p.p.m. solutions of carbofuran, aldicarb or terbufos for 15–30 min reduced infection (Prasad and Krishnappa, 1981). Seed treatment of okra with carbofuran at 3–6% w/w reduced populations of *R. reniformis* (Sivakumar, Meerzainuddin and Kuppuswamy, 1973). Seed dressing of some edible legumes (pulses) with carbofuran at 1.25% a.i. w/w in combination with field application at 4 kg a.i./ha (Panda and Seshadri, 1979) also gave good control, but the low yield potential of these crops makes field application uneconomical. Treatment of mungbean (*Phaseolus aureus*) seed with carbofuran at 1% a.i. w/w has also been found effective (Patel and

Thakar, 1986). Treatment of cotton seed with carbofuran at 0.7% a.i. w/w reduced *R. reniformis* populations by 82–87% and increased yield by 21–52% (Lordello and Brancalion, 1986). Application of fenamiphos, isozophos or oxamyl at 1–5 mg/kg to banana (*Musa* spp.) rhizome, or oxamyl to leaf axils or as a foliar spray, reduced reniform nematode numbers in banana (Gowen, 1977, 1978; Robalino, Román and Cordero, 1983). Use of FMC, an analogue of carbofuran, or fenitrothion at 0.05% concentration as a foliar spray on peanuts 20 days after sowing, combined with 1 kg a.i./ha of carbofuran, aldicarb or phorate, reduced the nematode population and increased yield (Mahapatra and Padhi, 1988).

In *in vitro* tests, some amino acids, such as glutamine and glycine (Naqvi and Saxena, 1976), and organic acids, such as acetic, butyric and propionic acids (Badra, Saleh and Oteifa, 1979), have been found to be toxic to *R. reniformis*. A growth regulator, ethephon, applied at 750–3500 p.p.m. to mango (*Mangifera indica* L.) reduced reniform nematode densities by 31–76% (Badra and Khattab, 1982). Application of aldicarb + indole-acetic acid also reduced *R. reniformis* on chickpea (Mahapatra and Padhi, 1986). Application of indole-acetic, -butyric or -propionic acids at 4–10 mg per two plants increased plant growth as well as multiplication of the reniform nematode (Mahmood and Saxena, 1986).

### CULTURAL METHODS

The use of most nematicides to control *R. reniformis* is precluded on economic and environmental grounds; cultural methods are the most practical means for managing these nematodes. Of the wide variety of cultural methods, crop rotation, organic amendments, summer ploughing, moisture management, soil solarization and good crop husbandry seem to be the most successful.

#### *Crop rotation*

The reniform nematodes decrease to very low numbers in moist soil within 1 year, thus theoretically permitting short rotations; however, the host range is very wide, extending to many dicotyledon and some monocotyledon crops and weeds. This makes selection of commercially acceptable non-host crops difficult and necessitates proper weed control (Haque and Gaur, 1985, 1986). Inclusion of marigold (*Tagetes erecta* L.), Pangolangrass, zinnia (*Zinnia elegans*, Jacq.), sugarcane, maize and mustard, for example, in cropping sequences caused reduction in populations (Alam, Saxena and Khan, 1981; Khan, Saxena and Mahmood, 1984). Some varieties of these crops may increase or maintain the population densities (Nakasono, 1973). In multiple cropping systems, restricting cultivation of susceptible crops to the winter season in the tropics helps in keeping populations low (Sethi and Gaur, 1990). Specific crops and their sequence would depend upon local factors.

#### *Organic amendments*

Organic amendment of soil with green manure, crop residues or de-oiled cakes reduced reniform nematode population densities (Khan, Khan and

Saxena, 1974; Lal, Yadav and Nandwana, 1977; Alam, Khan and Saxena, 1980). Addition of animal manures and plant leaves produces decomposition products in the soil which also reduced nematode population densities (Castillo, 1985). The high doses of organic amendments that are required for effective control are the biggest limitation to their use.

Ten tons/ha of organic amendments were as effective as 10 kg/ha of fenamiphos; application of fresh chicken dung and composted sawdust caused maximum reduction of *R. reniformis* (Duhaylongsod, 1988). De-oiled cakes of neem (*Azadirachta indica*) and karanj (*Pangamia glabra*) can cause significant reduction in nematode populations, as well as provide nutrients for the plants. In pot experiments, Padhi and Misra (1987) found that neem cake at a dose equivalent to 2 tons/ha reduced *R. reniformis* populations by 86%, and more than doubled the dry weight of French bean plants; karanj cake and prawn exoskeleton were also effective.

Rao, Padhi and Acharya (1987) suggested integration of aldicarb (1 kg a.i./ha) plus neem or karanj cake (0.5 ton/ha) plus urea (15 kg N/ha), which effectively reduced nematode populations and increased plant growth. Application of chopped weeds, such as water hyacinth [*Eichhornia crassipes* (Mart.) Solms], reduced root-knot and reniform nematodes; the water extracts of this weed also had nematicidal properties, reducing nematode multiplication when used as a root-dip treatment. Extracts of many other plants have been found toxic but to lesser extents. Neem and Persian lilac (*Melia azedarach*) contain compounds, such as nimbidine and thionemone and several phenolics such as hydroquinone, *p*-cresol, catechol, pyrogallol and gallic acid, which have been found toxic to *R. reniformis* (Vijayalakshmi, Gaur and Goswami, 1985). Even the leaf extracts of neem and Persian lilac, when used as root-dip treatment, reduced *R. reniformis* populations (Siddiqui and Alam, 1988a). Seed dressing with extracts of neem or Persian lilac seed, leaves, flowers, bark or gum decreased population growth of *Meloidogyne incognita* and *R. reniformis* (Siddiqui and Alam, 1987). Certain plants, such as species of *Calotropis* and *Euphorbia*, have latex which, when used as a seed dressing on tomato, egg-plant or okra seed, imparted some resistance/tolerance to the seedlings (Siddiqui and Alam, 1988b).

### *Summer ploughing and moisture management*

There is little published work on the effect of summer ploughing on the reniform nematodes, although it has been found highly effective against root-knot, cyst and other nematodes. Experience has shown that *R. reniformis* population densities are greatly reduced after 2–3 deep ploughings in a hot summer, although some nematodes survive in deeper layers or in a quiescent state.

Survival of *R. reniformis* in soil is inversely related to the rate of drying (Sehgal and Gaur, 1989b) and alternate wetting and drying causes greater decline (Gaur and Sehgal, 1988). Summer ploughing increases the rate of drying, and irrigation between ploughing results in alternate wetting and drying which may induce exsheathment of the young adults (Gaur and Perry,

1991) rendering them more prone to desiccation. Irrigated fallow has been found to cause a marked decline in populations (Rohrbach and Apt, 1986).

### *Soil solarization*

Mulching of moist soil with clear polyethylene sheet for 4–6 weeks in hot summer months (soil solarization) has been used to control *R. reniformis*. Using soil solarization techniques in Texas, Heald and Robinson (1987) found that a temperature of 47°C for 1 h or 41°C for 25 h would cause 90% mortality of *R. reniformis*, sufficient to reduce nematode population densities in the 0–15 cm soil layer. In Egypt, soil solarization depressed *R. reniformis* population densities for up to 60 days, which allowed good early establishment of broad bean, onion, tomato and clover (Abdel-Rahim *et al.*, 1988). Although reniform nematode populations decreased in solarized chickpea and pigeonpea fields (Sharma and Nene, 1985), the cost of polyethylene and labour makes solarization over whole fields expensive and, in intensive cropping systems, it is often impractical to take the land out of use for about 2 months. In transplantable host crops, the area used for nursery-beds can be conveniently solarized while the fields can be summer ploughed or used for growing non-host crops. Gaur and Dhingra (1991) obtained healthy and virtually nematode-free egg-plant seedlings by solarization of nursery-beds.

### *Crop husbandry*

Good crop husbandry practices such as field preparation, fertilizers and moisture management can improve crop tolerance to nematodes, including *R. reniformis*. Damage to the host is greater if the soil is deficient in nitrogen and potassium (Oteifa and Elgindi, 1974). Application of fertilizers can improve crop performance even in infested soil (Sivakumar and Meerzainuddin, 1975; Badra and Khattab, 1980, 1982; Mahmood and Saxena, 1980). The additional fertilizers increase tolerance by improving growth, and may also induce some resistance due to increased phenols (Mahmood and Saxena, 1980) or increased auxin–monophenol ratio (Badra and Khattab, 1982). Haque and Mukhopadhyaya (1983) noted that application of zinc or copper reduced reproduction while manganese or molybdenum increased it. Application of ammoniacal or phosphatic fertilizers increased the proportion of males (which do not feed) in populations of *R. reniformis*, and this was positively related to the level of phenylalanine in the root (Badra and Ismail, 1981). On the other hand, deficiency of nitrogen or potassium also increased male numbers.

Avoiding moisture stress in actively growing crops can reduce crop loss. Tillage, especially during the hot season, generally reduces nematode populations and improves crop growth. However, Hainlein and Black (1983) found better growth of pigeonpea in no-tillage conditions.

Many weeds not only compete with the crop but also support nematode populations during and between crop seasons (Ferraz, 1985). Control of weeds mechanically or with herbicides reduced population densities of *R.*

*reniformis* and other nematodes (Gaur and Haque, 1986; Ray and Patra, 1986); post-emergence herbicides caused greater decline of sedentary parasites, including *R. reniformis.*

HOST RESISTANCE

Developing commercially and agronomically suitable resistant cultivars is likely to be the best method of avoiding crop losses due to nematodes. A number of attempts have been made to screen for sources of resistance and to breed resistant cultivars. Varaprasad (1986) has listed resistant sources/varieties of several commercial crops, and subsequently other sources have been identified (*Table 2*). In many cases these sources have been identified on the basis of single-location pot tests. It is necessary to test resistance under field conditions and to examine the agronomic performance of the plants. Generally, the number of mature females on roots is taken as the criterion for assessing resistance. Gaur (1986) noted expression of resistance at different stages of nematode invasion and reproduction in cowpea varieties and suggested that total egg production would be a better criterion. The ideal period for assessment has been found to be 21–31 days after planting in soil infested with 1800–2800 infective females per 500 $cm^3$ soil (Williams *et al.*, 1979).

The genetics of resistance against *R. reniformis* has been studied in soybean (Williams *et al.*, 1981). Harville, Green and Birchfield (1985) have suggested that the resistance in soybean is quantitative and is controlled by two gene pairs with unequal effects, the genotypes of four varieties with different reactions being:

| | |
|---|---|
| Susceptible soybean cv. Davis | $Rn_1$ $Rn_1$ $Rn_2$ $Rn_2$ |
| Moderately susceptible soybean cv. Bragg | $Rn_1$ $Rn_1$ $rn_2$ $rn_2$ |
| Moderately resistant soybean cv. Dare | $rn_1$ $rn_1$ $Rn_2$ $Rn_2$ |
| Resistant soybean cv. Pickett-71 | $rn_1$ $rn_1$ $rn_2$ $rn_2$ |

Such understanding has helped in producing soybean cv. Gregg (= Bragg × Pickett-71), for example, which has a high degree of resistance against *R. reniformis*; it is also resistant to *Heterodera glycines* Ichinohe (Race 3), *Phytophthora megasperma* f. sp. *glycinea* and *Xanthomonas phaseoli* var. *sojensi*.

Attempts have been made to understand the biochemical nature of resistance, which seems to be related to the phenolic and amino-acid contents. Tomato and egg-plant cultivars contained more amino acids and *o*-dihydroxy phenols after inoculation with *R. reniformis* and nematode reproduction was inversely related to the amino-acid and phenolic concentration (Mahmood, 1985; Mahmood and Saxena, 1986). Resistance in pigeonpea cv. 556–1 was also related to increased phenolic levels (Thakar and Yadav, 1986).

**Table 2.** List of sources/varieties of crops reported as resistant against *Rotylenchulus reniformis* (updating Varaprasad, 1986)

| Crop | Resistant source/variety | Reference |
|---|---|---|
| Beans | Jack bean (*Canavalia ensiformis* L.), velvet bean (*Mucuna pruriens* L.) | Peacock (1956) |
| Blackgram (urdbean) (*Vigna mungo* L.) | Ratnapur–1, UG201, UG135 | Routaray, Sahoo and Das (1986) |
| Chickpea (*Cicer arietinum* L.) | PBNG34, BGM425, BGM426, BGM434, BG268, BG273, GNG–158 | Sahoo, Routaray and Das (1986) |
| | F–496 | Anver and Alam (1990) |
| Coffee (*Coffea* spp.) | *Coffea robusta* | Peacock (1956) |
| | *C. canephora* | Macedo (1974) |
| Cotton (*Gossypium* spp.) | Auburn 56, Plains, Delfos 425–920, Delfos 9169, Deltapine 6, Deltapine 6 × Delfos 6102, Cook 307–6 × Delfos 9169, Louisiana Hybrid 33 × 14–312, Louisiana 33, Roxie, Coker 100 wilt, Stoneville 2B strains | Sasser (1972) |
| | La 434–1031–4 | Beasley (1986) |
| | La RN 4–4, La RN 909, La RN 910, La RN 1032 | Jones *et al.* (1988) |
| Cotton (related species) | *Gossypium longicalyx, G. barbadense, G. stocksii, G. somalense, G. herbaceum* (PI 408775), *G. arboreum* (PI 41895, PI 417891, CB 3839), *G. hirsutum, Hibiscus* spp. | Yik and Birchfield (1984) |
| Cowpea (*Vigna unguiculata* L.) | V–16 | Thakar and Patel (1985) |
| | Puza-Phalguni, RC–48 | Gaur (1986) |
| | C152 | Haque and Padmavathy (1986); Makadia *et al.* (1987) |
| | S–488, EC–4213A, RC–8 | Khan and Husain (1988a, 1989a) |
| Horsegram (*Dolichos uniflorus* Lam.) | BGM–1, Kuruvai, K–42, DP–1–1236, Kalarahang | Nayak, Routaray and Das (1987) |
| Greengram (Mungbean) (*Phaseolus aureus* Roxb.) | ML62, ML80, PDM–14, TT8E, Pusa 103 | Routaray, Sahoo and Das (1986) |
| | ML70, EC–955 | Patel and Thakar (1986) |
| | IC–8924, IC–11312 | Patel, Chavda and Patel (1989) |
| Onion (*Allium cepa* L.) | Evergreen | Birchfield and Brister (1962) |
| Papaya (*Carica papaya* L.) | Solo, Washington, Coorg Honeydew | Patel, Patel and Patel (1989) |
| Pepper and related species (*Capsicum* spp.) | Sweet Bell, California Wonder, *Capsicum annuum* var. *fasciculatum* | Birchfield and Brister (1962) |
| | *C. frutescens* var. *microcarpum* | Peacock (1956) |
| | Chilli – Pusa Jwala | Routaray, Sahoo and Das (1988) |
| Pigeonpea [*Cajanus cajan* (L.) Mill] | 4658, 2387, 556–1, 4167, 3830–2 Norman | Thakar and Yadav (1985, 1986, 1987) |
| | ICP–12744, Basant, PDM 1 | Patel *et al.* (1987) |
| | AGS–522, GAUT–83–23, GAUT–82–74, GAUT–84–22 | Chavda, Patel and Patel (1988) |

*cont'd*

**Table 2.** (*cont'd*)

| Crop | Resistant source/variety | Reference |
|---|---|---|
| Potato (*Solanum* spp.) | La Rouge, Red La Soda, B7642–2 B7154–8, B7680–6, PI 377741 | Rebois and Webb (1979) |
| Rice (*Oryza sativa* L.) | Blue Bonnet | Birchfield and Brister (1962) |
| Sorghum [*Sorghum bicolor* (L.) Moench] | Sweet | Birchfield and Brister (1962) |
| Soybean [*Glycine max* (L.) Merr.] | Forrest | Hartwig and Epps (1973) |
| | Pickett, Pickett–71, D68–78, D68–80, D68–128, D68–180, D68–185, D68–201, D68–214, D68–216, D68–221 | Birchfield *et al.* (1971) |
| | Peking, Dyer, Cluster | Good (1972) |
| | Centennial | Hartwig and Epps (1977) |
| | TK5, Strain 12, Strain 99, PI 200451, GC 40078–2–40, EGSY 12, EGSY 6, GC 40142–066, UPL–SY2 | Lim and Castillo (1979) |
| | Gregg | Harville *et al.* (1988) |
| | Padre | Hartwig, Hinson and Scott (1988) |
| Sugarcane (*Saccharum* spp.) | CP 44–101 | Birchfield and Brister (1962) |
| Sweet potato [*Ipomoea batatas* (L.) Lam.] | Goldrush | Good (1972) |
| Tobacco (*Nicotiana tabacum* L.) | GT4, Anand 2, Anand 145 | Patel, Desai and Patel (1986b) |
| Tomato (*Lycopersicon* spp.) | Kalyanpur Sel–1, Kalyanpur Sel–III, LA 121, EC 118272, EC 118276, 11276 | Balasubramanian and Ramakrishnan (1983) |
| | Peto–108, Peto–95, Nema–1400, Ace–55, Petopride | Montasser (1986) |
| *L. pimpinellifolium* (Tusl.) Mill. | PI 375937 | Rebois *et al.* (1973) |
| Turnip (*Brassica rapa* L.) | Purple Top, White Globe | Birchfield and Brister (1962) |

## BIOLOGICAL CONTROL

A number of soil inhabiting or artificially applied antagonistic micro-organisms are potential biocontrol agents (Kerry, 1987), although very few antagonists have been tried specifically against the reniform nematodes. Sitaramaiah and Sikora (1981, 1982) found suppression of the invasion and development of *R. reniformis* on tomato, bushbean (*Phaseolus vulgaris* L.), cucumber (*Cucumis sativus* L.) and muskmelon (*Cucumis melo* L.) in the presence of the endomycorrhizal fungi, *Glomus fasciculatum* and *Glomus mosseae*, and an unidentified nematophagous fungus was found trapping *R. reniformis* (Alcantara and De Azevedo, 1981). Recent tests in India showed that *Paecilomyces lilacinus* (Thom) Samson parasitized the eggs and mature females of *R. reniformis* and reduced its multiplication on cowpea (Khan and

Husain, 1988b), tomato and egg-plant (Reddy and Khan, 1988, 1989), at inoculation levels of about 240 million spores per plant, alone or in combination with carbofuran; the proportion of males in the population was increased (Reddy and Khan, 1988). The metabolites in the cultures of certain fungi, for example *Aspergillus niger, Helminthosporium* sp. and *Fusarium* sp., have been found to inhibit the activity of *R. reniformis* within 12 h (Alcantara and De Azevedo, 1980).

Some species of higher plants are antagonistic to nematodes, and use of such plants as soil amendments, intercrops or as alternative crops in rotation can reduce population densities. Intercropping of neem or Persian lilac seedlings with tomato and egg-plant reduced root-knot and reniform nematode populations (Siddiqui and Saxena, 1987) but, in the field, neem and Persian lilac are not suitable intercrops. Marigold species (*Tagetes patula, T. erecta, T. minuta* and *T. tenuifolia*) have been found to reduce reniform nematode populations when grown as intercrops (Siddiqui and Alam, 1987, 1988b) or alone (Sahoo and Padhi, 1986), since the root exudates are toxic. Intercropping with zinnia (Tiyagi *et al.*, 1986) or its inclusion in cropping sequences (Alam, Saxena and Khan, 1981) could also reduce reniform nematode populations. Leaf extracts of an insect-repellent plant, *Cymbopogon flexuosus*, have been found to be highly toxic to *M. incognita* and *R. reniformis*.

The use of biological control methods, especially those involving antagonistic microflora, is in its infancy. There is a need to study the edaphic factors that may encourage the build-up of such microflora to become a major component of the antagonistic potential of the soil, without the need for repeated introductions. The flowers and natural pigments of some antagonistic crops such as *Tagetes* spp. have commercial value, especially near cities. Eventually it may be possible to find antagonistic crops suitable for inclusion in cropping sequences; some kinds of mustard, such as *Brassica campestris, B. juncea* and *Eruca sativa*, have such potential (Haque and Gaur, 1985; Sahoo and Padhi, 1986).

### NON-CONVENTIONAL CONTROL METHODS

Irradiation at 7500–8000 erg/cm$^2$/s for 6–10 days at 28–31°C inhibited the development and moulting of J2, and at 11000 erg/cm$^2$/s irradiation was lethal (Nakasono, 1972; Nakasono and Katsura, 1976). Ultra-high frequency (UHF) electromagnetic energy at 366–732 J/cm$^2$ reduced *R. reniformis* population densities in the upper 10 cm soil layer (Heald and Wayland, 1975) and increased pear yield, although application of 1,3-dichloropropene at 75 l/ha was better singly and in combination with UHF (Heald, Menges and Wayland, 1974). At present, these methods appear of academic interest only.

### INTEGRATED CONTROL

Combinations of more than one chemical or non-chemical control methods have been tried against the reniform nematodes, and have generally given better control than individual methods. Gaur (1975) found that combinations of nursery-bed treatment with nematicide, crop rotation including wheat, application of organic amendments (de-oiled neem cake) or application of additional NPK fertilizers gave better control of *M. incognita* and *R. reniformis* than each treatment alone. Growing a non-host crop, such as chilli, and application of a nematicide (carbofuran, aldicarb, CGA12223, phorate, disulfoton) at 1 kg a.i./ha gave about 90% reduction in nematode numbers (Sivakumar, Lakshmanan and Balasubramanian, 1979). Similarly, application of aldicarb and poultry droppings gave enhanced control and increased yields of tomato (Badra and Mohamed, 1980). Chelliah *et al.* (1976a,b) suggested integrated multiple pest and disease control systems for tomato and okra, to include control of viral and fungal diseases, insects, and the reniform and lesion nematodes. These systems included fungicide treatment of seed or nursery-beds, application of aldicarb at 1 kg a.i./ha 1 week after transplanting tomato or sowing okra, and insecticide and fungicide sprays 1 and 2 months after planting.

For the control of *M. javanica, R. reniformis* and *Pratylenchus brachyurus* (Godfrey) Filipjev and Schuurmans Steckhoven in pineapple fields in Hawaii, Rohrbach and Apt (1986) proposed a scheme including:

1. avoidance of monoculture;
2. increasing soil pH above 6.0 and avoiding ammonium sulphate or other acidic fertilizers;
3. irrigated fallow, as these nematodes prefer low moisture regimes; and
4. application of 1,3-dichloropropene at 224–336 l/ha (preplant under plastic mulch), or fenamiphos at 0.56–3.36 kg a.i./ha at various crop growth intervals (not exceeding a total for the plant crop of 44.7 kg a.i./ha), or oxamyl spray at 1.12–4.48 kg a.i./ha at various intervals (not exceeding a total of 53.8 kg a.i./ha).

The information available on the reniform nematodes shows that, depending upon local conditions and crops, it should be possible to evolve specific methods to manage these nematodes by certain combinations of practices. These include: crop rotation with non-host or antagonistic crops; growing resistant cultivars; summer ploughing (preferably incorporating alternate wetting and drying) or moist fallow; soil solarization; organic amendments; restricting susceptible crops to the autumn/winter season; use of nematicides as seed-dressing, root-dip or nursery-bed treatments; or, for high-value crops, promoting natural enemies or biocontrol agents by modifying the soil environment.

The bulk of the information collected so far has been from greenhouse, microplot or small-scale field experiments. There is an urgent need to carry out integrated control trials under field conditions at diverse locations to develop practical, economic and environmentally safe methods to combat the reniform nematodes. Detailed studies on the biology, especially moulting behaviour and survival strategies, are likely to help by identifying susceptible phases of their life cycle. Understanding the biochemical nature and genetic basis of host resistance may help in evolving resistant germplasm and commercial cultivars. The genetic variability among the various populations or races of *R. reniformis* requires investigation, especially in relation to work on host resistance.

## References


ABDEL-RAHIM, M. F., SATOUR, M. M., MICKAIL, K. Y., EL-ERAKI, S. A., GRINSTEIN, A., CHEN, Y. AND KATAN, J. (1988). Effectiveness of soil solarization in furrow-irrigated Egyptian soils. *Plant Disease* **72**, 143–146.

ACHARYA, A. AND PADHI, N. N. (1987). Population threshold of *Rotylenchulus reniformis* in relation to its disease intensity on betelvine. *International Nematology Network Newsletter* **4**, 21–23.

ACOSTA, N. AND AYALA, A. (1975). Pathogenicity of *Pratylenchus coffeae, Scutellonema bradys, Meloidogyne incognita* and *Rotylenchulus reniformis* on *Dioscorea rotundata. Journal of Nematology* **7**, 1–6.

ACOSTA, N., CRUZ, C. AND ABREU, E. (1984). Chemical control of nematodes and insects on tomatoes. *Journal of Agriculture of the University of Puerto Rico* **68**, 79–86.

AHMAD, S., TIYAGI, S. A. AND KHAN, S. (1987). Interaction between root-knot and reniform nematode on chilli. *Pakistan Journal of Nematology* **5**, 19–25.

ALAM, M. M., KHAN, A. M. AND SAXENA, S. K. (1980). Mechanism of control of plant parasitic nematodes as a result of the application of organic amendments to the soil. V. Role of phenolic compounds. *Indian Journal of Nematology* **9**, 136–142.

ALAM, M. M., SAXENA, S. K. AND KHAN, A. M. (1981). Soil population of plant parasitic nematodes under various cropping sequences. *Nematologia Mediterranea* **9**, 49–56.

ALCANTARA, V. S. B. AND DE AZEVEDO, J. L. (1980). Toxicidade de filtrados de culturas de fungos ao nematóide *Rotylenchulus reniformis. Fitopatologia Brasileira* **5**, 201–206.

ALCANTARA, V. S. B. AND DE AZEVEDO, J. L. (1981). Isolamento e selação de fungos predadores de nematoides. *Revista de Agricultura Piracicaba Brazil* **56**, 135–146.

ANVER, S. AND ALAM, M. M. (1990). Susceptibility of chickpea accessions to the reniform nematode *Rotylenchulus reniformis. Tests of Agrochemicals and Cultivars 11. Annals of Applied Biology* **116**, (Suppl.), 94–95.

APT, W. J. (1976). Survival of reniform nematodes in desiccated soils. *Journal of Nematology* **8**, 278.

APT, W. J. (1979). Control of reniform nematode on pineapple in Hawaii with phenamiphos. *Abs. No. 612 Department of Plant Pathology, University of Hawaii, Honolulu, USA.*

AYALA, A. (1962). Pathogenicity of the reniform nematode on various hosts. *Journal of Agriculture of University of Puerto Rico* **46**, 73–82.

AYALA, A. AND ROMÁN, J. (1963). Distribution and host range of the burrowing nematode on Puerto Rican soils. *Journal of Agriculture of University of Puerto Rico* **47**, 28–37.

BADRA, T. AND ISMAIL, S. A. (1981). Accelerating the role of phenylalanine in shunting sex differentiation of *Rotylenchulus reniformis* by means of some nutrient formulae. *Revue de Nématologie* **4**, 209–212.

BADRA, T. AND KHATTAB, M. M. (1980). The effect of nitrogen fertilizers on the growth of olive and in relation to infestations of *Rotylenchulus reniformis. Nematologia Mediterranea* **8**, 67–72.

BADRA, T. AND KHATTAB, M. M. (1982). Chemically-induced resistance to *Rotylenchulus reniformis* by ethephon growth regulant and relevant pathometabolites in mango seedlings. *Nematologia Mediterranea* **10**, 49–56.

BADRA, T. AND MOHAMED, M. K. (1980). Influence of combinations of organic amendments and nematicides on tomato infected with *Rotylenchulus reniformis* and associated populations of predacious and saprophytic arthropods and microphagous nematodes. *Indian Journal of Nematology* **9**, 101–110.

BADRA, T., SALEH, M. A. AND OTEIFA, B. A. (1979). Nematicidal activity and composition of some organic fertilizers and amendments. *Revue de Nématologie* **2**, 29–36.

BALASUBRAMANIAN, P. AND RAMAKRISHNAN, C. (1983). Resistance to the reniform nematode *Rotylenchulus reniformis* in tomato. *Nematologia Mediterranea* **11**, 203–204.

BEASLEY, J. P. (1986). Evaluation of cotton *Gossypium hirsutum* L., genotypes for resistance to the reniform nematode, *Rotylenchulus reniformis* Linford and Oliveira. *Dissertation Abstracts International, B (Sciences and Engineering)* **47**, 856B.

BIRCHFIELD, W. (1962). Host-parasite relations of *Rotylenchulus reniformis* on *Gossypium hirsutum. Phytopathology* **52**, 862–865.

BIRCHFIELD, W. AND BRISTER, L. R. (1962). New hosts and non hosts of reniform nematode. *Plant Disease Reporter* **46**, 683–685.

BIRCHFIELD, W. AND MARTIN, W. J. (1967). Reniform nematodes survival in air-dried soil. *Phytopathology* **57**, 804.

BIRCHFIELD, W. AND MARTIN, W. J. (1976). Effects of some organophosphate and carbamate nematicides on reniform nematodes and sweet potato yields. *Journal of Nematology* **8**, 279–280.

BIRCHFIELD, W. AND WILLIAMS, C. (1974). Effects of nematicides and resistant soybean varieties on reniform nematode populations and soybean yields. *Proceedings of the American Phytopathological Society* **1**, 70.

BIRCHFIELD, W., WILLIAMS, C., HARTWIG, E. E. AND BRISTER, L. R. (1971). Reniform nematode resistance in soybeans. *Plant Disease Reporter* **55**, 1043–1045.

BIRD, A. F. (1983). Growth and moulting in nematodes: changes in the dimensions and morphology of *Rotylenchulus reniformis* from start to finish of moulting. *International Journal for Parasitology* **13**, 201–206.

BIRD, A. F. (1984). Growth and moulting and development of the hatched larva of *Rotylenchulus reniformis. Parasitology* **89**, 107–119.

BISHNOI, S. P. AND YADAV, B. S. (1989). Effect of some host plants on the biology of *Rotylenchulus reniformis* Linford and Oliveira, 1940. *International Nematology Network Newsletter* **6**, 10–11.

BOST, S. C. (1984). Evaluation of nematicides for control of the reniform nematode on cotton 1984. *Fungicide and Nematicide Tests. American Phytopathological Society* **40**, 97–98.

BRODIE, B. B. AND COOPER, W. E. (1964). Relation of parasitic nematodes to postemergence damping-off of cotton. *Phytopathology* **54**, 1023–1027.

BURROWS, P. R. (1990). The use of DNA to identify plant parasitic nematodes. *Nematological Abstracts* **59**, 1–8.

CARTER, W. W. (1980). Interaction of *Rotylenchulus reniformis* and *Macrophomina phaseolina* in charcoal rot of cantaloupe. *Journal of Nematology* **12**, 217.

CASTILLO, M. B. (1985). Some studies on the use of organic soil amendments for nematode control. *Philippine Agriculturist* **68**, 76–93.

CASTILLO, M. B., ARCEO, M. B. AND LITSINGER, J. A. (1978). Nematodes in cropping patterns. IV. Populations of plant parasitic nematodes associated with cropping patterns under different rice-growing environments in Manaoag Pangasinan. *Philippine Phytopathology* **12**, 24–29.

CHAVDA, J. C., PATEL, B. A. AND PATEL, D. J. (1988). Screening of pigeonpea lines to *Rotylenchulus reniformis. International Nematology Network Newsletter* **5**, 28–29.

CHELLIAH, S., MURUGESAN, S., SIVAKUMAR, V. C. AND RAMAKRISHNAN, L.

(1976a). Combination treatments for the control of insect pests, mite, virus vector, nematodes, fungal and viral diseases of Ghendi *Abelmochus esculentus* (L.) Moench. *Madras Agricultural Journal* **63**, 345–349.

CHELLIAH, S., MURUGESAN, S., SIVAKUMAR, C. V. AND RAMAKRISHNAN, L. (1976b). Combination treatments for the control of virus vectors, nematodes, fungal and viral diseases of tomato. *Madras Agricultural Journal* **63**, 350–353.

CHITWOOD, B. G. AND CHITWOOD, M. B. (1950). *An Introduction to Nematology. Section 1, Anatomy* (revised). Monumental Printing Co., Baltimore, Maryland.

CLARK, C. A. AND WRIGHT, V. L. (1983). Effect and reproduction of *Rotylenchulus reniformis* on sweet potato selections. *Journal of Nematology* **15**, 198–203.

CLARK, C. A., STRAND, C. AND WATSON, B. (1983). Comparison of soil fumigant with ethoprop for reniform nematode control in sweet potato 1982. *Fungicide and Nematicide Tests* **38**, 12–13.

COHN, E. (1976). Cellular changes induced in roots by two species of the genus *Rotylenchulus. Nematologica* **22**, 169–172.

DASGUPTA, D. R. AND SESHADRI, A. R. (1971). Races of the reniform nematode, *Rotylenchulus reniformis* Linford and Oliveira, 1940. *Indian Journal of Nematology* **1**, 21–24.

DASGUPTA, D. R., RASKI, D. J. AND SHER, S. A. (1968). A revision of the genus *Rotylenchulus* Linford and Oliveira, 1940 (Nematoda: Tylenchidae). *Proceedings of the Helminthological Society of Washington* **35**, 169–192.

DUHAYLONGSOD, R. D. (1988). Populations of root-knot and reniform nematodes in soil amended with fresh and composted organic materials. *International Nematology Network Newsletter* **5**, 24–26.

EGUNJOBI, O. A., AKONDE, P. T. AND CAVENESS, F. E. (1986). Interaction between *Pratylenchus sefaensis, Meloidogyne javanica* and *Rotylenchulus reniformis* in soil and mixed crops of maize and cowpea. *Revue de Nématologie* **9**, 61–70.

ELGINDI, A. Y., OTEIFA, B. A. AND KHADR, A. S. (1974). Interrelationships of *Rotylenchulus reniformis, Fusarium oxysporum f. vasinfectum* and potassium nutrition of cotton, *Gossypium barbedense. Potash Review* **5**, Subject 23, 43rd suite.

FERRAZ, L. C. C. B. (1985). Comportamento de diversas plantas daninhas de occurrência comum no Estado de São Paulo em relação ao nematóide reniforme. *Nematologia Brasileira* **9**, 143–152.

FORTUNER, R. (1984). List and status of the genera and families of plant parasitic nematodes. *Helminthological Abstracts, B. Plant Nematology* **53**, 87–133.

FORTUNER, R. (1987). A reappraisal of Tylenchina (Nemata). 8. The family Hoplolaimidae Filip'ev, 1934. *Revue de Nématologie* **10**, 219–232.

GAUR, H. S. (1975). *Crop damage in relation to the density of nematode population and an integrated approach to nematode population management.* Ph.D. Thesis, Indian Agricultural Research Institute, New Delhi.

GAUR, H. S. (1986). Invasion and reproduction of the reniform nematode on some cultivars of cowpea and green gram. *International Nematology Network Newsletter* **3**, 8–10.

GAUR, H. S. (1989). Dissemination and mode of survival of nematodes in dust storms. *Indian Journal of Nematology* **18**, 94–98.

GAUR, H. S. AND DHINGRA, A. (1991). Management of *Meloidogyne incognita* and *Rotylenchulus reniformis* in nurserybeds by soil solarization and organic soil amendment. *Revue de Nématologie* **14**, 189–196.

GAUR, H. S. AND HAQUE, M. M. (1986). Interspecific relations between concomitant populations of *Tylenchorhynchus* spp., *Rotylenchulus reniformis* and *Hoplolaimus indicus* under different crops. *Indian Journal of Nematology* **16**, 241–246.

GAUR, H. S. AND HAQUE, M. M. (1987). Effect of irrigation and stage of moisture stress on the growth of nematode populations under pea. *Indian Journal of Nematology* **17**, 189–192.

GAUR, H. S. AND PERRY, R. N. (1991). The role of the moulted cuticles in the desiccation survival of adults of *Rotylenchulus reniformis*. *Revue de Nématologie*, in press.

GAUR, H. S. AND SEHGAL, M. (1988). A comparative study of the effect of moisture stress and period of storage on the survival of *Meloidogyne incognita, Rotylenchulus reniformis* and *Tylenchulus semipenetrans* in soil. *International Nematology Network Newsletter* **5**, 5–8.

GAUSEMAN, H. W., HEALD, C. M., JR AND ESCOBAR, D. E. (1975). Effect of *Rotylenchulus reniformis* on reflectance of cotton plant leaves. *Journal of Nematology* **7**, 368–374.

GERMANI, G. (1978a). Tests preliminaires de sensibilité de deux cultivars de tomate et d'un cultivar d'arachide a deux souches de *Rotylenchulus reniformis* (Nematoda: Tylenchida). *Revue de Nématologie* **1**, 111–112.

GERMANI, G. (1978b). Caracteres morpho-biometriques de trois espèces ouest-africaines de *Rotylenchulus* Linford and Oliveira, 1940 (Nematoda: Tylenchida). *Revue de Nématologie* **1**, 241–250.

GOOD, J. M. (1972). Management of plant parasitic nematode populations. *Proceedings Tall Timbers Conference on Ecological Animal Control by Habitat Management, Tallahassee, Florida, 24–25 February*, pp. 109–127.

GOWEN, S. R. (1977). Nematicidal effects of oxamyl applied to leaves of banana seedlings. *Journal of Nematology* **9**, 158–161.

GOWEN, S. R. (1978)., Efecto de diferentes nematicides sobre el rendimiento del banano en las islas de Barlovento. *Nematropica* **8**, 9–13.

HAINLEIN, M. AND BLACK, I. D. (1983). The use of fallows, planting techniques and nematicide application for control of *Rotylenchulus reniformis* in pigeon peas. *Fiji Agricultural Journal* **45**, 35–42.

HAQUE, M. M. AND GAUR, H. S. (1985). Effect of multiple cropping on the dynamics of nematode populations and crop performance. *Indian Journal of Nematology* **15**, 262–263.

HAQUE, M. M. AND GAUR, H. S. (1986). Population behaviour and build-up of *Rotylenchulus reniformis* and other nematodes in some selected cowpea cultivars. *National Nematology Conference on Plant Parasitic Nematodes, India: Problems and Progress, 17–20 December 1986, IARI, New Delhi, India.*

HAQUE, M. M. AND PADMAVATHY, P. V. (1986). On the multiplication of *Rotylenchulus reniformis* in susceptible and resistant cultivars of cowpea. *Indian Journal of Nematology* **15**, 223–225.

HAQUE, M. S. AND MUKHOPADHYAYA, M. C. (1983). Influence of some micronutrients on *Rotylenchulus reniformis*. *Indian Journal of Nematology* **13**, 115–116.

HARTWIG, E. E. AND EPPS, J. M. (1973). Registration of Forrest soybeans. *Crop Science* **13**, 287.

HARTWIG, E. E. AND EPPS, J. M. (1977). Registration of Centennial soybeans (Reg. no. 114). *Crop Science* **17**, 979.

HARTWIG, E. E., HINSON, K. AND SCOTT, A. (1988). Registration of 'Padre' soybean. *Crop Science* **28**, 1025.

HARVILLE, B. G., GREEN, A. AND BIRCHFIELD, W. (1985). Genetic resistance to reniform nematodes in soybeans. *Plant Disease* **69**, 587–589.

HARVILLE, B. G., ELINS, W. M., HALL, W. AND GREEN, M. A. (1988). Registration of 'Gregg' soybean. *Crop Science* **28**, 196.

HEALD, C. M. (1975). Pathogenicity and histopathology of *Rotylenchulus reniformis* infecting cantaloupe. *Journal of Nematology* **7**, 149–152.

HEALD, C. M. (1980). Effect of reniform nematode on cantaloup yields. *Plant Disease* **64**, 282–283.

HEALD, C. M. AND ROBINSON, A. F. (1987). Effects of soil solarization on *Rotylenchulus reniformis* in the lower Rio Grande Valley of Texas. *Journal of Nematology* **19**, 93–103.

HEALD, C. M. AND WAYLAND, J. R. (1975). Factors affecting the control of

*Rotylenchulus reniformis* with UHF electromagnetic energy. *Journal of Nematology* **7**, 171–174.

HEALD, M. G., INSERRA, R. N. AND VOVLAS, N. (1988). Parasitism and reproduction of *Meloidogyne incognita* and *Rotylenchulus reniformis* on cantaloupe in two soils. *Nematropica* **18**, 53–58.

HEALD, C. M., MENGES, R. M. AND WAYLAND, J. R. (1974). Efficacy of ultra-high frequency (UHF) electromagnetic energy and soil fumigation on the control of the reniform nematode and common purslane among southern peas. *Plant Disease Reporter* **58**, 985–987.

HUSAIN, S. I. AND KHAN, A. M. (1967). A new subfamily, a new subgenus and eight new species of nematodes from India belonging to superfamily Tylenchoidea. *Proceedings of the Helminthological Society of Washington* **34**, 175–186.

HUTTON, D. G. AND HAMMERTON, J. L. (1975). Investigating the role of *Rotylenchulus reniformis* in a decline of pigeon pea. *Nematropica* **5**, 24.

ISMAIL, W. AND SAXENA, S. K. (1980). Potassium nutrition of castor in relation to infestations by *Meloidogyne incognita* and *Rotylenchulus reniformis. Indian Journal of Nematology* **10**, 1–8.

JIMENEZ, G. AND LOPEZ, R. (1987). Seasonal fluctuation of the spatial distribution of *Meloidogyne incognita* and *Rotylenchulus reniformis* on papaya (*Carica papaya* L.). *Turrialba – Revista Interamericana de Ciencias Agricolas* **37**, 165–170.

JOHNSON, A. W. AND FASSULIOTIS, G. (1984). Nematode parasites of vegetable crops. In *Plant and Insect Nematodes* (W. R. Nickle, Ed.), pp. 323–372. Marcel Dekker, New York.

JONES, J. E., BEASLEY, J. P., DICKSON, J. I. AND CALDWELL, W. D. (1988). Registration of four cotton germplasm lines with resistance to reniform and root-knot nematodes. *Crop Science* **28**, 199–200.

JONES, M. G. K. AND DROPKIN, V. H. (1975). Scanning electron microscopy of syncytial transfer cells induced in roots by cyst nematodes. *Physiological Plant Pathology* **7**, 259–263.

JONES, M. G. K. AND PAYNE, H. L. (1978). Early stages of nematode-induced giant cell formation in roots of *Impatiens balsamina. Journal of Nematology* **10**, 70–84.

KERRY, B. R. (1987). Biological control. In *Principles and Practice of Nematode Control in Crops* (R.H. Brown and B. R. Kerry, Eds), pp. 233–263. Academic Press, Australia.

KHAN, A. H., SAXENA, S. K. AND MAHMOOD, I. (1984). Effect of different cropping sequences on the population of plant parasitic nematodes. *Pakistan Journal of Nematology* **2**, 29–36.

KHAN, F. A. (1985). Hatching response of *Rotylenchulus reniformis* to root leachates of certain hosts and nonhosts. *Revue de Nématologie* **8**, 391–393.

KHAN, F. A. AND KHAN, A. M. (1974). Studies on the reniform nematode *Rotylenchulus reniformis*. I. Host range and population changes. *Indian Journal of Nematology* **3**, 24–30.

KHAN, M. W., KHAN, A. M. AND SAXENA, S. K. (1974). Rhizosphere fungi and nematodes of eggplant as influenced by oilcake amendments. *Indian Phytopathology* **27**, 480–484.

KHAN, R. M. AND KHAN, M. W. (1985). Histopathology of cauliflower roots infected with reniform nematode. *Pakistan Journal of Nematology* **3**, 83–85.

KHAN, R. M., KHAN, A. M. AND KHAN, M. W. (1985a). Cellular changes in tomato roots following infection of reniform nematode. *Pakistan Journal of Nematology* **3**, 73–82.

KHAN, R. M., KHAN, M. W. AND KHAN, A. M. (1985b). Cohabitation of *Meloidogyne incognita* and *Rotylenchulus reniformis* in tomato roots and effect on multiplication and plant growth. *Nematologia Mediterranea* **13**, 153–159.

KHAN, T. A. AND HUSAIN, S. I. (1988a). Response of cowpea cultivars to *Rotylenchulus reniformis. Nematropica* **18**, 159–162.

KHAN, T. A. AND HUSAIN, S. I. (1988b). Studies on the efficacy of *Paecilomyces lilacinus* as biocontrol agent against a disease complex caused by the interaction of *Rotylenchulus reniformis, Meloidogyne incognita* and *Rhizoctonia solani* on cowpea. *Nematologia Mediterranea* **16**, 229–231.

KHAN, T. A. AND HUSAIN, S. I. (1989a). Relative resistance of six cowpea cultivars as affected by the concomitance of two nematodes and a fungus. *Nematologia Mediterranea* **17**, 39–41.

KHAN, T. A. AND HUSAIN, S. I. (1989b). Studies of the pathogenicity of *Meloidogyne incognita* and *Rotylenchulus reniformis* on cowpea (*Vigna unguiculata*) var. Pusa Garsati. *Pakistan Journal of Nematology* **7**, 109–113.

LAL, A., YADAV, B. S. AND NANDWANA, R. P. (1977). Effect of chopped leaves of various plants and sewage on the plant growth and reniform nematode, *Rotylenchulus reniformis. Indian Journal of Mycology and Plant Pathology* **7**, 68–69.

LELE, V. C., DURGAPAL, J. C., ATARWAI, D. K. AND SETHI, C. L. (1978). Crown and root gall of grape (*Vitis vinifera* L.) in Andhra Pradesh. *Current Science* **47**, 280.

LIM, B. K. AND CASTILLO, M. B. (1979). Screening soybeans for resistance to reniform nematode disease in the Philippines. *Journal of Nematology* **11**, 275–282.

LINFORD, M. B. AND OLIVEIRA, J. M. (1940). *Rotylenchulus reniformis*, nov. gen. sp., a nematode parasite of roots. *Proceedings of the Helminthological Society of Washington* **7**, 35–42.

LOOF, P. A. A. AND OOSTENBRINK, M. (1962). *Rotylenchulus borealis* n. sp. with a key to the species of *Rotylenchulus. Nematologica* **7**, 83–90.

LORDELLO, L. G. E. AND BRANCALION, A. M. (1986). Avaliaçáo de nematicida Furadan 350 FMC no controle do nematóide *Rotylenchulus reniformis* atacando algocloeira. *Nematologica Brasileira* **10**, 103–107.

LUC, M., SIKORA, R. A. AND BRIDGE, J. (1990). *Plant Parasitic Nematodes in Subtropical and Tropical Agriculture*, p. 629. CAB International, Wallingford, UK.

MACEDO, M. C. M. (1974). Susceptibility of coffee to the reniform nematode. *Plant Breeding Abstracts* **47**, 1472.

MCSORLEY, R. (1987). Plot size and design for acquisition of field data in nematology. In *Vistas on Nematology* (J. A. Veech and D. W. Dickson, Eds), pp. 52–58. Society of Nematologists, DeLeon Springs, Florida.

MCSORLEY, R. AND PARRADO, J. L. (1983). Control of nematodes on white-fleshed sweet potatoes with Vydate, 1982. *Fungicide and Nematicide Tests* **38**, 13.

MCSORLEY, R., MCMILLAN, R. T., JR AND PARRADO, J. L. (1986). Soil fumigants for tomato production on Rockdale soils. *Proceedings of the Florida State Horticultural Society* **98**, 232–237.

MCSORLEY, R., DANKERS, W. H., PARRADO, J. L. AND REYNOLDS, J. S. (1985). Spatial distribution of the nematode community on Perrine marl soils. *Nematropica* **15**, 77–92.

MAGGENTI, A. R., LUC, M., RASKI, D. J., FORTUNER, R. AND GERAERT, E. (1988). A reappraisal of Tylenchina (Nemata). II. List of generic and supra-generic taxa, with their junior synonyms. *Revue de Nématologie* **11**, 177–188.

MAHAPATRA, B. C. AND PADHI, N. N. (1986). Pathogenicity and control of *Rotylenchulus reniformis* on *Cicer arietinum. Nematologia Mediterranea* **14**, 287–290.

MAHAPATRA, S. N. AND PADHI, N. N. (1988). Chemical control of nematodes in peanut crops. *Nematologia Mediterranea* **16**, 121–122.

MAHMOOD, I. (1985). Changes in amino acid content in different cultivars of tomato and eggplant as a result of infection with *Rotylenchulus reniformis* Linford and Oliveira 1940. *Bangladesh Journal of Botany* **14**, 1–6.

MAHMOOD, I. AND SAXENA, S. K. (1980). Effect of different doses of ammonium sulphate and urea on the plant growth and resulting biochemical changes in eggplant cv. Pusa purple long infected with *Rotylenchulus reniformis. Acta Botanica Indica* **8**, 171–174.

MAHMOOD, I. AND SAXENA, S. K. (1986). Effect of certain plant growth regulators on the growth of tomato inoculated with *Rotylenchulus reniformis*, nematode multiplication and resulting changes in phenolics. *Indian Journal of Plant Pathology* **4**, 139–142.

MAKADIA, B. M., PATEL, D. J., JOGANI, D. K. AND SHAH, H. M. (1987). Reactions of root-knot nematode resistant and susceptible cowpea varieties to reniform nematode, *Rotylenchulus reniformis*. *Gujarat Agricultural University Research Journal* **12**, 62–64.

MISHRA, S. D. AND GAUR, H. S. (1981). Pathogenicity of *Meloidogyne incognita* and *Rotylenchulus reniformis* to mothbean, *Vigna aconitifolius* Jacq. *Indian Journal of Nematology* **11**, 228–230.

MONTASSER, S. A. (1986). Resistance in tomato cultivars to the reniform nematode *Rotylenchulus reniformis*. *Pakistan Journal of Nematology* **4**, 79–82.

MUKHOPADHYAYA, M. C. AND HAQUE, M. D. S. (1980). Influence of soil types on *Rotylenchulus reniformis* on castor. *Indian Journal of Nematology* **10**, 88–91.

NAKASONO, K. (1966). An instance of male and young female development of the reniform nematode (*Rotylenchulus* sp.) within the egg-shell (Tylenchida: Hoplolaimidae). *Applied Entomology and Zoology* **1**, 49–50.

NAKASONO, K. (1972). Developmental retardation and morphological abnormalities observed in moulting larvae of the reniform nematode, *Rotylenchulus reniformis*. *Applied Entomology and Zoology* **7**, 243–245.

NAKASONO, K. (1973). Relationship between cultivation of Tagetes plants and changes in population of the reniform nematode, *Rotylenchulus reniformis*. *Japanese Journal of Nematology* **3**, 38–41.

NAKASONO, K. (1977). Sexual attraction of *Rotylenchulus reniformis* females and some differences in the attractiveness between amphimictic and parthenogenetic populations (Nematoda: Nacobbidae). *Applied Entomology and Zoology* **12**, 154–161.

NAKASONO, K. (1978). Effect of soil temperature on development, reproduction and sex ratio of *Rotylenchulus reniformis*: differences between the amphimictic and parthenogenic populations. *Japanese Journal of Nematology* **8**, 32–42.

NAKASONO, K. (1980). Studies on post embryonic development of the reniform nematode, *Rotylenchulus reniformis* Linford and Oliveira (Nematoda: Rotylenchulidae). II. Development of the reproductive system in larvae of the parthenogenic population. *Applied Entomology and Zoology* **15**, 425–432.

NAKASONO, K. (1983). Studies on morphological and physio-ecological variations of the reniform nematode, *Rotylenchulus reniformis* Linford and Oliveira, 1940 with an emphasis on differential geographical distribution of amphimictic and parthenogenetic populations in Japan. *Bulletin of the National Institute of Agricultural Sciences, Japan C* **38**, 1–67.

NAKASONO, K. AND KATSURA, N. (1976). Inhibitory effect of blue light on the moulting and development of larvae of *Rotylenchulus reniformis*. *Japanese Journal of Nematology* **6**, 84–88.

NANJAPPA, C. K., PANDA, M. AND SESHADRI, A. R. (1978). Effects of *Meloidogyne incognita* and *Rotylenchulus reniformis* on the emergence and survival of seedlings. *Proceedings of the 3rd International Congress of Plant Pathology München, W. Germany, 16–23 August*, p. 150.

NAQVI, Q. A. AND HASEEB, A. (1978). Influence of marigold mottle virus (MMV), the reniform nematode, *Rotylenchulus reniformis* and the stunt nematode on the growth of chilli and on nematode multiplication. *Acta Botanica Indica* **6**, 174–175.

NAQVI, Q. A. AND SAXENA, S. K. (1976). Studies on the influence of amino acids on the mortality of *Rotylenchulus reniformis*. *Comparative Physiology and Ecology* **1**, 93–96.

NATH, R. P., HAIDAR, M. G., AKHTER, S. W. AND PRASAD, H. (1979). Studies on the nematodes of vegetables in Bihar. I. Effect of the reniform nematode,

*Rotylenchulus reniformis*, on *Trichosanthes dioica. Indian Journal of Nematology* **6**, 175–177.

NAYAK, D. K., ROUTARAY, B. N. AND DAS, S. N. (1987). Relative susceptibility of some horsegram cultivars against reniform nematode, *Rotylenchulus reniformis. International Nematology Network Newsletter* **4**, 19–20.

NEAL, D. C. (1954). The reniform nematode and its relationship to the incidence of fusarium wilt of cotton at Baton Rouge, Louisiana. *Phytopathology* **44**, 447–450.

OTEIFA, B. A. AND ELGINDI, A. Y. (1974). Host nutrition in relation to infection by the reniform nematode *Rotylenchulus reniformis*. In *Simposio Internacional (XII) de Nematologia Sociedad Europea de Nematologos 1–7 Septembre, 1974, Granada, Spain.*

PADHI, N. N. AND MISRA, R. P. (1987). Pathologic reactions of french bean to the reniform nematode at varying levels of inoculation. *Indian Journal of Nematology* **17**, 128–129.

PANDA, M. AND SESHADRI, A. R. (1979). Effect of initial inoculum level of *Rotylenchulus reniformis* on different growth attributes of cowpea, mungbean and urid. *Indian Journal of Nematology* **9**, 59–60.

PATEL, B. A., CHAVDA, J. C. AND PATEL, D. J. (1989). Screening of certain greengram lines against reniform nematode, *Rotylenchulus reniformis. International Nematology Network Newsletter* **6**, 25–26.

PATEL, B. A., CHAVDA, J. C., PATEL, S. T. AND PATEL, D. J. (1987). Reaction of some pigeonpea lines to reniform nematode (*Rotylenchulus reniformis*). *International Pigeonpea Newsletter* **6**, 57–59.

PATEL, D. J. (1986). Histopathology of tobacco roots infected by *Rotylenchulus reniformis. Pakistan Journal of Nematology* **4**, 75–78.

PATEL, D. J., DESAI, M. V. AND PATEL, D. B. (1986a). Effect of *Rotylenchulus reniformis* on tobacco seed germination and seeding development. *International Nematology Network Newsletter* **3**, 6–7.

PATEL, D. J., DESAI, M. V. AND PATEL, D. B. (1986b). Susceptibility of tobacco varieties to *Rotylenchulus reniformis. Tobacco Research* **12**, 202–203.

PATEL, D. S. AND THAKAR, N. A. (1986). Reaction of some mungbean varieties to the reniform nematode, *Rotylenchulus reniformis* Linford and Oliviera, 1940. *Indian Journal of Nematology* **15**, 239.

PATEL, H. V., PATEL, D. J. AND PATEL, B. A. (1989). Reaction of papaya varieties to *Rotylenchulus reniformis. International Nematology Network Newsletter* **6**, 24.

PATEL, O. S. AND THAKAR, N. A. (1986). Efficacy of some systemic chemicals in control of *Rotylenchulus reniformis* infecting mungbean. *Indian Journal of Nematology* **16**, 281–282.

PATEL, S. K., VYAS, K. N. AND PATEL, D. J. (1988). Histological changes due to combined inoculation of two root-knot and reniform nematodes in tobacco. *Pakistan Journal of Nematology* **6**, 87–92.

PATHAK, K. N., NATH, R. P. AND HAIDER, M. G. (1986). Effect of initial inoculum levels of *Meloidogyne incognita* and *Rotylenchulus reniformis* on pigeonpea and their interrelationships. *Indian Journal of Nematology* **15**, 177–179.

PEACOCK, F. C. (1956). The reniform nematode in the Gold Coast. *Nematologica* **1**, 307–310.

PERRY, R. N. (1977a). Desiccation survival of larval and adult stages of the plant parasitic nematodes, *Ditylenchus dipsaci* and *D. myceliophagus. Parasitology* **74**, 139–148.

PERRY, R. N. (1977b). The water dynamics of stages of *Ditylenchus dipsaci* and *D. myceliophagus* during desiccation and rehydration. *Parasitology* **75**, 45–70.

PRASAD, K. S. K. AND KRISHNAPPA, K. (1981). Effect of bare root-dip treatments with pesticides on *Rotylenchulus reniformis* affecting brinjal. *Indian Journal of Nematology* **11**, 154–158.

PRASAD, K. S. K. AND PADAGNUR, G. M. (1980). Observations on the association of

*Rotylenchulus reniformis* with Verticillium wilt of cotton. *Indian Journal of Nematology* **10**, 91–92.

RADEWALD, J. D. AND TAKESHITA, G. (1964). Desiccation studies on five species of plant-parasitic nematodes in Hawaii. *Phytopathology* **54**, 903–904.

RAJENDRAN, G. AND NAGANATHAN, T. G. (1979). Nematicidal control of *Rotylenchulus reniformis* in grapevine var. anab-e-shahi. 87 (En) *All India Co-ordinated Fruits Improvement Project, College of Horticulture, Tamil Nadu Agricultural University, Coimbatore, India.*

RAO, A. B. V. K., PADHI, N. N. AND ACHARYA, A. (1987). Effect of different nematicides, oilcakes and urea in the control of *Rotylenchulus reniformis* on okra. *Indian Journal of Nematology* **17**, 171–173.

RAO, B. H. K. AND PRASAD, K. S. K. (1971). Population studies on *Meloidogyne javanica* and *Rotylenchulus reniformis* occurring together and separately, and their effect on the host. *Indian Journal of Entomology* **32**, 194–200.

RAY, S. AND PATRA, P. C. (1986). Effect of weedicides on population dynamics of *Rotylenchulus reniformis* and *Tylenchorhynchus cuticaudatus* affecting rabi groundnut. *Indian Phytopathology* **39**, 485–488.

RAZAK, A. R. AND EVANS, A. A. F. (1976). An intracellular tube associated with feeding by *Rotylenchulus reniformis* on cowpea root. *Nematologica* **22**, 182–189.

REBOIS, R. V. (1973). Effect of soil water on infectivity and development of *Rotylenchulus reniformis* on soybean, *Glycine max*. *Journal of Nematology* **5**, 246–249.

REBOIS, R. V. (1980). Ultrastructure of a feeding peg and tube associated with *Rotylenchulus reniformis* in cotton. *Nematologica* **26**, 396–405.

REBOIS, R. V. AND WEBB, R. E. (1979). Reniform nematode resistance in potato clones. *American Potato Journal* **56**, 313–319.

REBOIS, R. V., ELDRIDGE, B. J. AND WEBB, R. E. (1978). *Rotylenchulus reniformis* parasitism of potatoes and its effect on yield. *Plant Disease Reporter* **62**, 520–523.

REBOIS, R. V., MADDEN, P. A. AND ELDRIDGE, B. J. (1975). Some ultrastructural changes induced in resistant and susceptible soybean roots following infection by *Rotylenchulus reniformis*. *Journal of Nematology* **7**, 122–139.

REBOIS, R. V., ELDRIDGE, B. J., GOOD, J. M. AND STONER, A. K. (1973). Tomato resistance and susceptibility to the reniform nematode. *Plant Disease Reporter* **57**, 169–172.

REDDY, P. P. AND KHAN, R. M. (1988). Evaluation of *Paecilomyces lilacinus* for the biological control of *Rotylenchulus reniformis* infecting tomato, compared with carbofuran. *Nematologia Mediterranea* **16**, 113–115.

REDDY, P. P. AND KHAN, R. M. (1989). Evaluation of biocontrol agent *Paecilomyces lilacinus* and carbofuran for the management of *Rotylenchulus reniformis* infecting brinjal. *Pakistan Journal of Nematology* **7**, 55–59.

REDDY, D. D. R. AND SESHADRI, A. R. (1974). Studies on some systemic nematicides. II. Further studies on the action of thionazin and aldicarb on *Meloidogyne incognita* and *Rotylenchulus reniformis*. *Indian Journal of Nematology* **2**, 182–190.

RIDDLE, D. L. AND BIRD, A. F. (1985). Responses of the plant parasitic nematodes *Rotylenchulus reniformis, Anguina agrostis* and *Meloidogyne javanica* to chemical attractants. *Parasitology* **91**, 185–195.

ROBALINO, G., ROMÁN, J. AND CORDERO, M. (1983). Efecto del nematicida-insecticida oxamil aplicado al suelo y a las axilas de las hojas del bananero. *Nematropica* **13**, 135–143.

ROBINSON, A. F. (1989). Thermotactic adaptation in two foliar and two root-parasitic nematodes. *Revue de Nématologie* **12**, 125–131.

ROBINSON, A. F. AND HEALD, C. M. (1989). Accelerated movement of nematodes from soil in Baermann funnels with temperature gradients. *Journal of Nematology* **21**, 370–378.

ROBINSON, A. F., HEALD, C. M., FLANAGAN, S. L., THAMES, W. H. AND AMADOR,

J. (1987). Geographical distribution of *Rotylenchulus reniformis, Meloidogyne incognita* and *Tylenchulus semipenetrans* in the Lower Rio Grande Valley as related to soil texture and land use. *Annals of Applied Nematology* **1**, 20–25.

RODRIGUEZ-FUENTES, M. E. (1976). Einige Untersuchungen über die Entwicklung von *Rotylenchulus reniformis* Linford and Oliveira, 1940. In *2 Vortragstagung zu Aktuellen Problemen der Phytonematologie am 27–5–1976 in Rostock Manuskrips – druck ver Vorträge*, pp. 126–136. Biologische Gesellschaft der Deutschen Demokratischen Republik, Rostock. Rostock DDR.

RODRIGUEZ-FUENTES, M. E. (1980). Supervivencia de *Rotylenchulus reniformis* Linford and Oliveira 1940 sin la influencia del cultivo. *Centro Agricola Cuba* **7**, 67–74.

ROHRBACH, K. G. AND APT, W. J. (1986). Nematode and disease problems of pineapple. *Plant Disease* **70**, 81–87.

ROUTARAY, B. N., SAHOO, H. AND DAS, S. N. (1986). Evaluation of greengram and blackgram varieties against reniform nematode *Rotylenchulus reniformis. Indian Journal of Nematology* **16**, 27–29.

ROUTARAY, B. N., SAHOO, H. AND DAS, S. N. (1988). Evaluation of some chilli cultivars against *Meloidogyne incognita* and *Rotylenchulus reniformis. Journal of the Indian Botanical Society* **67**, 220–221.

SAHOO, H. AND PADHI, N. N. (1986). Susceptibility of plants to the reniform nematode, *Rotylenchulus reniformis. Indian Journal of Nematology* **16**, 97–98.

SAHOO, H., ROUTARAY, B. N. AND DAS, S. N. (1986). Evaluation of bengalgram varieties against reniform nematode, *Rotylenchulus reniformis. Journal of the Indian Botanical Society* **65**, 86–89.

SASSER, J. N. (1972). Nematode diseases of cotton. In *Economic Nematology* (J. M. Webster, Ed.), pp. 187–214. Academic Press, London.

SEHGAL, M. AND GAUR, H. S. (1989a). Survival and infectivity of the reniform nematode, *Rotylenchulus reniformis* in relation to moisture stress in soil without host. *Indian Journal of Nematology* **18**, 49–54.

SEHGAL, M. AND GAUR, H. S. (1989b). Effect of the rate of soil moisture loss on the survival, infectivity and development of *Rotylenchulus reniformis,* the reniform nematode. *Pakistan Journal of Nematology* **7**, 83–90.

SEINHORST, J. W. (1965). The relation between nematode density and damage to plants. *Nematologica* **11**, 137–154.

SETHI, C. L. AND GAUR, H. S. (1990). Population dynamics of plant parasitic nematodes in relation to cropping systems. In *Progress in Plant Nematology* (S. K. Saxena, M. Wajid Khan, Abdul Rashid and Rashid M. Khan, Eds), pp. 223–262. CBS Publishers and Distributors, PVT Ltd, Delhi, India.

SHARMA, S. B. AND NENE, Y. L. (1985). Effect of pre-sowing solarization on plant parasitic nematodes in chickpea and pigeonpea fields. *Indian Journal of Nematology* **15**, 277–278.

SHER, S. A. (1961). Revision of the *Hoplolaiminae* (Nematoda) I. Classification of nominal genera and nominal species. *Nematologica* **6**, 155–169.

SIDDIQI, M. R. (1972). Rotylenchulus reniformis. *C.I.H. Descriptions of Plant-parasitic Nematodes* Set 1 (5). Commonwealth Institute of Helminthology, St Albans, UK.

SIDDIQI, M. R. (1986). *Tylenchida: parasites of plants and insects.* Commonwealth Agricultural Bureaux, Farnham Royal, Slough.

SIDDIQUI, M. A. AND ALAM, M. M. (1987). Efficacy of seed dressing with extracts of neem and Persian lilac against *Meloidogyne incognita* and *Rotylenchulus reniformis. Nematologia Mediterranea* **15**, 399–403.

SIDDIQUI, M. A. AND ALAM, M. M. (1988a). Control of root-knot and reniform nematodes by bare-root dip in leaf extracts of margosa and Persian lilac. *Zeitschrift für Pflanzenkrankheiten und Pflanzenschutz* **95**, 138–142.

SIDDIQUI, M. A. AND ALAM, M. M. (1988b). Effect of latex seed dressing on

*Rotylenchulus reniformis* and plant growth of some vegetables. *Nematologia Mediterranea* **16**, 129–130.

SIDDIQUI, M. A. AND SAXENA, S. K. (1987). Effect of interculture of margosa and Persian lilac with tomato and eggplant on root-knot and reniform nematodes. *International Nematology Network Newsletter* **4**, 5–8.

SINGH, N. D. (1975). Effects of chemicals and plastic mulch on nematode populations and yield of tomato. *Nematropica* **5**, 6–10.

SINGH, R. V. AND KHERA, S. (1978). Culturing and life history studies of *Rotylenchulus reniformis* Linford and Oliveira on brinjal (Nematoda). *Bulletin of the Zoological Survey of India* **1**, 115–128.

SITARAMAIAH, K. AND SIKORA, R. A. (1981). Influence of the endomycorrhizal fungus *Glomus mosseae* on *Rotylenchulus reniformis* penetration and development on bush bean, cucumber and muskmelon. *Mededelingen van de Faculteit Landbouwwetenschappen Rijksuniversiteit Gent* **46**, 695–702.

SITARAMAIAH, K. AND SIKORA, R. A. (1982). Effect of mycorrhizal fungus *Glomus fasciculatus* on the host-parasite relationship of *Rotylenchulus reniformis* in tomato. *Nematologica* **28**, 412–419.

SIVAKUMAR, C. V. AND MEERZAINUDDIN, M. (1973). The effect of yellow vein mosaic disease of okra on the reniform nematode, *Rotylenchulus reniformis*. *South Indian Horticulture* **21**, 68–69.

SIVAKUMAR, C. V. AND MEERZAINUDDIN, M. (1975). Influence of N, P and K on the reniform nematode *Rotylenchulus reniformis* and its effect on the yield of okra. *Indian Journal of Nematology* **4**, 243–244.

SIVAKUMAR, C. V. AND SESHADRI, A. R. (1971). Life history of the reniform nematode *Rotylenchulus reniformis* Linford and Oliveira, 1940. *Indian Journal of Nematology* **1**, 7–20.

SIVAKUMAR, C. V. AND SESHADRI, A. R. (1974). Histopathology of infection by the reniform nematode *Rotylenchulus reniformis* Linford and Oliveira 1940 on castor, papaya and tomato. *Indian Journal of Nematology* **2**, 173–181.

SIVAKUMAR, C. V. AND SESHADRI, A. R. (1976). Longevity of the reniform nematode, *Rotylenchulus reniformis* in host-free soil. *Indian Journal of Nematology* **6**, 138–144.

SIVAKUMAR, C. V., LAKSHMANAN, P. AND BALASUBRAMANIAN, M. (1979). Interaction of a non-host plant *Capsicum annuum* and certain systemic nematicides in population decline of *Rotylenchulus reniformis*. *Revue de Nématologie* **2**, 165–168.

SIVAKUMAR, C. V., MEERZAINUDDIN, M. AND KUPPUSWAMY, S. (1973). Evaluation of certain systemic nematicides for the control of the root-knot nematode, *Meloidogyne incognita* and the reniform nematode *Rotylenchulus reniformis* on okra. *Madras Agricultural Journal* **60**, 530–533.

SRIVASTAVA, A. N. AND SETHI, C. L. (1986). Seasonal population fluctuations of *Heterodera zeae, Rotylenchulus reniformis* and *Tylenchorhynchus vulgaris* in maize–cowpea–wheat rotation. *Indian Journal of Nematology* **16**, 16–18.

SUD, U. C., VARAPRASAD, K. S., SESHADRI, A. R. AND KHER, K. K. (1985). Relationship between initial densities of *Rotylenchulus reniformis* and damage to cotton with a fit to Seinhorst's curves. *Indian Journal of Nematology* **14**, 148–151.

SUNDARAM, R., VELAYUTHAM, B. AND VADIVELU, S. (1987). Seasonal fluctuations of nematodes affecting garden bean. *International Nematology Network Newsletter* **4**, 22–24.

TAHA, A. H. Y. AND KASSAB, A. S. (1980). Interrelations between *Meloidogyne javanica, Rotylenchulus reniformis* and *Rhizobium* sp. on *Vigna sinensis*. *Journal of Nematology* **12**, 57–62.

TAHA, A. H. Y. AND SULTAN, S. A. (1979). Cellular response of grape seedlings to *Rotylenchulus reniformis* and *Tylenchulus semipenetrans* infections. *Nematologia Mediterranea* **7**, 45–50.

TAYLOR, L. R. (1961). Aggregation, variance and the mean. *Nature* **189**, 732–735.

TCHATCHOUA, J. AND SIKORA, R. A. (1978). Untersuchungen über die Wechselbeziehungenzwischen *Rotylenchulus reniformis* Linford and Oliviera 1940 und *Verticillium dahliae* an Baumwollpflanzen. *Mededelingen van de Faculteit Landbouwwetenschappen, Rijkuniversiteit Gent* **43**, 757–764.

THAKAR, N. A. AND PATEL, C. C. (1985). Screening of few cowpea varieties against reniform nematode, *Rotylenchulus reniformis* Linford and Oliveira, 1940. *Indian Journal of Nematology* **14**, 204.

THAKAR, N. A. AND PATEL, C. C. (1988). Action of few systemic nematicides on egg hatching of reniform nematode, *Rotylenchulus reniformis* Linford and Oliveira, 1940. *Indian Journal of Entomology* **48**, 520–521.

THAKAR, N. A. AND YADAV, B. S. (1985). Screening pigeonpeas for their resistance to reniform nematode. *International Pigeonpea Newsletter* **4**, 42–43.

THAKAR, N. A. AND YADAV, B. S. (1986). Role of total phenols in pigeonpea resistance to reniform nematode. *Indian Journal of Nematology* **16**, 261–263.

THAKAR, N. A. AND YADAV, B. S. (1987). Evaluation of pigeonpea varieties/lines for resistance against the reniform nematode. *Indian Journal of Nematology* **17**, 132–133.

THOMAS, R. J. AND CLARK, C. A. (1983). Population dynamics of *Meloidogyne incognita* and *Rotylenchulus reniformis* alone and in combination and their effects on sweet potato. *Journal of Nematology* **15**, 204–211.

THORNE, G. (1949). On the classification of the Tylenchida, new order (Nematoda: Phasmida). *Proceedings of the Helminthological Society of Washington* **16**, 37–73.

TIYAGI, S. A., ANVER, S., BANO, M. AND SIDDIQUI, M. A. (1986). Feasibility of growing zinnia as a mix crop along with tomato for control of root-knot and reniform nematodes. *International Nematology Network Newsletter* **3**, 6–7.

TRIANTAPHYLLOU, A. C. AND HIRSCHMANN, H. (1964). Reproduction in plant and soil nematodes. *Annual Review of Phytopathology* **2**, 57–80.

TSAI, B. Y. AND APT, W. J. (1979). Anhydrobiosis of the reniform nematode: survival and coiling. *Journal of Nematology* **11**, 316.

UPADHYAYA, R. S., KHAN, A. M. AND SAXENA, S. K. (1974). Multiplication of *Rotylenchulus reniformis* around the roots of castor in autoclaved and naturally infested soil. *Indian Journal of Nematology* **2**, 191–193.

VAN DEN BERG, E. AND SPAULL, V. W. (1981). *Rotylenchulus* species found associated with sugar cane in South Africa with a description of *R. sacchari* sp.n. (Nematoda: Rotylenchulinae). *Phytophylactica* **13**, 43–47.

VARAPRASAD, K. S. (1986). *Rotylenchulus reniformis* Linford and Oliveira, 1940 – A comparative account of systematics, biology and management. In *Plant Parasitic Nematodes of India, Problems and Progress* (G. Swarup and D. R. Dasgupta, Eds), pp. 194–210. Indian Agricultural Research Institute, New Delhi, India.

VARAPRASAD, K. S., SUD, U. C. AND SWARUP, G. (1987). Relationship between *Rotylenchulus reniformis* densities and host damage with reference to tolerance levels. *Indian Journal of Nematology* **17**, 11–16.

VERMA, S. K. AND PRASAD, S. K. (1969). The reniform nematode *Rotylenchulus reniformis* Linford and Oliveira, 1940. I Bio-ecological studies. *Indian Journal of Entomology* **31**, 36–47.

VIJAYALAKSHMI, K., GAUR, H. S. AND GOSWAMI, B. K. (1985). Neem for the control of plant-parasitic nematodes. *Neem Newsletter* **2**, 35–42.

VOVLAS, N. AND EKANAYAKE, H. M. R. K. (1985). Histological alterations induced by *Rotylenchulus reniformis* alone or simultaneously with *Meloidogyne incognita* on banana roots. *Nematropica* **15**, 9–17.

WILLIAMS, C., GILMAN, D. F., FONTENOT, D. S. AND BIRCHFIELD, W. (1979). A rapid technique for screening soybeans for reniform nematode resistance. *Plant Disease Reporter* **63**, 827–829.

WILLIAMS, C., GILMAN, D. F., FONTENOT, D. S. AND BIRCHFIELD, W. (1981). Inheritance of reaction to the reniform nematode in soybean. *Crop Science* **21**, 93–94.

WILLIAMS, J. R. (1960). Studies on the nematode soil fauna of sugarcane fields in Mauritius. Tylenchoidea (Partim). *Mauritius Sugar Industry Research Institute* **4**, 1–30.

WOMERSLEY, C. AND CHING, C. (1989). Natural dehydration regimes as a prerequisite for the successful induction of anhydrobiosis in the nematode *Rotylenchulus reniformis. Journal of Experimental Biology* **143**, 359–372.

YIK, C. P. AND BIRCHFIELD, W. (1979). Host studies and reactions of rice cultivars to *Meloidogyne graminicola. Phytopathology* **69**, 497–499.

YIK, C. P. AND BIRCHFIELD, W. (1984). Resistant germplasm in *Gossypium* species and related plants to *Rotylenchulus reniformis. Journal of Nematology* **16**, 146–153.

YING, K. C. (1985). Morphological character of different stages of *Rotylenchulus reniformis. Chinese Journal of Zoology* **20**, 29–32.

# 8
# Resistance in Plants to Cyst and Root-Knot Nematodes

ROGER COOK

*AFRC Institute of Grassland and Environmental Research, WPBS, Plas Gogerddan, Aberystwyth, Dyfed SY23 3EB, UK*

Introduction
Occurrence and use of resistance
Definition and specificity of resistance
Resistance to cereal cyst nematode
Resistance phenomena
    Before invasion—Root invasion—After invasion
Non-host resistance
Multiple resistance
Plant–nematode recognition
    Genetic interactions—Elicitors and defence
Novel resistance
Concluding remarks
References

## Introduction

Cyst-forming nematodes (species of *Heterodera* and *Globodera*) and root-knot nematodes (*Meloidogyne* spp.) are specialized plant parasites with highly developed interactions with their hosts. Resistance that prevents nematode reproduction is widely available in crop plants and their relatives. Resistant cultivars of some crops have been used alone or as one component of integrated systems for nematode control. In other cases, available resistance has not been used (Cook and Evans, 1987). Nematode resistance has been authoritatively reviewed in recent years (Huang, 1985; Kaplan and Davis, 1987). The present article compares knowledge of mechanisms that contribute to cyst and root-knot nematode resistance with resistance to other pathogens and pests, and considers the prospects for improved exploitation.

## Occurrence and use of resistance

Resistance of plants to nematodes, like that to other pathogens, is the normal

*Agricultural Zoology Reviews* — Vol. 4, February 1991
0269–0543 $20.00 + $0.00 

rather than the exceptional situation. Few crops have more than 10 major nematode pathogens, and most are non- or poor hosts of many other parasitic nematodes. None the less, nematodes are major pests, especially of tropical crops. Damage to crops is estimated to range from 7% (for wheat) to 20% (for bananas), throughout the world each year, with average losses of 40 major crops equivalent to 12.3% of annual production (Sasser and Freckman, 1987). It has been calculated that expenditure on nematological research and development (R & D) amounts to less than 0.2% of these losses. In developing countries, the proportion of R & D expenditure is significantly less than this world average, although losses from nematodes are no less severe.

Nematode resistance has been reported in most crops or in related germplasm, although it may not have been used in plant breeding programmes to develop resistant cultivars. Armstrong and Jensen (1978) listed resistance in some 111 crop genera, and Fassuliotis (1976) noted at least 235 cultivars of 15 crops with root-knot resistance. Sasser and Kirby (1979) listed root-knot-resistant cultivars of major crops, and sources of seed. In a recent world-wide survey of nematologists, it was recorded that nematode-resistant germplasm was available in 48 crops (Sasser and Freckman, 1987). Availability was greatest in South America (28 crops) and least in Africa (four crops). Undoubtedly, literature reviews overestimate the number of resistant cultivars by inclusion of references to experimental and source materials. Similarly, surveys of scientists may not reflect the true scene. Luc, Bridge and Sikora (1990) have emphasized that, in the tropics where nematode control is much needed, resistant cultivars have not been a real solution. This view is confirmed by non-nematological literature: a review of pest and disease control of more than 50 tropical food crops noted nematode pests of 20, but recorded use of resistant cultivars for nematode control only in Lima bean (*Phaseolus lunatus* L.) and tomato (*Lycopersicon esculentum* Mill.) (Martin, 1984). It seems clear that nematode resistance could be exploited in many areas where it is not.

Cook and Evans (1987) have discussed some reasons why nematode resistance breeding programmes have not been developed. The need for resistance to several nematodes in one crop is a major reason for non-exploitation in the tropics (Bridge, 1987; Noe, 1988). Resistance-breaking pathotypes have been a concern in breeding programmes in temperate crops, and must be expected in crops bred for the tropics. Often other factors (agronomic, consumer preferences or tolerance of other stresses) have to be given priority. In diverse and complex cropping systems, the production and introduction of nematode resistance needs joint application of the specialized skills of nematologist and breeder. These difficulties could be overcome given sufficient R & D input. Consideration of the costs and benefits of alternative investment strategies for crop improvement are beyond the scope of this review. The new techniques for genetic engineering, which promise to revolutionize plant breeding, may offer improved means of developing nematode-resistant crops but, for this, a thorough understanding of the nature and mechanisms of resistance will be essential.

**Definition and specificity of resistance**

Resistance is the ability of a plant to prevent reproduction of a nematode. This definition is important as it is independent of the effects of nematodes on plant growth. The classic description of plant–nematode interactions provided by Dropkin (1955) recognized that four combinations were possible between nematode reproduction (much v. little) and plant growth (good v. poor). Some have attempted to confine the descriptors of these interactions to a single word for each (Canto-Saenz, 1985). Such semantic debate obscures the clear central concepts. The four cells of the simple two-by-two interaction represent the extremes of continuously variable phenomena (Cook, 1974). A two-dimensional representation, as given by Evans and Haydock (1990), enables the effects of environmental influences, including variation in initial nematode population density, to be visualized as a third dimension. In practice, resistance ranges from complete, in which no nematode reproduction occurs, through degrees of resistance in which there is less reproduction than on a susceptible control. Other plant genotypes may be more susceptible than commonly used control cultivars, which may have some resistance (Cook and York, 1982). This review is concerned with those features of plant–nematode genotype interactions that lead to cyst and root-knot nematode reproduction being suppressed or prevented.

The primitive condition of host–nematode relationships may be resistance, as has been postulated for plant interactions with fungi and insects (Browning, 1980; Harris and Frederiksen, 1984), from which basic compatibility evolved by genetic changes in parasite and host (Heath, 1982). Co-evolutionary selection pressures have resulted in more or less high degrees of specificity between host and nematode genotypes, especially in cyst nematodes where gene-for-gene interactions are identifiable.

Plants devote many genes to the recognition of pathogens and to defence (Goodell, Deangelis and Ayres, 1985). Biotrophic pathogens must devote a similar genetic effort to avoid or overcome defences. The genetic basis of plant–nematode interactions should therefore appear complex. Agriculture and plant breeding has made it easier to recognize major effects of some plant genotypes. Likewise, nematode populations may have simple virulence phenotypes where only a small proportion of a nematode genepool has been introduced into a region, for example potato cyst nematode, *Globodera rostochiensis* (Woll.) Behrens in Europe (Stone, 1985); or where a crop has exerted major selection pressure for virulence, for example barley, *Hordeum vulgare* L., on *Heterodera avenae* Wollenweber in Britain (Cook and York, 1982). Extensive tests with many host genotypes would be necessary to determine the variation in virulence within an unselected nematode population. Cyst nematodes are more host-specific than the major tropical root-knot nematodes, and pathotypes have been more readily defined in cyst nematodes than in root-knot nematodes. This may be because those *Meloidogyne* spp. with wide host ranges (more than 700 species in many plant families) reproduce by parthenogenesis, whereas major temperate cyst nematode parasites are amphimictic. Parthenogenesis associated with polyploidy fixes heterozygosity, allowing very wide host ranges. This is advantageous to parasites since it prevents temporary selective forces acting against long-term beneficial adaptations (Price, 1980).

Within the species of *Meloidogyne*, host races are recognized by tests with selected crop genera (Sasser, 1979). When defined resistant genotypes are available and used, it is likely that variation at the pathotype level will be recognized. The term 'pathotype' has been used in nematology to describe intraspecific variation (Dropkin, 1988). Pathotypes are recognized by their virulence phenotype on selected plant differentials. The same pathotype phenotype could therefore be expected to occur in different host races of root-knot, or in related species of cyst nematode.

## Resistance to cereal cyst nematode

Some of the consequences of resistance are exemplified by studies of temperate cereals and cereal cyst nematode (*H. avenae*). Oats (*Avena sativa* L.) are the most susceptible hosts of *H. avenae* in much of north-western Europe with quite a range in the reaction of different cultivars (Cook and York, 1982). Observation of root invasion and subsequent rates of nematode development showed the effects of selected cultivars (*Tables 1* and *2*). When individual root tips of four cultivars were each inoculated with a single nematode, similar proportions of nematodes invaded. When experiments were done in pots, applying hatched juveniles to seedlings, the most resistant cv., Nelson, had fewest juveniles in its roots 10 days after inoculation. In the next 10 days, the number of nematodes in the less resistant line, 06334CnR, declined. In three other cultivars, nematode invasion was related to host status, as defined by numbers of females developing. Nematode development was also different in cv. Nelson, with few juveniles advancing beyond the invasive stage. In the other cultivars, development advanced more quickly in the more susceptible plants. At the final assessment, neither of the two more resistant cultivars supported females, in spite of their earlier differences. Nematodes clearly failed to establish feeding sites in cv. Nelson and many emerged from the roots. Those that remained probably died. When emergence of invaded nematodes from resistant cv. Panema was compared with

**Table 1.** Invasion of roots of five genotypes of oats (*Avena sativa*) by juveniles of cereal cyst nematode (*Heterodera avenae*)

| Oat cv. | *H. avenae* reaction | 1 nematode/root tip: Number inoculated | 1 nematode/root tip: % invaded at 3 days | 4000 nematodes/plant: Number invaded at 10 days | 4000 nematodes/plant: Number invaded at 20 days |
|---|---|---|---|---|---|
| Nelson | R | 60 | 58 | 163 | 124 |
| 06334CnR | R | – | – | 424 | 189 |
| Silva | pR | 111 | 63 | 415 | 266 |
| Sun II | pS | 88 | 50 | 292 | 383 |
| Milford | S | 182 | 58 | 385 | 418 |

R, resistant; S, susceptible; p, partially.
Means of six replicates.

**Table 2.** Development of juveniles of cereal cyst nematode (*Heterodera avenae*) in roots of five genotypes of oat (*Avena sativa*)

| Oat cv. | *H. avenae* reaction | Time (days after inoculation) 10 | | 20 | | | 100 |
|---|---|---|---|---|---|---|---|
| | | development stage (% all nematodes) | | | | | females/plant |
| | | %j2 | %j3 | %j2 | %j3 | %j4 | |
| Nelson | R | 99 | 1 | 78 | 9 | 13 | 0 |
| 06334CnR | R | 99 | 1 | 32 | 29 | 39 | 0 |
| Silva | pR | 73 | 27 | 27 | 14 | 59 | 4 |
| Sun II | pS | 64 | 36 | 7 | 18 | 75 | 24 |
| Milford | S | 60 | 40 | 9 | 13 | 78 | 34 |

R, resistant; S, susceptible; p, partially.
Means of six replicates.

that from susceptible cv. Peniarth, the proportions that left the roots were 40 and 18%, respectively (Cook, unpublished).

In these oat genotypes resistance is influenced by several genes; in cv. Nelson a single dominant gene was introduced from *A. sativa* CI 3444. Selection 06334CnR has a dominant gene introduced from wild oat, *A. sterilis* L. I 376, with two or three complementary dominant genes. The genes in cultivars Nelson and 06334CnR are the same, allelic or closely linked. There are additional genes in cv. Nelson which (in segregating progeny of the cross Nelson × Sun II) increased the susceptibility of plants lacking the dominant resistant gene. In cv. Sun II, opposing alleles at these modifying loci acted as minor resistance genes (Chew, Cook and Thomas, 1981; Cook and York, 1982).

Although there is some evidence that cereal cyst nematode populations differ in their virulence on oats, major pathotype groupings have been identified on barley. There is evidence from the pattern of the barley–*H. avenae* interactions for a gene-for-gene system. Resistant barleys have been widely grown in north-western Europe, and the nematode populations have responded to selection so that pathotypes are more or less homogeneous for virulence on barley genes *Rha1* and *Rha2* (Cook and York, 1982). In areas where a diversity of cereal germplasm exists (e.g. North Africa and India) it seems that many different genes for virulence on barley may be present in nematode populations. The pathotype schemes developed in response to barley breeding programmes in Scandinavia and Britain were relatively simple and, perhaps, atypical of the full range of possible interactions. Positive selection for major gene nematode resistance is relatively recent. Older European cultivars, such as oat cv. Sun II and many, older spring-sown barley cultivars, have good levels of background resistance. Where nematodes have been significant influences during the domestication or breeding of other crops, it is likely that similar levels of background resistance could be present.

## Resistance phenomena

The mechanisms expressed in these observations and those encountered in other host–nematode systems, form the subject of the next part of this review.

It is convenient to examine these according to the stage of the nematode–plant encounter at which their effects are apparent.

## BEFORE INVASION

Invasion of roots is influenced by plant characteristics. Host root exudates stimulate hatching of invasive juveniles of some cyst nematodes (Perry, 1987). It is not clear whether there are differences between cultivars in the effectiveness of their root diffusates. The identification of hatching factors has received varying levels of attention over the years, with the objective of synthesizing analogues to stimulate hatch in the absence of susceptible hosts (Perry, 1987). The hatching factors (glycinoeclepins) from hosts of soybean cyst nematode, *H. glycines* Ichinohe, have been structurally identified and analogues produced (Miwa *et al.*, 1987).

Diffusates and other emissions contribute to the attractiveness of roots to nematodes; genotypic differences might result in fewer nematodes at root tips. More *M. hapla* Chitwood juveniles were attracted to roots of susceptible than of resistant alfalfa, *Medicago sativa* L., when these were grown together (Griffin and Waite, 1971). This differential attractiveness was not so apparent in studies with each plant grown separately. Various root exudates repel nematodes, for example, cucurbits and *M. incognita* (Kofoid and White) Chitwood (Haynes and Jones, 1976); sweet potato, *Ipomoea batatas* (L.) Lam., and *M. incognita* (Jatala and Russell, 1972). Other roots produce toxic exudates, for example *Eragrostis curvula* Nees and various Compositae to *Meloidogyne* spp. (Scheffer, Kickuth and Visser, 1962; Gommers, 1981).

At the root surface, mutual recognition phenomena seem to involve nematode cuticular proteins and amphidial carbohydrates. Variations in these characteristics have been shown between life stages of cyst and root-knot nematodes (Reddigari *et al.*, 1986; Forrest, Robertson and Milne, 1989; Robinson, Delgado and Parkhouse, 1989), between pathotypes of cyst nematodes (Forrest and Robertson, 1986) and between both species and host races of root-knot nematodes (McClure and Stynes, 1988). Continuing research into the possible binding mechanisms between these molecules and plant root cells or exudates may identify varietal differences.

Less attractive roots might cause juveniles to deplete energy reserves, before reaching and recognizing a suitable root. Plants with such roots would have increased resistance through the reduced invasion and decreased fitness of invaded juveniles. Root-knot and cyst nematode invasive juveniles are functionally non-biotrophic. They have sufficient food reserves (albeit obtained biotrophically by their parent) for penetration of roots and for migration to suitable feeding sites. Some can emerge from a root and reinvade a second one, at least in experimental circumstances (Mullin and Brodie, 1988b).

## ROOT INVASION

Nematode energy reserves may also affect the ability of invasive juveniles to penetrate pre-formed barriers in roots. Thus older roots with secondary thickening or those with wound periderm are usually less invaded. Such roots may also be less attractive than root tip regions. Resistance of rice (*Oryza sativa* L.) to *M. graminicola* Golden and Birchfield was associated with the physical characteristics of roots, including a highly sclerotized exoderm (Jena and Rao, 1974). In an analogous interaction, tall fescue infected by the endophytic fungus (*Acremonium coenophialum* Morgan-Jones and Gams) was less susceptible to a number of nematodes, including *M. marylandi* Jepson and Golden. This has been attributed to thickening of root cell walls as an indirect effect of the endophyte infection (Bernard, Gwinn and Kimmons, 1990).

In many other crops, resistant cultivars are fully susceptible to invasion by *Meloidogyne* spp., for example, rice (Roy, 1975), alfalfa (Griffin and Elgin, 1977), cotton (*Gossypium hirsutum* L.) (McClure, Ellis and Nigh, 1974), tomato (Riggs and Winstead, 1959) and bean (*Phaseolus vulgaris* L.) (Fassuliotis, Deakin and Hoffman, 1970). Similar observations have been made with cyst-nematode-resistant plants, for example cereals (Cotten, 1969; O'Brien and Fisher, 1978a), potato (*Solanum tuberosum* L.) (Forrest, Trudgill and Cotes, 1986) and soybean [*Glycine max* (L.) Merril] (Riggs, Kim and Gibson, 1973).

Those resistant plants in which nematode invasion has been studied have been identified originally by their effects on nematode reproduction, generally after inoculation with abundant juveniles. This may have obscured differences in invasion suitability. It cannot be assumed that mechanisms affecting varietal resistance to invasion are absent from plants. There are a number of plant–nematode associations where nematode invasion is relatively less than in susceptible interactions. Non-hosts were shown not to be penetrated by *M. hapla* (Sasser, 1954). There are other examples of reduced invasion relative to susceptible controls: *Lycopersicon peruvianum* (L.) Mill. by root-knot (Liao and Dunlop, 1950), wild *Solanum* species by *Globodera* spp. (Roberts and Stone, 1983) and wild legumes by *H. glycines* (Riggs, 1987). Oat roots are invaded by more *H. avenae* than the poorer host, barley (Spaull and Hague, 1978).

Invading juveniles may not remain in roots if they fail to establish a feeding site and, as with oat and *H. avenae*, this may lead to observations of reduced invasion of resistant roots. Some reports of less invasion of resistant roots may be of this type. In some of these examples there is evidence that a rapid host response was responsible for reduced invasion by root-knot nematodes. Invasion of *L. peruvianum* was arrested when nematodes had only partially penetrated (Liao and Dunlop, 1950); in *Capsicum*, less penetration of two resistant cultivars was associated with an immediate necrotic response (Hendy, Dalmasso and Cardin, 1985). Such a post-infectional host response induced by first-invading juveniles, causing severe necrosis, may decrease the attractiveness of a root to other nematodes. In those parts of the tropics where preplanting root-knot populations are small, it could be a useful characteristic, unless associated with undesirable concentrations of pre-formed compounds. Screening individual root responses to individual nematodes would be necessary to select for such a mechanism. Price and Hague (1981) demonstrated differential invasion of oat and barley roots by *H. avenae* in this way.

## AFTER INVASION

Once nematodes have penetrated into the cortex, they migrate towards the stele and induce the development of a feeding site in susceptible roots. In suitable roots, juveniles migrate using the mechanical power of the stylet and cellulase, pectinase and other wall-degrading enzymes produced from oesophageal glands. Cyst nematode juveniles move through root cells. In some cases, root-knot nematodes seem to push plant cells apart without wall breakdown, whereas in others there is intracellular migration (Endo, 1987). Some juveniles migrate within roots for considerable distances; *H. schachtii* Schmidt juveniles moved for up to five times their body length through host roots (Yu and Steele, 1981). During this migration, juveniles continue to be functionally non-biotrophic; plant root differences requiring greater expenditure of juvenile energy reserves seem likely to impart resistance.

### *Feeding-site induction*

After invasion of roots, juveniles probe at cells of the endodermis or pericycle or other cells adjacent to or within the vascular tissue of the stele. At this stage there are changes in the activities of the oesophageal glands of both cyst and root-knot nematodes. The dorsal glands appear to be involved in feeding-cell induction, whereas the sub-ventral gland secretions are involved in digestion. The secretions from the two sets of glands differ (Hussey, 1987; Wyss, 1987). It is not known whether the plant genome affects this change in secretory activity, or whether it is directly and solely under nematode regulation.

Root-knot nematodes feed at several adjacent giant cells and cyst nematodes at syncytial feeding sites. These modified plant cells function as transfer cells for solutes (Jones and Northcote, 1972) and seem to be sites of intense metabolic activity. There is an increasing amount of information on the changes that occur during the development of both giant cells and syncytia. Their ultrastructure and biochemistry in susceptible hosts have been reviewed (Huang, 1985; Endo, 1987). In the susceptible plant, proliferation of cell organelles and smooth endoplasmic reticulum is consistent with the functions of secretion and transport. A theoretical model advanced by Premachandran and Dasgupta (1983) invokes induced transcription of host genes by nematode secretions that lead to derepression of parts of the plant genome. Recently, induction of feeding cells by *M. incognita* in tomato has been associated with the secretion into the initial feeding cell, via the stylet, of a high molecular weight glycoprotein produced in juvenile dorsal oesophageal glands (Hussey, Paguio and Seabury, 1990).

None the less, we are far from a complete understanding of their functioning in susceptible hosts and, therefore, not best able to recognize phenomena that may lead to breakdown of the plant–nematode interaction in resistant plants. Root-knot nematodes induce susceptible plants to produce galls, surrounding them and their feeding sites. This process is separate from giant cell formation (Fassuliotis, 1976). The continuing effects of nematodes on plant tissues surrounding them is shown by the observations that production of the gelatinous matrix into which eggs are deposited within root galls by *M. javanica* was associated with activity of pectolytic, cellulytic and proteolytic enzymes (Orion and Franck, 1990). Cyst nematode infested roots are frequently more branched than healthy roots, with proliferation of lateral roots at infection sites.

Insertion of the nematode stylet into the feeding cells is not continuous. *Heterodera schachtii* juveniles withdrew stylets after periods of secretion into the syncytium, and ingestion from it (Wyss, 1987). Root-knot juveniles feed on separate giant cells in turn. The interface between nematode stylet and host cell cytoplasm is probably bound by host plasmalemma, although the complete continuity of this has not been confirmed (Wyss, Stender and Lehmann, 1984). Feeding is interrupted for longer periods during moulting (three times between four feeding periods for cyst nematodes, but only once for root-knot nematodes, which only feed as second-stage juveniles and adult females). Feeding tubes, formed from nematode secretions, remain in feeding cells after each stylet withdrawal. They may function as filters to avoid the stylet aperture being blocked, or simply to increase the area through which solutes can pass (Rumpenhorst, 1984; Wyss, Stender and Lehmann, 1984).

### *Ultrastructural responses*

The responses of the initial feeding cell in resistant plants varies. In general, cyst nematodes induce development of feeding cells in resistant plants, but these subsequently degenerate; root-knot nematode initial feeding cells commonly react hypersensitively and die before feeding begins (Wyss, Stender and Lehmann, 1984). There are some examples to which this generalization does not apply. Ultrastructural study of degenerating feeding cells shows the accumulation of electron-dense vacuolated cytoplasm, with thickening of the cell walls. It has been hypothesized that such responses may lead to, or may result from, a breakdown of the transfer functions into and out of the cell. The immobile nematode dies either as a result of starvation or through the effects of an accumulation of toxic compounds within the feeding cell (Riggs, Kim and Gibson, 1973). In cyst-nematode-induced syncytia in resistant plants, there is more rough endoplasmic reticulum than in susceptible plants, and increased vacuolation of the cytoplasm.

### *Biochemical changes*

A hypersensitive response is a common plant reaction to wounding and infection by nematodes. Necrosis, typical of hypersensitivity, may occur in response to cell penetration by juveniles or by their stylets (Kaplan and

Davis, 1987). Nematodes may be affected during migration and also when initiating feeding at the stele. In *Capsicum*, immediate necrosis occurred when resistant roots were penetrated by root-knot (Hendy, Dalmasso and Cardin, 1985) and, in tomato resistant to *M. incognita*, invaded juveniles were walled off by the necrotic response (Hung and Rohde, 1973). Resistance to *Meloidogyne* in a number of crops has been associated with plant cell necrosis and death of juveniles. This often occurs before feeding cell induction, in contrast to resistance to cyst nematodes which has been associated with death of initially normal feeding cells (Wyss, Stender and Lehmann, 1984). Some cyst nematode juveniles leave roots of resistant plants without inducing feeding cells and, in oat and potato, this emigration is associated with hypersensitive responses (Cook, 1974; Rice, Leadbeater and Stone, 1985).

Nematodes within roots are subject to the effects of pre-formed as well as induced compounds within the roots. Gommers (1981) concluded that the presence in plants of constitutive nematicidal compounds does not indicate that they are effective *in vivo*. In cotton, peroxidase activity was greater in root-knot-resistant cultivars (Noel and McClure, 1978); in tomato, the concentration of chlorogenic acid localized in the endodermis was highest in plants most resistant to *Meloidogyne* (Hung and Rohde, 1973). In *M. incognita*-resistant tomato roots, catalase, peroxidase and polyphenoloxidase levels were all higher than in susceptible cultivars. Concentrations of the glycoalkaloid, alpha-tomatine, were related to resistance (Zhuchenko, Balashova and Andryushchenko, 1975; Okopnyi and Sadykin, 1976). Tobacco (*Nicotiana repanda*) resistant to root-knot had more chlorogenic acid and higher chlorogenic acid oxidase activity than susceptible roots (Milne, Boshoff and Buchan, 1965). Singh and Choudhury (1973) correlated total phenolics in tomato with resistance to *M. incognita, M. javanica* (Treub) Chitwood and *M. arenaria* (Neal) Chitwood. Phenolic concentrations were higher and increased more in response to infection in a barley cv. resistant to *M. incognita* race 1 than in susceptible cultivars (Rezk, Ibrahim and Ibrahim, 1987).

The pre-formed substances implicated in resistance, whether enzymes, glycoalkaloids or phenolics, are also encountered in responses occurring after infection. Many of the enzyme systems involved in phenolic metabolism and wound responses are activated rapidly in response to nematode infection (Kaplan and Davis, 1987). Nicotine increased in infected roots of both resistant and susceptible cultivars of tobacco (*N. tabacum* L.) to concentrations toxic to *Meloidogyne* juveniles *in vitro*. Davis and Rich (1987) considered that its greater *in vivo* toxicity in resistant plants resulted from localization at infection sites. The increase in resistant plants was greater than in susceptible ones. In root-knot-resistant cotton the activities of peroxidase and 6-phosphogluconate dehydrogenase increased within 6 days of inoculation, more than in susceptible roots (Noel and McClure, 1978). Terpenoid aldehydes decreased in susceptible but increased in resistant cotton (Veech and McClure, 1977), in response to infection by *M. incognita. M. javanica* and *M. arenaria* resistant tomato roots produced phenolics which were toxic to juveniles (Macaron, 1975). *Globodera rostochiensis* juveniles that emerged from resistant potato roots were less fit than ones emerging from susceptible roots (Mullin and Brodie, 1988b). Zacheo, Molinari and Dacoda (1988) found that in resistant tomato, *M. incognita* triggered enhanced hydroxyproline activities in cell walls. Infection of resistant roots increased activity of peroxidase, related to lignification around necrotic sites. In susceptible roots increased peroxidase activity was implicated in giant cell wall development. In *Prunus* rootstocks, invaded juveniles of *M. incognita* and *M. javanica* had deposits of lignin around their heads (Fresno, 1975). Similar changes after invasion of cowpea (*Vigna unguiculata* Walp.) resistant to *M. incognita* were related to enhanced peroxidase activity and production of polyphenols and lignin (Raja and Dasgupta, 1986).

Some of the responses induced in nematode-resistant plants result in the production of substances with phytoalexin activity (Kaplan and Keen, 1980). Phytoalexins are low molecular weight antimicrobial compounds, often isoflavenoids, polyacetylenes, terpenoids or steroids. Glyceollin has been associated with soybean resistance to *M. incognita* in such a way as to satisfy the criteria (developed from fungus-resistance studies) which establish linkage between phytoalexin induction and effects of resistance (Kaplan, Keen and Thomason, 1980a, 1980b).

### *Temperature effects*

Resistance to plant parasitic nematodes may be lost at temperatures above those at which plant and nematode usually interact. This temperature-sensitive resistance is reported for root-knot nematodes on a variety of crops of the warm temperate and subtropical regions. Among experimental treatments affecting the expression of resistance, higher than normal temperatures provide some interesting information on the mechanisms of plant response. Resistance of some bean (*P. vulgaris*) lines but not others is temperature sensitive, being eroded at 30°C (Omwega, Thomason and Roberts, 1990a). There is also evidence that resistance of cereals to cereal cyst nematode is less effective at higher temperatures (25–30°C) (Cook and McLeod, 1980). The general resistance of oats to the root-knot nematode (*M. naasi* Franklin) is ineffective at temperatures above 25°C (Cook, 1972; Gooris and D'Herde, 1972).

Loss of resistance of tomato to *M. incognita* at 34°C compared with 27°C, was associated with suppression of polyphenoloxidase and peroxidase activities and with reduced superoxide generation and lipid peroxidation. The activities of superoxide dismutase and catalase were enhanced. The latter two enzymes 'scavenge' the toxic superoxide and hydrogen peroxide molecules produced in response to wounding. This may lead to increased susceptibility by preventing their toxicity to both plant and nematode (Zacheo, Pricolo and Bleve-Zacheo, 1988). The reactions are quick, being detectable within hours (Kaplan and Davis, 1987). In resistant tomato, necrotic cells had high concentrations of phenolics 3 days after inoculation; carboxylesterase activities in adjacent cells had increased within 24 h after infection. Carboxylesterases are involved in secondary wall changes, and their increased activity preceded suberization or phenolic production, both changes correlated with the hypersensitive response (Melillo *et al.*, 1989).

### *Plant growth substances*

The role of plant growth substances in resistance has also been explored experimentally. Changes in plant hormone activity related to resistance have been reported. In oats, development of *H. avenae* females was favoured by a high ratio of cytokinins to auxins and that of males by the opposite; resistant cereals had low endogenous cytokinin activity and susceptible ones high levels (Lung, 1987). Tomato, resistant to *M. incognita*, became more susceptible after exogenous application of cytokinin. In treated roots, increased juvenile growth was associated with less necrosis and more galling (Dropkin, Helgeson and Upper, 1969). Both auxin (NAA) and cytokinin (kinetin) affected the response of resistant tomato to *M. incognita*; galls, but not giant cells, were produced in treated roots (Sawhney and Webster, 1975). Susceptible tomato roots infected by *M. incognita* had more auxins and cytokinin-like substances than healthy roots; abscisic acid levels were lower than in healthy roots, perhaps because the high auxin levels prohibited its transport from the shoot (Ganguly, Rajan and Dasgupta, 1989). Gibberellin applied to *Prunus* root stocks resistant to *M. incognita* and *M. javanica*, led to loss of resistance (Fresno, 1975). Kinetin and NAA applied exogenously to root stocks encouraged both giant cell and gall formation, reducing resistance to *M. javanica*. Infected susceptible cultivars had higher endogenous auxins and cytokinins than infected resisters or non-infected susceptibles (Kochba and Samish, 1971, 1972). Maleic hydrazide applied to tomato increased numbers of males of *M. incognita* and *M. javanica* (Davide and Triantaphyllou, 1968). A similar antimitotic compound, morphactin, had similar effects (Orion and Minz, 1971). It is not clear whether endogenous plant growth substances have a direct role in plant nematode interactions or whether changes in their activity are a consequence of cell responses to nematodes.

### *Fate of nematodes*

The initial feeding cell can be affected by pre-existing or induced mechanisms of resistance. After death of this cell or an early resistant response, the juvenile may probe a second cell, leave the root or die.

The stage of nematode development at which the feeding cell degeneration is completed has marked effects on nematode development; in some cases cyst nematodes develop as males in resistant plants but not as females. It is possible for cyst nematode females removed from their feeding cell to produce some eggs (Cook, 1977). Similarly, it is likely that juvenile cyst nematodes may obtain sufficient reserves from early feeding to continue development for a time after degeneration of their feeding cell, if the only effect is of starvation and not toxicity. Other factors that may limit food availability, such as the size of feeding cell and vascular contacts, have not been investigated; there is some evidence to suggest that such phenomena may affect fecundity and hence population reproduction [for example, potatoes with *Globodera* resistance (Dellaert *et al.*, 1988)]. In any plant–nematode interaction various and differing degrees of expression of these responses may be encountered (Kaplan and Davis, 1987), leading to different and characteristic expressions of resistance.

In cyst nematodes resistance has been shown to result from breakdown of initially normal feeding cells (Wyss, Stender and Lehmann, 1984). This occurred at different stages in different resistant cultivars of soybean and *H. glycines* (Endo, 1965; Acedo, Dropkin and Luedders, 1984); of potato and *Globodera* spp. (Rice, Leadbeater and Stone, 1985); and in various hosts of *H. schachtii* (Yu and Steele, 1981; Stender, 1987). Nematodes at such cells may suffer one of several fates resulting from insufficient food for normal development. In some cases, nematodes may obtain sufficient food to undergo moults into adult males. In resistant potatoes fewer *G. rostochiensis* males developed than on susceptibles (Mullin and Brodie, 1988b) whereas in *Cucumis* spp. resistant to *M. incognita* more males than normal were produced (Fassuliotis, 1970). In many cases, nematode death may result, probably from starvation.

Feeding site size may also affect nematode development and hence degree of plant resistance (Wyss and Zunke, 1987). Size of feeding cells may be influenced by constitutive factors, such as root size, as well as by induced mechanisms limiting feeding-cell expansion. Syncytial or giant cell initiation may break down at different stages after induction. Non-functioning feeding cells probably result from the thickening of adjacent cell walls, preventing passage of solutes into and out of the cell(s). The phenolic metabolism discussed in relation to induced mechanisms, and which is associated with death of feeding cells, may also directly affect juveniles at these cells.

Occasionally one or a few females mature in roots of cyst-nematode-resistant plants (Cook and York, 1982). These may have virulent genotypes, but this is not easily demonstrated since, if the virulence gene is recessive, outbreeding with avirulent males would mask virulence in the progeny (Jones and Perry, 1978). Alternatively, O'Brien and Fisher (1978b) hypothesized, albeit without convincing experimental evidence, that the first invading females of *H. avenae* induced a systemic resistance to subsequent nematodes. The discovery that potato cyst nematodes induced the production of pathogenesis-related proteins systemically (Hammond-Kosack, Atkinson and Bowles, 1989) adds some credence for the hypothesis. The progenies of the few females that developed on resistant potato from inoculations with an avirulent pathotype, have been shown to be less fit than those from susceptible plants (Mullin and Brodie, 1988a). This suggests that syncytia in resistant roots may offer nutrition that is deficient in quality or quantity. Progress in understanding such effects demands more knowledge of the nutritional requirements of the nematodes and of the changes that take place in their feeding cells throughout their activity.

## Non-host resistance

A promising area for research into resistance mechanisms is that shown by non-host plants. Past studies have focused on plants selected, on the basis of suppression of nematode reproduction, from the germplasm of genetically well-understood crops. Some potentially valuable, but incompletely effective, resistance mechanisms may be absent from such selections. Studies of resistance to fungi in non-hosts tend to confirm the view of Wood (1982) that genes promoting resistance in natural populations may have high costs in terms of fitness in other respects. Where these costs affect crop agronomic or production attributes, such genes may be lost. When further selection of improved crops is done in the absence of the pathogen, it is wholly likely that many defence genes will be lost. Broader-based resistance could be combined with agronomically desirable traits if nematode-resistance selection criteria were based upon the effects of defined mechanisms.

There are few studies of nematode–plant interactions comparing non-host with resistant host responses, although *M. hapla* juveniles were shown not to invade non-hosts (Sasser, 1954). There is also evidence that non-host roots inoculated with abundant juveniles are invaded, albeit by fewer juveniles than good hosts (*Table 3*). Detailed study of plant responses in such cases would be of interest in comparison with those of resistant and susceptible hosts. Where observations have been made on wild or primitive relatives of cultivated plants, a number of expressions of resistance have been identified. These tests identified resistance affecting hatch, cysts per plant and eggs per cyst (Dellaert *et al.*, 1988). Similar observations have been made with potatoes resistant to *Meloidogyne* spp. (Mendoza and Jatala, 1985).

**Table 3.** Numbers of juveniles of cereal cyst nematode (*Heterodera avenae*) in roots of host and non-host plants at four times after inoculation of 10-day-old seedlings with 450 juveniles

| Plant | Host status | Nematodes per plant Time (days after inoculation) | | | |
|---|---|---|---|---|---|
| | | 1 | 2 | 3 | 7 |
| Oat | ++ | 165 | 239 | 190 | 190 |
| Barley | + | 80 | 105 | 170 | 105 |
| Red clover | – | 3 | 4 | 20 | 5 |
| Tomato | – | 12 | 10 | 9 | 11 |

++, Good host; +, host; –, non-host.

## Multiple resistance

A frequently cited problem with the introduction of nematode-resistant crops is the need for resistance to more than one nematode pest. Usually, conscious selection has been for a single species. However, some selections have proved resistant to more than one species, especially with root-knot. Alfalfa cv. Vernal was resistant to *M. javanica* and immune to *M. hapla* (Goplen and Stanford, 1959). There are a number of other examples of cultivars with multiple resistance. Tomatoes with the *Mi* resistance gene were effective against *M. incognita, M. javanica* and *M. arenaria*; sweet potato was resistant to *M. incognita* and *M. javanica*; and pepper (*Capsicum annuum* L.) to *M. incognita, M. javanica* and *M. arenaria* but not to *M. hapla* (Fassuliotis, 1979). Similarly, four of 28 accessions of pigeon pea (*Cajanus cajani* Mill.) were resistant to both *M. incognita* and *M. javanica* (Moore *et al.*, 1988). Resistance of maize (*Zea mays* L.) to *M. incognita* and *M. arenaria* had different uncorrelated mechanisms (Windham and Williams, 1988). Potatoes with resistance to all races of *M. incognita* and *M. arenaria*, as well as to *M. javanica* and *M. hapla*, have been selected. After four generations of recurrent selection with a single nematode isolate, a small percentage of seedlings had multiple resistance. These tended to be of poor agronomic type, needing further improvement (Mendoza and Jatala, 1985). Resistance in the source(s) was probably broadly based, compared with that exploited in those crops where resistance is effective only against particular root-knot races. In contrast, bean (*P. vulgaris*) accessions with resistance to *M. incognita* race 1, *M. javanica* and *M. arenaria*, proved to have a single dominant gene controlling resistance to all three nematodes. This is probably vulnerable to the emergence of resistance-breaking genotypes of nematode (Omwega, Thomason and Roberts, 1990b).

Resistance to cyst nematodes tends to be more specific, and pathotypes have been identified in many field populations. Even so, in tests of wild and primitive potato seedlings, about 50% of 1284 accessions were resistant to a single *G. pallida* pathotype, and of these four had high resistance to many isolates of both *G. pallida* and *G. rostochiensis* (Dellaert *et al.*, 1988). There are also reports of cyst nematode resistance combined with resistance to another nematode; for example, root-knot-resistant tomato with *G. rostochiensis* resistance (Choleva, Georgiev and Solirova, 1987); soybean with resistance to *H. glycines* races 1 and 3 and also to *M. incognita* and *M. javanica*, as well as moderate resistance to *M. arenaria* (Boerma *et al.*, 1988). Soybeans resistant to both *H. glycines* and another sedentary endoparasitic nematode, *Rotylenchulus reniformis* Linford and Oliveira, had apparently identical hypersensitivity to each nematode, leading to poorly developed feeding cells (Rebois, Epps and Hartwig, 1970). Combined resistance to *M. incognita* and *R. reniformis* was found also in one of 12 accessions of *Capsicum annuum* (Routaray, Sahoo and Das, 1988). These are significant and encouraging observations, worthy of more detailed study of their mechanisms.

## Plant–nematode recognition

### GENETIC INTERACTIONS

Jones (1974) speculated that, since all pathotypes of cyst nematode can induce feeding cells (in a susceptible plant), and since all plants can develop them (when induced by a virulent nematode), the specificity of the nematode–plant interactions must derive from a recognition system separate from that involved with the induction and production of the feeding cells. This recognition system would determine whether the nematode induction/plant response produced a feeding cell or led to its breakdown or non-induction. There is no evidence for abnormal host physiology or biochemistry in the feeding cell, other than differences in time, position and quantity of expression (Jones, 1980).

It has been hypothesized that genes conditioning avirulence of a (bacterial) pathogen are highly mutable and dispensable, in so far as they have no effect on intrinsic pathogenic ability. Such genes have no selective value, being ephemerally involved in recognition events between host and pathogen (Gabriel, 1989). In contrast, genes directly involved in host species interactions (determining host range) have high selective value and are unlikely to change. Similar observations have been made in respect of the interactions between gall-forming insects and their plant hosts (Rohfritsch and Shorthouse, 1982).

The interaction at the nematode feeding site is likely to involve products of plant and nematode genes of fundamental importance to the organisms. Conflicting selection pressures means that there is little scope for modifying their function in feeding cells because of adverse effects on their primary action at other times or sites. The probability of resistance being due to the reaction of surrounding cells also indicates that processes other than direct interactions in the feeding cell are involved. Where the feeding cell degenerates, this may be due to wall changes around that cell rather than to the direct interaction of nematode and plant gene products. Modifications of nematode enzymes or host substrates might lead to quantitative or qualitative variations in the resistance response. Hypersensitivity is a common plant response to wounding, and may be the primitive resistance from which basic compatibility has arisen.

Other biotrophs have to avoid causing the cell death which leads to the release or production of the elicitors of phytoalexins (Bailey, 1982). In non-biotrophic phases the defence response may be avoided by speed of growth (rate of action) or by the evolution of some way to negate its effects. In a similar way, cyst and root-knot nematodes must suppress defence reactions in the feeding cell while expressing plant genes for food production. In their non-biotrophic phases, nematode movement may avoid expressed defence products. The glycocalyx secreted by many nematodes (Bird, Bonig and Bacic, 1988) may be a means of protecting the nematode from host defence products. Feeding cyst nematodes also secrete a subcrystalline layer (Wyss and Zunke, 1987) which may play a role in protection at the feeding site from general defence mechanisms.

## ELICITORS AND DEFENCE

Extensive studies on other systems indicate that the apoplast is the site of signals which elicit defence responses (Bowles, 1990). The apoplast (plant cell walls and intercellular spaces) provides a continuous network between plant tissues and organs. Defence responses can be grouped either as local ones, involving strengthening of cell walls and deterrence and antibiosis; or as systemically induced responses of unknown function (Bowles, 1990).

Nematodes, in their passage through roots, are subject to the first two classes of response; timing and degree of expression are crucial to the effectiveness of plant defence to pathogens and the same is likely to be true with nematodes. The role of induced resistance is not clear; there seem to be no good examples proven for nematodes. Cereal primary root axes, infested by *H. avenae* females, frequently have some proliferation of lateral roots which are often the site of feeding cells at which males develop. Plant growth substances and syncytial size may determine the sex of developing cyst nematodes, but it is possible that this is the indirect effect of an induced resistance mechanism. The induction of pathogenesis-related proteins throughout potato plants whose roots are invaded by cyst nematodes has not been shown to be related to decreased susceptibility to nematodes (Hammond-Kosack, Atkinson and Bowles, 1989).

The substances involved in plant defence responses function *in vivo* as a co-ordinated series of events with the array of products spatially and temporally regulated. There are three core enzymes of phenyl propanoid metabolism, responsible for three core reactions in plant defence. Precise characterization of genes encoding the core enzymes of phenylpropanoid metabolism shows that the regulation of the pathways is complex: the first enzyme, phenyl ammonium lyase (PAL), is present in multiple isomeric forms in all systems so far studied (Bowles, 1990). PAL has been shown to increase in resistant roots following infection by *M. incognita*. This increase did not occur at the elevated temperatures at which resistance was lost (Brueske, 1980). These pathways and their products are also involved in normal plant growth and development, leading to the production of structural compounds and plant growth substances as well as a variety of defence compounds, including suberin, lignin and other wall-bound phenolics (Hahlbrock and Scheel, 1989). Some of the products of these pathways are compounds with phytoalexin activity (flavenoids, isoflavenoids, coumarins and soluble esters). More than 12 000 secondary products are known from plants, where they are often physically or chemically compartmentalized in harmless forms within cells (Conn, 1985). In plants, most phenolics are of low activity *in vivo*, being bound as glycosides. Root-knot nematode homogenates contain β-glycosidase and β-glucosidase, but breakdown of plant glycoside to phenolics has not been demonstrated *in vivo* (Huang, 1985).

The *in vivo* responses of plants are likely to involve a highly co-ordinated network of events involving continuing exchange and recognition of signals leading to local, and possibly systemic, gene activation. Each signal may trigger more than one response, in order to co-ordinate phenomena of containment, repair and defence. Nematode enzyme systems change during development and parasitism: some of the gland secretions have been characterized (Hussey, 1987) and their roles are likely to be illuminated by the use of monoclonal antibodies for *in situ* labelling studies (Atkinson *et al.*, 1988; Hussey, 1989). Experimental work with these pathways needs precisely controlled systems. The identification of what may be the inducer of giant cell development in *M. incognita* is just such a precision tool, opening up numerous experimental opportunities (Hussey, Paguio and Seabury, 1990). Analyses of nematode–plant interactions in whole plants are almost certain to obscure the highly specific and local effects of elicitors (Dixon and Lamb, 1990).

The most rapid defence response in organized plant tissue is the deposition of callose in response to membrane damage. Membrane potentials and their transport functions are immediately impaired by some elicitors (Bowles, 1990). In interactions with fungi, maintenance of the integrity of the plasmalemma between the two organisms seems to be important in maintaining biotrophy (Gay and Woods, 1987). The precise nature of this interface between the nematode and plant is not yet clear. The feeding tubes produced in feeding cells may be bound by a host membrane. The reasons why cyst nematodes regularly withdraw their feeding stylet from syncytia is an important question, which may be related to maintenance of biotrophy (that is, in avoiding resistance). The deposition of callose may have effects on the ability of nematodes to withdraw and reinsert their stylets. Wyss has observed problems in stylet reinsertion in his video filming of *H. schachtii* feeding and development (Wyss, 1990). It may be that regular withdrawal of the stylet and development of new feeding tubes on reinsertion are necessary to avoid triggering plant defences. Alternatively, it may be simply an essential part of the feeding cycle. Other plant parasitic nematodes, for example the cortical endoparasite *Pratylenchus penetrans* Cobb, feeding for prolonged periods cause the death of the cell on which they feed, although in shorter periods of feeding this does not occur (Zunke, 1990).

In other plant defence responses, pectic fragments from plant cell walls have elicitor activity, presumably portending harmful circumstances (Anderson, 1989). Nematodes in their non-biotrophic passage through or between cells may stimulate, avoid or negate defence responses, which may or may not be different in quality or quantity from those affecting induction, maintenance and use of a feeding cell. The importance of cell wall elicitor activity emphasizes the need for further study of the penetration path of nematodes through roots, and the factors affecting this.

## Novel resistance

Characterization of mechanisms resulting in the observed resistance phenomena has concentrated on relatively few nematode–plant interactions. Observations on the biochemistry of resistance have focused on even fewer; none the less, these have proven very complex. More knowledge of feeding cell biochemistry and metabolism is needed so that these can be compared. Wyss and Zunke (1987) have indicated that amino-acid composition varies between syncytia in resistant and susceptible plants. More knowledge of the biosynthetic pathways induced by nematodes, as well as of the elicitors and recognition systems, is likely to lead nearer to the goal of introducing broad-based mechanisms of resistance without affecting other plant characteristics.

Plant biotechnology, using techniques of nucleic acid hybridization and genetic engineering, has great potential for improving the use of resistance for nematode control. Although molecular understanding of host–nematode interactions is one goal of such techniques (Burrows, 1990), the prospect of identifying and cloning a nematode-resistance gene of wide effectiveness remains a distant one. The possibility has been discussed of using the technologies of genetic engineering for gene transfer without associated agronomic weaknesses, etc. The technology to clone genes for transfer into crops from other sources can be used only when plant gene products can be identified and related to their effects on nematodes. At present, such techniques may be more readily applied to introduce alien genes into crops. This could provide nematode resistance of widespread effectiveness and durability.

Both cyst and root knot nematodes have been shown to complete normal life cycles *in vitro* on root cultures transformed by *Agrobacterium rhizogenes* (*H. schachtii* by Paul *et al.*, 1987; *M. javanica* by Verdejo, Jaffee and Mankau, 1988; *G. rostochiensis* by Kumar and Forrest, 1990). *Agrobacterium rhizogenes* can be used to incorporate genetic material into differentiated tissue. This is a useful test system for determining the effects on nematodes of novel genes from alien sources or from genetically engineered biosynthetic pathways. Alternative investigation on more defined conditions is dependent upon the development of alternative culture systems. Recently, the plant-parasitic migratory nematode *Pratylenchus scribneri* Steiner has been shown to ingest substances from artificial media (Tsai and Van Gundy, 1990). Nevertheless, the use of defined media for culturing obligate plant pathogens seems a long way off.

Some alien genes have been introduced into plants for insect pest control. The endotoxin genes from *Bacillus thuringiensis* have been cloned and expressed in transgenic plants (Vaeck *et al.*, 1987). Another class of Bt toxins, the exotoxins, have been shown to have nematicidal activity (Ignoff and Dropkin, 1977; Osman, Salem and Ghattas, 1988; Devidas, Cibulsky and Rehberger, 1989). Animal-parasitic nematodes are killed by the effects of bacterial endotoxins on membrane permeability (Bone, Bottjer and Gill, 1987). Activation of the endotoxins in insects depends upon the high pH of their gut. It is not clear whether ingested Bt toxins might affect nematodes. Similarly, the cowpea trypsin inhibitor gene, which in transgenic plants confers resistance to insect pests (Hilder *et al.*, 1987) requires activation by alkaline pH. Nevertheless, these developments demonstrate possibilities for nematode control using genetically engineered plants.

There are numerous opportunities for genetic engineering to produce plants with stable nematode resistance. Increasing knowledge of the biosynthesis by plants of attractants and hatching agents may lead to their exploitation in this way. Perry (1987) has suggested that such compounds could be synthesized for field application to induce hatch in the absence of hosts. Transgenic non-host plants producing a hatching stimulus would be a useful 'application' system. Alternatively, it may be possible to modify or regulate genes controlling enzymes involved in hatching factor production so as to produce ineffective compounds in host plants. Further candidates for genetic engineering include mimics or inhibitors of nematode moulting hormones or sex pheromones (Endo, 1987), or of nematode-specific metabolic processes. Hussey, Paguio and Seabury (1990) have suggested that their selection of a monoclonal antibody specific for a *M. incognita* secretory glycoprotein possibly involved in feeding cell induction could lead to the development of novel resistance. They suggest that transgenic plants, producing antibodies to the nematode glycoprotein, might inhibit nematode development.

## Concluding remarks

The creation of novel forms of resistance and the development of stable resistance requires a more complete understanding of the biochemistry of both nematode metabolism and of plant–nematode interactions. The potential of molecular genetic techniques both as tools for investigation as well as for potential exploitation of this knowledge is an additional spur. Research into the genome of the bacterial feeding nematode *Caenorhabditis elegans* (Kenyon, 1988), is likely to provide tools and knowledge of relevance to the development of novel resistance mechanisms.

The existence of good levels of resistance in many crops should not be ignored. For example, in barleys in north-west Europe Andersen and Andersen (1987) have recorded that as many as 95% of invading female *H. avenae* die in those widely grown cultivars used as susceptible controls. Many tropical crops may have good resistance. The separate phenomenon of tolerance, depending upon other mechanisms, may also be present in cultivars selected under stress conditions (Wallace, 1987). Clearly, there is evidence that resistance is naturally available and that alternative selection criteria could contribute to more stable resistance. The use of nucleic acid technologies to map plant genomes will provide ways of selecting for quantitative traits associated with resistance.

It seems that in compatible interactions there are numerous genes of nematode and plant involved, creating opportunities for non-specific gene interactions involving substrate effects and enzymes as well as more specific recognition systems. Resistance may be derived from the modification of these gene products: understanding of mechanisms would be improved by studies of closely comparable plant genotypes, including isogenic lines and defined nematode pathotypes. Comparison of these with the resistance of non-hosts or with that conferred by novel mechanisms would then be possible.

Vasil (1990) has indicated the need for continuing close interactions between biotechnology and traditional breeding so that the achievements of molecular genetics can be translated into practice; in the meantime there is much to be done in characterizing model plant–nematode interactions on a broader base than has so far been done, as well as undertaking the more detailed analysis of presently well-defined systems.

## References

ACEDO, J. R., DROPKIN, V. H. AND LUEDDERS, V. D. (1984). Nematode population attrition and histopathology of *Heterodera glycines*–soybean associations. *Journal of Nematology* **16**, 48–57.

ANDERSEN, K. AND ANDERSEN, S. (1987). *Heterodera avenae*: virulence and resistance. In *Cyst Nematodes* (F. Lamberti and C. E. Taylor, Eds), pp. 227–286. Plenum Press, London.

ANDERSON, A. J. (1989). The biology of glycoproteins as elicitors. In *Molecular and Genetic Perspectives, Volume 3, Plant–microbe interactions* (T. Kosuge and E. W. Nester, Eds), pp. 87–130. McGraw Hill, New York.

ARMSTRONG, J. M. AND JENSEN, H. J. (1978). *Indexed bibliography of nematode resistance in plants*. Bulletin 639, Agricultural Experiment Station, Oregon State University, Corvallis.

ATKINSON, H. J., HARRIS, P. D., HALK, E. J., NOVITSKI, C., LEIGHTON-SANDS, J., NOLAN, P. AND FOX, P. C. (1988). Monoclonal antibodies to the soybean cyst nematode, *Heterodera glycines*. *Annals of Applied Biology* **112**, 459–469.

BAILEY, J. A. (1982). Mechanisms of phytoalexin accumulation. In *Phytoalexins* (J. A. Bailey and J. A. Mansfield, Eds), pp. 289–318. Blackie, London.

BERNARD, E. C., GWINN, K. D. AND KIMMONS, C. A. (1990). Interrelationships of endophyte-infected tall fescue and plant parasitic nematodes. *Nematologica* **36**, in press.

BIRD, A. F., BONIG, I. AND BACIC, A. (1988). A role for the 'excretory' system in Secernentean nematodes. *Journal of Nematology* **20**, 493–497.

BOERMA, H. R., HUSSEY, R. S., PHILLIPS, D. W. AND WOOD, E. D. (1988). Registration of 'Twiggs' soybean. *Crop Science* **28**, 375.

BONE, L. W., BOTTJER, K. P. AND GILL, S. S. (1987). Alteration of *Trichostrongylus colubriformis* egg permeability by *Bacillus thuringiensis israelensis* toxin. *Journal of Parasitology* **73**, 295–299.

BOWLES, D. J. (1990). Defense-related proteins in higher plants. *Annual Review of Biochemistry* **59**, 873–907.

BRIDGE, J. (1987). Control strategies in subsistence agriculture. In *Principles and Practice of Nematode Control in Crops* (R. H. Brown and B. R. Kerry, Eds), pp. 389–420. Academic Press, Sydney.

BROWNING, J. A. (1980). Genetic protective mechanisms of plant–pathogen populations: their coevolution and use in breeding for resistance. In *Biology and Breeding for Resistance to Arthropods and Pathogens of Cultivated Plants* (M. K. Harris, Ed.), pp. 52–75. Miscellaneous publication 1451, Texas Agricultural Experiment Station.

BRUESKE, C. H. (1980). Phenylalanine ammonia lyase activity in tomato roots infected and resistant to the root-knot nematode, *Meloidogyne incognita*. *Physiological Plant Pathology* **16**, 409–414.

BURROWS, P. R. (1990). The use of DNA to identify plant parasitic nematodes. *Nematological Abstracts* **59**, 1–8.

CANTO-SAENZ, M. (1985). The nature of resistance to *Meloidogyne incognita* (Kofoid and White, 1919) Chitwood, 1949. In *An advanced treatise on* Meloidogyne. *Volume 1, Biology and control* (J. N. Sasser and C. C. Carter, Eds), pp. 225–231. North Carolina State University and United States Agency for International Development, North Carolina.

CHEW, B. H., COOK, R. AND THOMAS, H. (1981). Investigations on resistance of oats to *Heterodera avenae*: location of resistance genes. *Euphytica* **30**, 669–673.

CHOLEVA, B., GEORGIEV, H. AND SOLIROVA, V. (1987). Study of the resistance of lines and hybrids of tomato to root-knot nematodes (*Meloidogyne* sp.) and potato cyst nematode (*Globodera rostochiensis* (Woll.) Mulvey and Stone). *Biotekhnologiya and Biotekhnika* **2**, 30–33. (In *Helminthological Abstracts, Series B* **57**, 1427).

CONN, E. E. (1985). Chemical conjugation and compartmentalization: plant adaptations to toxic natural products. In *Cellular and Molecular Biology of Plant Stress* (J. L. Key and T. Kosuge, Eds), pp. 351–365. A. R. Liss, New York.

COOK, R. (1972). Reaction of some oat cultivars to *Meloidogyne naasi. Plant Pathology* **21**, 41–43.

COOK, R. (1974). Nature and inheritance of nematode resistance in cereals. *Journal of Nematology* **6**, 165–174.

COOK, R. (1977). The relationship between feeding and fecundity of females of *Heterodera avenae. Nematologica* **23**, 403–410.

COOK, R. AND EVANS, K. (1987). Resistance and tolerance. In *Principles and Practice of Nematode Control in Crops* (R. H. Brown and B. R. Kerry, Eds), pp. 179–232. Academic Press, Sydney.

COOK, R. AND MCLEOD, R. W. (1980). Resistance in wheat to *Heterodera avenae* in Australia and Britain. *Nematologica* **26**, 274–277.

COOK, R. AND YORK, P. A. (1982). Resistance of cereals to *Heterodera avenae*: methods of investigation, sources and inheritance of resistance. *European and Mediterranean Plant Protection Organisation Bulletin* **12**, 423–434.

COTTEN, J. (1969). Cereal varieties resistant to *Heterodera avenae* and *Ditylenchus dipsaci. Proceedings of the 5th British Insecticide and Fungicide Conference* **1**, 164–168.

DAVIDE, R. G. AND TRIANTAPHYLLOU, A. C. (1968). Influence of the environment on development and sex differentiation of root-knot nematodes III. Effects of foliar application of maleic hydrazide. *Nematologica* **14**, 37–46.

DAVIS, E. L. AND RICH, J. R. (1987). Nicotine content of tobacco roots and toxicity to *Meloidogyne incognita. Journal of Nematology* **19**, 23–29.

DELLAERT, L. M. W., VINKE, H., MEIJER, K., WOUTERS, D. AND VAN KEULEN, H. (1988). Resistentie tegen de aardappelcystenaaltjes in wilde en primitieve Solanaceae. *Prophyta* January, 90–94.

DEVIDAS, P., CIBULSKY, R. J. AND REHBERGER, L. (1989). Evaluation of *Bacillus thuringiensis* exotoxin for nematode control. *Nematologica* **35**, 265.

DIXON, R. A. AND LAMB, C. J. (1990). Molecular communication in interactions between plants and microbial pathogens. *Annual Review of Plant Physiology and Molecular Biology* **41**, 339–367.

DROPKIN, V. H. (1955). The relationships between nematodes and plants. *Experimental Parasitology* **4**, 282–322.

DROPKIN, V. H. (1988). The concept of race in phytonematology. *Annual Review of Phytopathology* **26**, 145–161.

DROPKIN, V. H., HELGESON, J. P. AND UPPER, C. D. (1969). The hypersensitive reaction of tomatoes resistant to *Meloidogyne incognita*: reversal by cytokinins. *Journal of Nematology* **1**, 55–61.

ENDO, B. Y. (1965). Histological responses of resistant and susceptible soybean varieties and backcross progeny to entry and development of *Heterodera glycines. Phytopathology* **55**, 375–381.

ENDO, B. Y. (1987). Histopathology and ultrastructure of crops invaded by certain sedentary endoparasitic nematodes. In *Vistas on Nematology* (J. A. Veech and D. W. Dickson, Eds), pp. 196–210. Society of Nematologists, Maryland.

EVANS, K. AND HAYDOCK, P. P. J. (1990). A review of tolerance by potato plants of cyst nematode attack, with consideration of what factors may confer tolerance and methods of assaying and improving it in crops. *Annals of Applied Biology* **117**, 703–740.

FASSULIOTIS, G. (1970). Resistance of *Cucumis* spp. to the root-knot nematode, *Meloidogyne incognita acrita. Journal of Nematology* **2**, 174–178.

FASSULIOTIS, G. (1976). Progress, problems and perspectives in breeding food crops for root-knot resistance. In *Proceedings of the Research Planning Conference on Root-Knot Nematodes* (Meloidogyne *spp.*), pp. 81–93. North Carolina State University.

FASSULIOTIS, G. (1979). Plant breeding for nematode resistance. In *Root-knot nematodes* (Meloidogyne *spp.*): *systematics, biology and control* (F. Lamberti and C. E. Taylor, Eds), pp. 425–453. Academic Press, London.

FASSULIOTIS, G., DEAKIN, J. R. AND HOFFMAN, J. C. (1970). Root-knot nematode resistance in snap beans: breeding and nature of resistance. *Journal of the American Society of Horticultural Science* **95**, 640–645.

FORREST, J. M. S. AND ROBERTSON, W. M. (1986). Characterization and localization of saccharides on the head region of four populations of the potato cyst nematode *Globodera rostochiensis* and *G. pallida. Journal of Nematology* **18**, 23–25.

FORREST, J. M. S., ROBERTSON, W. M. AND MILNE, E. W. (1989). Changes in the structure of amphidial exudate and the nature of lectin labelling on freshly hatched, invaded and emigrant second stage juveniles of *Globodera rostochiensis. Nematologica* **34**, 422–431.

FORREST, J. M. S., TRUDGILL, D. L. AND COTES, L. M. (1986). The fate of juveniles of *Globodera rostochiensis* Ro1 in roots of susceptible and resistant potato cultivars with gene H1. *Nematologica* **32**, 106–114.

FRESNO, A. P. (1975). Resistance of *Prunus persica* to *Meloidogyne* spp. *Nematological Society of Southern Africa Newsletter* **7**, 12–13.

GABRIEL, D. W. (1989). Genetics of plant–parasite populations and host–parasite specificity. In *Molecular and genetic perspectives, Volume 3, Plant–microbe interactions* (T. Kosuge and E. W. Nester, Eds), pp. 343–380. McGraw Hill, New York.

GANGULY, A. K., RAJAN, S. AND DASGUPTA, D. R. (1989). Development of auxin, cytokinin and abscisic acid-like substances in the root-knot nematode, *Meloidogyne incognita* infected tomato root. *Indian Journal of Nematology* **18**, 158–159.

GAY, J. L. AND WOODS, A. M. (1987). Induced modifications in the plasma membranes of infected cells. In *Fungal Infection of Plants* (G. F. Pegg and P. G. Ayres, Eds), pp. 79–91. Cambridge University Press, Cambridge.

GOMMERS, F. J. (1981). Biochemical interactions between nematodes and plants and their relevance to control. *Helminthological Abstracts Series B* **50**, 9–24.

GOODELL, J. J., DEANGELIS, P. L. AND AYRES, A. R. (1985). Immunochemical identification of antigens involved in plant–pathogen interactions. In *Cellular and Molecular Biology of Plant Stress* (J. L. Key and T. Kosuge, Eds), pp. 447–457. A. R. Liss, New York.

GOORIS, J. AND D'HERDE, C. J. (1972). Le cycle de développement de *Meloidogyne naasi* Franklin sur cereales de printemps et d'hiver et sur betteraves. *Revue de l'Agriculture (Brussels)* **25**, 651–657.

GOPLEN, B. P. AND STANFORD, E. H. (1959). Studies on the nature of resistance in alfalfa to two species of root-knot nematodes. *Agronomy Journal* **51**, 486–488.

GRIFFIN, G. D. AND ELGIN, J. H. (1977). Penetration and development of *Meloidogyne hapla* in resistant and susceptible alfalfa under differing temperatures. *Journal of Nematology* **9**, 51–56.

GRIFFIN, G. D. AND WAITE, W. W. (1971). Attraction of *Ditylenchus dipsaci* and *Meloidogyne hapla* by resistant and susceptible alfalfa seedlings. *Journal of Nematology* **3**, 215–219.

HAHLBROCK, K. AND SCHEEL, D. (1989). Physiology and molecular biology of phenylpropanoid metabolism. *Annual Review of Plant Physiology and Molecular Biology* **40**, 347–369.

HAMMOND-KOSACK, K. E., ATKINSON, H. J. AND BOWLES, D. J. (1989). Systemic accumulation of novel proteins in the apoplast of the leaves of potato plants following root invasion by the cyst nematode *Globodera rostochiensis. Physiological and Molecular Plant Pathology* **35**, 495–506.

HARRIS, M. K. AND FREDERIKSEN, R. A. (1984). Concepts and methods regarding host plant resistance to arthropods and pathogens. *Annual Review of Phytopathology* **22**, 247–272.

HAYNES, R. L. AND JONES, C. M. (1976). Effects of the Bi locus in cucumber on reproduction, attraction and response of the plant to infection by the southern root-knot nematode. *Journal of the American Society for Horticultural Science* **101**, 422–424.

HEATH, M. C. (1982). Absence of active defence mechanisms in compatible host–pathogen interactions. In *Active Defence Mechanisms in Plants* (R. K. S. Wood, Ed.), pp. 143–146. Plenum Press, London.

HENDY, H., DALMASSO, A. AND CARDIN, M. C. (1985). Differences in resistant *Capsicum annuum* attacked by different *Meloidogyne* species. *Nematologica* **31**, 72–78.

HILDER, V. A., GATEHOUSE, A. M. R., SHEERMAN, S. E., BARKER, R. F. AND BOULTER, D. (1987). A novel mechanism of insect resistance engineered into tobacco. *Nature* **300**, 160–163.

HUANG, J. (1985). Mechanisms of resistance to root-knot nematodes. In *An Advanced Treatise on* Meloidogyne, *Volume I, Biology and Control* (J. N. Sasser and C. C. Carter, Eds), pp. 165–172. North Carolina State University and United States Agency for International Development, North Carolina.

HUNG, C. L. AND ROHDE, R. A. (1973). Phenol accumulation related to resistance in tomato to infection by root-knot and lesion nematodes. *Journal of Nematology* **5**, 253–258.

HUSSEY, R. S. (1987). Secretions of esophageal glands of tylenchid nematodes. In *Vistas on Nematology* (J. A. Veech and D. W. Dickson, Eds), pp. 221–228. Society of Nematologists, Maryland.

HUSSEY, R. S. (1989). Monoclonal antibodies to secretory granules in esophageal glands of *Meloidogyne* species. *Journal of Nematology* **21**, 392–398.

HUSSEY, R. S., PAGUIO, O. R. AND SEABURY, F. (1990). Localization and purification of a secretory protein from the oesophageal glands of *Meloidogyne incognita* with a monoclonal antibody. *Phytopathology* **80**, 709–714.

IGNOFF, C. M. AND DROPKIN, V. H. (1977). Deleterious effects of the thermostable toxin of *Bacillus thuringiensis* on species of soil inhabiting myceliophagous and plant parasitic nematodes. *Journal of the Kansas Entomological Society* **50**, 394–398.

JATALA, P. AND RUSSELL, C. C. (1972). Nature of sweet potato resistance to *Meloidogyne incognita* and the effects of temperature on parasitism. *Journal of Nematology* **4**, 1–7.

JENA, R. N. AND RAO, Y. S. (1974). Root-knot nematode resistance in rice. *Indian Journal of Genetics and Plant Breeding* **34**, 433–449.

JONES, F. G. W. (1974). Host parasite relationships of potato cyst nematodes: a speculation arising from the gene-for-gene hypothesis. *Nematologica* **20**, 437–443.

JONES, F. G. W. AND PERRY, J. N. (1978). Modelling populations of cyst nematodes (Nematoda: Heteroderidae). *Journal of Applied Ecology* **15**, 349–371.

JONES, M. G. K. (1980). Microgel electrophoretic examination of soluble proteins in giant transfer cells and associated root-knot nematodes (*Meloidogyne javanica*) in balsam roots. *Physiological Plant Pathology* **16**, 359–367.

JONES, M. G. K. AND NORTHCOTE, D. H. (1972). Nematode induced syncytium – a multinucleate transfer cell. *Journal of Cell Science* **10**, 789–809.

KAPLAN, D. T. AND DAVIS, E. L. (1987). Mechanisms of plant incompatibility with nematodes. In *Vistas on Nematology* (J. A. Veech and D. W. Dickson, Eds), pp. 267–276. Society of Nematologists, Maryland.

KAPLAN, D. T. AND KEEN, N. T. (1980). Mechanisms conferring plant incompatibility to nematodes. *Revue de Nématologie* **3**, 123–134.

KAPLAN, D. T., KEEN, N. T. AND THOMASON, I. J. (1980a). Association of glyceollin with the incompatible response of soybean roots to *Meloidogyne incognita. Physiological Plant Pathology* **16**, 309–318.

KAPLAN, D. T., KEEN, N. T. AND THOMASON, I. J. (1980b). Studies on the mode of action of glyceollin in soybean incompatibility to the root-knot nematode *Meloidogyne incognita. Physiological Plant Pathology* **16**, 319–325.

KENYON, C. (1988). The nematode *Caenorhabditis elegans*. *Science* **240**, 1448–1453.

KOCHBA, J. AND SAMISH, R. M. (1971). Effect of kinetin and 1-napthylacetic acid on root-knot nematodes in resistant and susceptible peach root stocks. *Journal of the American Society for Horticultural Science* **96**, 458–461.

KOCHBA, J. AND SAMISH, R. M. (1972). Levels of endogenous cytokinins and auxins in roots of nematode resistant and susceptible peach root stocks. *Journal of the American Society for Horticultural Science* **97**, 115–119.

KUMAR, A. AND FORREST, J. M. S. (1990). Reproduction of *Globodera rostochiensis* on transformed roots of *Solanum tuberosum* cv. Désirée. *Journal of Nematology* **22**, 395–398.

LIAO, S. C. AND DUNLOP, A. A. (1950). Arrested invasion of *Lycopersicon peruvianum* roots by the root-knot nematode. *Phytopathology* **40**, 216–218.

LUC, M., BRIDGE, J. AND SIKORA, R. A. (1990). Reflections on nematology in subtropical and tropical agriculture. In *Plant Parasitic Nematodes in Subtropical and Tropical Agriculture* (M. Luc, R. A. Sikora and J. Bridge, Eds), pp. xi–xvii. CAB International, Wallingford.

LUNG, G. (1987). Wirtspflanzen-parasit-interaktion bei zystenbildende nematoden. Neue einblicke mit ubersicht uber die verschiedenen resistenzmechanismen. *Mededelingen van de Faculteit Landbouwwetenschappen Rijksuniversiteit Gent* **52**, 593–606.

MACARON, J. (1975). Etude de la resistance de deux variétés de tomate aux *Meloidogyne* spp. et au *Phytophthora parasitica*. *Nematologia Mediterranea* **3**, 35–41.

MCCLURE, M. A. AND STYNES, B. A. (1988). Lectin binding sites on the amphidial exudates of *Meloidogyne*. *Journal of Nematology* **20**, 321–326.

MCCLURE, M. A., ELLIS, K. C. AND NIGH, E. L. (1974). Post-infection development and histopathology of *Meloidogyne incognita* in resistant cotton. *Journal of Nematology* **6**, 21–26.

MARTIN, F. W. (1984). *Handbook of tropical food crops*. CRC Press, Florida.

MELILLO, M. T., BLEVE-ZACHEO, T., ZACHEO, G. AND GAHAN, P. B. (1989). Histochemical localisation of carboxylesterases in roots of *Lycopersicon esculentum* in response to *Meloidogyne incognita* infection. *Annals of Applied Biology* **114**, 325–330.

MENDOZA, H. A. AND JATALA, P. (1985). Breeding potatoes for resistance to the root-knot nematode *Meloidogyne* spp. In *An Advanced Treatise on* Meloidogyne, *Volume I, Biology and Control* (J. N. Sasser and C. C. Carter, Eds), pp. 165–172. North Carolina State University and United States Agency for International Development, North Carolina.

MILNE, D. L., BOSHOFF, D. N. AND BUCHAN, P. W. W. (1965). The nature of resistance of *Nicotiana repanda* to the root-knot nematode *Meloidogyne javanica*. *South African Journal of Agricultural Science* **8**, 557–570.

MIWA, A., NII, Y., OKAWARA, H. AND SAKAKIBARA, M. (1987). Synthetic study on hatching stimulus for the soybean cyst nematode. *Agricultural and Biological Chemistry* **51**, 3459–3461.

MOORE, K. M., RICH, J. R., BUHR, K. L. AND KNAUFT, D. A. (1988). Evaluation of pigeon pea *Cajanus cajani* accessions for resistance to *Meloidogyne incognita* race 3 and *M. javanica*. *Nematropica* **18**, 129–136.

MULLIN, B. A. AND BRODIE, B. B. (1988a). Effects of host resistance on the fecundity of *Globodera rostochiensis*. *Journal of Nematology* **20**, 109–112.

MULLIN, B. A. AND BRODIE, B. B. (1988b). Effects of resistance on second stage juveniles and adult males of *Globodera rostochiensis*. *Journal of Nematology* **20**, 335–339.

NOE, J. P. (1988). Theory and practice of the cropping systems approach to reducing nematode problems in the tropics. *Journal of Nematology* **20**, 204–213.

NOEL, G. R. AND MCCLURE, M. A. (1978). Peroxidase and 6-phosphogluconate dehydrogenase in resistant and susceptible cotton infected by *Meloidogyne incognita. Journal of Nematology* **10**, 34–39.

O'BRIEN, P. C. AND FISHER, J. M. (1978a). Factors influencing the number of larvae of *Heterodera avenae* within susceptible wheat and barley seedlings. *Nematologica* **24**, 295–304.

O'BRIEN, P. C. AND FISHER, J. M. (1978b). Studies on the mechanism of resistance of wheat to *Heterodera avenae. Nematologica* **24**, 463–471.

OKOPNYI, N. S. AND SADYKIN, A. V. (1976). The resistance of tomatoes to the gall nematodes. *Izadatel'stvo Shtiintsa*, pp. 108–109.

OMWEGA, C. O., THOMASON, I. J. AND ROBERTS, P. A. (1990a). Effect of temperature on expression of resistance to *Meloidogyne* spp. in common bean (*Phaseolus vulgaris*). *Journal of Nematology* **22**, 446–451.

OMWEGA, C. O., THOMASON, I. J. AND ROBERTS, P. A. (1990b). A single dominant gene in common bean conferring resistance to three root-knot nematode species. *Phytopathology* **80**, 745–748.

ORION, D. AND FRANCK, A. (1990). An electron microscope study of cell wall lysis by *Meloidogyne javanica* gelatinous matrix. *Revue de Nématologie* **13**, 105–107.

ORION, D. AND MINZ, G. (1971). The influence of morphactin on the root-knot nematode *Meloidogyne javanica* and its galls. *Nematologica* **17**, 107–112.

OSMAN, G. Y., SALEM, F. M. AND GHATTAS, A. (1988). Bio-efficacy of two bacterial insecticide strains of *Bacillus thuringiensis* as a biological control agent in comparison with a nematicide, Nemacur, on certain parasitic Nematoda. *Anz Schaedlingskd Pflanzenschutz Umweltschutz* **61**, 35–37.

PAUL, H., ZIJLSTRA, C., LEEUWANGH, J. E., KRENS, F. A. AND HUIZING, H. J. (1987). Reproduction of the beet cyst nematode *Heterodera schachtii* Schm. on transformed root cultures of *Beta vulgaris* L. *Plant Cell Reports* **6**, 379–381.

PERRY, R. N. (1987). Physiology of hatching. In *Cyst Nematodes* (F. Lamberti and C. E. Taylor, Eds), pp. 119–132. Plenum Press, London.

PREMACHANDRAN, D. AND DASGUPTA, D. R. (1983). A theoretical model for plant nematode interaction. *Revue de Nématologie* **6**, 311–314.

PRICE, N. S. AND HAGUE, N. G. M. (1981). The invasion of root tips of cereals by the cereal cyst nematode *Heterodera avenae. Annals of Applied Biology* **99**, 301–306.

PRICE, P. W. (1980). *Evolutionary Biology of Parasites*. Princeton University Press, New Jersey.

RAJA, A. AND DASGUPTA, D. R. (1986). Enhanced synthesis of messenger RNA in relation to resistance expression in cowpea (*Vigna unguiculata*) infected with root-knot nematode, *Meloidogyne incognita. Revue de Nématologie* **9**, 35–48.

REBOIS, R. V., EPPS, J. M. AND HARTWIG, E. E. (1970). Correlation of resistance in soybeans to *Heterodera glycines* and *Rotylenchulus reniformis. Phytopathology* **60**, 695–700.

REDDIGARI, S. R., JANSMA, P. L., PREMACHANDRAN, D. AND HUSSEY, R. S. (1986). Cuticular collagenous proteins of second stage juveniles and adult females of *Meloidogyne incognita*: isolation and partial characterization. *Journal of Nematology* **18**, 294–302.

REZK, M. A., IBRAHIM, I. K. A. AND IBRAHIM, A. A. M. (1987). Effect of root-knot nematodes on the phenolic contents of barley and wheat. *Nematologia Mediterranea* **15**, 259–263.

RICE, S. L., LEADBEATER, B. S. C. AND STONE, A. R. (1985). Changes in cell structure in roots of resistant potatoes parasitised by potato cyst nematodes. I. Potatoes with resistance gene H1 derived from *Solanum tuberosum* ssp. *andigena. Physiological Plant Pathology* **27**, 219–234.

RIGGS, R. D. (1987). Non-host root penetration by soybean cyst nematode. *Journal of Nematology* **19**, 251–254.

RIGGS, R. D. AND WINSTEAD, N. N. (1959). Studies on resistance in tomato to root-knot nematodes and on the occurrence of pathogenic biotypes. *Phytopathology* **49**, 716–724.

RIGGS, R. D., KIM, K. S. AND GIBSON, I. (1973). Ultrastructural changes in Peking soybeans infected with *Heterodera glycines*. *Phytopathology* **63**, 76–84.

ROBERTS, P. A. AND STONE, A. R. (1983). Comparisons of invasion and development of *Globodera* spp. and European potato cyst-nematode pathotypes in roots of resistant *Solanum* sg. *Leptostemonum* spp. *Nematologica* **29**, 95–108.

ROBINSON, M. P., DELGADO, J. AND PARKHOUSE, R. M. E. (1989). Characterization of stage-specific cuticular proteins of *Meloidogyne incognita* by radio-iodination. *Physiological and Molecular Plant Pathology* **35**, 135–140.

ROHFRITSCH, O. AND SHORTHOUSE, J. D. (1982). Insect galls. In *Molecular Biology of Plant Tumors* (G. Kahl and N. Schell, Eds), pp. 131–151. Academic Press, New York.

ROUTARAY, B. N., SAHOO, H. AND DAS, S. N. (1988). Evaluation of some chilli cultivars against *Meloidogyne incognita* and *Rotylenchulus reniformis*. *Journal of the Indian Botanical Society* **67**, 220–221.

ROY, A. K. (1975). Studies on resistance of rice to the attack of *Meloidogyne graminicola*. *Zeitschrift für Pflanzenkrankheiten und Pflanzenschutz* **82**, 384–387.

RUMPENHORST, H. J. (1984). Intracellular feeding tubes associated with sedentary plant parasitic nematodes. *Nematologica* **30**, 77–85.

SASSER, J. N. (1954). Identification and host parasite relationships of the soybean cyst nematode in resistant soybean roots. *Phytopathology* **48**, 578–579.

SASSER, J. N. (1979). Pathogenicity, host ranges and variability in *Meloidogyne* species. In *Root-knot nematodes* (Meloidogyne *spp.*): *systematics, biology and control* (F. Lamberti and C. E. Taylor, Eds), pp. 257–268. Academic Press, London.

SASSER, J. N. AND FRECKMAN, D. W. (1987). A world perspective on nematology: the role of the Society. In *Vistas on Nematology* (J. A. Veech and D. W. Dickson, Eds), pp. 7–12. Society of Nematologists, Maryland.

SASSER, J. N. AND KIRBY, M. F. (1979). *Crop cultivars resistant to root-knot nematodes* Meloidogyne *spp. with information on seed sources.* Cooperative publication of the North Carolina State University and the United States Agency for International Development.

SAWHNEY, R. AND WEBSTER, J. M. (1975). The role of plant growth hormones in determining the resistance of tomato plants to the root-knot nematode *Meloidogyne incognita*. *Nematologica* **21**, 95–103.

SCHEFFER, F., KICKUTH, R. AND VISSER, J. H. (1962). Die wurzelausscheidungen von *Eragrostis curvula* (Schrad.) Nees und ihr einfluss auf Wurzeknoten-nematoden. *Zeitschrift für Pflanzenernaehrung Duengung Bodenkunde* **98**, 114–120.

SINGH, B. AND CHOUDHURY, B. (1973). The chemical characteristics of tomato cultivars resistant to root-knot nematode (*Meloidogyne* spp.). *Nematologica* **19**, 447–448.

SPAULL, A. M. AND HAGUE, N. G. M. (1978). Influence of cereal cultivar on the population dynamics of the cereal cyst nematode *Heterodera avenae*. *Nematologica* **24**, 376–383.

STENDER, C. (1987). Cell responses to *Heterodera schachtii* feeding on susceptible and resistant cultivars of oil radish. In *Cyst Nematodes* (F. Lamberti and C. E. Taylor, Eds), pp. 251–254. Plenum Press, London.

STONE, A. R. (1985). Co-evolution of potato cyst nematodes and their hosts: implications for pathotypes and resistance. *European and Mediterranean Plant Protection Organisation Bulletin* **15**, 131–138.

TSAI, B. Y. AND VAN GUNDY, S. D. (1990). Artificial feeding systems for plant parasitic nematodes. *Journal of Nematology* **22**, 327–332.

VAECK, M., REYNAERTS, A., HOFTE, H., JANSENS, S., DE BEUCKELEER, M., DEAN, C., ZABEAU, M., VAN MONTAGU, M. AND LEEMANS, J. (1987). Transgenic plants protected from insect attack. *Nature* **328**, 33–37.

VASIL, I. K. (1990). The realities and challenges of plant biotechnology. *Biotechnology* **8**, 296–301.

VEECH, J. A. AND MCCLURE, M. A. (1977). Terpenoid aldehydes in cotton roots susceptible and resistant to the root-knot nematode *Meloidogyne incognita*. *Journal of Nematology* **9**, 225–229.

VERDEJO, S., JAFFEE, A. AND MANKAU, R. (1988). Reproduction of *Meloidogyne javanica* on plant roots genetically transformed by *Agrobacterium rhizogenes*. *Journal of Nematology* **20**, 599–604.

WALLACE, H. R. (1987). A perception of tolerance. *Nematologica* **33**, 419–432.

WINDHAM, G. L. AND WILLIAMS, W. P. (1988). Resistance of maize inbreds to *Meloidogyne incognita* and *M. arenaria*. *Plant Disease* **72**, 67–69.

WOOD, R. K. S. (1982). Man-made and natural resistance. *Proceedings of the International Conference on Plant Protection in the Tropics, Malaysia*, pp. 49–63.

WYSS, U. (1987). Video assessment of root cell responses to dorylaimid and tylenchid nematodes. In *Vistas on Nematology* (J. A. Veech and D. W. Dickson, Eds), pp. 211–220. Society of Nematologists, Maryland.

WYSS, U. (1990). Behaviour patterns of *Heterodera schachtii* throughout development. *Nematologica* **36**, in press.

WYSS, U. AND ZUNKE, U. (1987). Gnotobiology of cyst nematodes – its potential in basic research. In *Cyst Nematodes* (F. Lamberti and C. E. Taylor, Eds), pp. 147–162. Plenum Press, London.

WYSS, U., STENDER, C. AND LEHMANN, H. (1984). Ultrastructure of feeding sites of the cyst nematode *Heterodera schachtii* Schmidt in roots of susceptible and resistant *Raphanus sativa* var *oleiformis* Pers. *Physiological Plant Pathology* **25**, 21–37.

YU, M. H. AND STEELE, A. E. (1981). Host parasitic interaction of resistant sugar beet and *Heterodera schachtii*. *Journal of Nematology* **13**, 206–212.

ZACHEO, G., MOLINARI, S. AND DACODA, D. (1988). Hydroxyproline-rich proteins and peroxidases in tomato roots infested by root-knot nematode. *Nematologia Mediterranea* **16**, 235–237.

ZACHEO, G., PRICOLO, G. AND BLEVE-ZACHEO, T. (1988). Effect of temperature on resistance and biochemical changes in tomato inoculated with *Meloidogyne incognita*. *Nematologia Mediterranea* **16**, 107–112.

ZHUCHENKO, A. A., BALASHOVA, N. N. AND ANDRYUSHCHENKO, V. K. (1975). The role of the glycoalkaloid alphatomatine in the resistance of tomatoes to diseases and pests. *Sel'skokhozyaistvennaya Biologiya* **10**, 451–453.

ZUNKE, U. (1990). Observations on the invasion and endoparasitic behaviour of the root lesion nematode *Pratylenchus penetrans*. *Journal of Nematology* **22**, 309–320.

9

# The Utilization of Organic Wastes by Earthworms

K. E. FLETCHER*

Ex *Entomology Department, Horticultural Research International, Wellesbourne, Warwicks CV35 9EF, UK*

Introduction
The problems of organic waste disposal
The origins of vermiculture
Soil improvement and reclamation
Vermicomposting and the stabilization of organic wastes
Earthworms as animal feed protein
Types of earthworms used in vermiculture
Methods of production in vermiculture
Conclusions

## Introduction

It has long been established that the activities of earthworms play a major role in the maintenance of soil fertility (Darwin, 1881; Russel, 1910; Salisbury, 1925). However, more recently there has been considerable interest in using numerous earthworm species to reclaim and improve agricultural soils, to accelerate the decay of noxious organic wastes to produce composts and to provide an alternative source of animal feed protein by the mass production of earthworm dry-matter.

All earthworms are classified within the order Oligochaeta, but the aquatic species within the order are not strictly defined as earthworms and use of this terminology may cause confusion. Although there are thought to be in excess of 3000 species of oligochaetes, only about 1800 of these are classified as earthworms. In temperate regions of the Palaearctic zone the Lumbricidae dominate the earthworm fauna, but lumbricids also occur in the tropics where they have been introduced inadvertently by man. Tropical regions are usually dominated by species from different families, e.g. the Glosso-

**Present address 'Ecocare', 15 Grange Gardens, Wellesbourne, Warwicks. CV35 9RL, UK*

*Agricultural Zoology Reviews* — Vol. 4, February 1991
0269–0543 $20.00 + $0.00 © Intercept Ltd, PO Box 716, Andover, Hampshire SP10 1YG, UK

scolecidae in Central and South America and the Megascolecidae in South-East Asia and Australia (Michaelson, 1903; Stephenson, 1930).

There is surprisingly little known about the basic biology and ecology of most earthworm species and, of the 150 species about which much is known, only very few have been used in ways that would benefit mankind (Edwards and Lofty, 1977; Lee, 1985). There are comprehensive bibliographies of earthworm research published by Satchell and Martin (1981, 1984), and Lofs-Holmin (1985) summarized recent literature on earthworm farming.

A synopsis of the British earthworm fauna (Sims and Gerard, 1985) shows the diversity of form and function within the Lumbricidae of the United Kingdom. Attempts to classify the various ecological strategies adopted by earthworms have been made (Lee, 1959; Bouché, 1971). There are a variety of ecological niches occupied by earthworms; few species actually construct permanent burrow systems and even fewer produce surface casts (Lofty, 1974). The ability of some earthworms to improve soil structure, drainage and fertility to the benefit of crops and grassland, and the possibility of using other species as a cost-effective treatment of some organic waste problems, has stimulated both scientific and commercial interest (Appelhof and Worden, 1981; Satchell, 1983; Pagliai and Omodeo, 1987; Edwards and Neuhauser, 1988). The growing and multiplying of stocks of earthworms for various purposes is called vermiculture, and the use of earthworms to accelerate the decomposition of various organic materials is called vermicomposting or vermistabilization (Neuhauser, Hartenstein and Kaplan, 1979; Loehr *et al.*, 1984). The full potential of these relatively new techniques has yet to be realized (Sabine, 1988).

## The problems of organic waste disposal

Modern farming techniques and the food processing industry typical of modern, highly industrialized societies, generate increasingly large quantities of organic wastes (*Table 1*), currently disposed of by either incineration, land-spreading or land-filling. These methods may cause environmental damage and represent large economic losses of primary nutrients and metabolic energy, unless more efficient waste management systems can be found (Orskov, 1980; Larkin, *et al.*, 1981; Russel, 1982).

As costs of disposal have risen and legislation becomes increasingly restrictive, there has been renewed interest in exploiting organic wastes as fertilizers and soil conditioners or ameliorators. The increasingly large quantities involved makes the search for new and environmentally acceptable methods ever more urgent. Many aspects of modern farm waste management are dealt with in Ministry of Agriculture, Fisheries and Food (MAFF) publications (1980, 1982, 1983, 1986, 1987).

Only a relatively small proportion of the organic wastes produced would be suitable for use in vermiculture and vermicomposting. Few animal wastes contain a suitably balanced spectrum of essential nutrients for plant growth. Many contain heavy metals, e.g. the high copper content of some pig manures, and they may even contain herbicides, applied to the growing

**Table 1.** Estimated quantities of organic waste produced each year and the energy they contain (adapted from Edwards, 1988)

| Material source | Quantity (million tonnes per year) | Metabolizable energy (MJ/kg) |
|---|---|---|
| Livestock excreta | | |
| Cattle | 83.2 | 3.0 |
| Pigs | 9.1 | 4.5 |
| Poultry | 4.6 | 7.0 |
| Sheep | 22.7 | 3.0 |
| Arable crop waste | | |
| Straw | 9.3 | 6.5 |
| Sugarbeet | 0.5 | 12.0 |
| Other | 0.9 | 10.0 |
| Factory processing waste | | |
| Brewing and distilling | 0.6 | 11.0 |
| Paper | 0.3 | 10.8 |
| Potato | 0.2 | 12.0 |
| Spent mushroom compost | 0.2 | 6.5 |
| Canning and freezing | 0.05 | 11.0 |

fodder, that have passed through the animal relatively unchanged (Edwards, 1983). Domestic garbage contains a mixture of materials which would require laborious sorting if the organic waste content were to be used (Appelhof, 1982) and sewage treatment plants usually produce sewage sludge contaminated by industrial effluents (Hartenstein, Neuhauser and Collier, 1980). The application of vermicomposting to organic waste disposal will encounter many such problems requiring many different solutions.

## The origins of vermiculture

Earthworms are an important link in the food chain of many vertebrate and invertebrate animals (Macdonald, 1983). Not surprisingly, early man used earthworms both as food and as a bait (Bouché, 1987). The use of earthworms from dungheaps to feed domestic animals is documented in the eighteenth century (Milocco, 1782) and for fishing by Izaak Walton (1653), but it was not until the beginning of the twentieth century in California, USA that the commercial production of earthworms for fishing bait began. These enterprises used the most cost-effective materials to grow worms, and animal manures were enriched with high protein additives to produce large angling worms.

A much more profitable approach was introduced when Oliver (1937) and Barrett (1942) suggested that the same earthworms and the materials on which they had fed could be used by farmers and growers to improve agricultural soils. The industry then expanded rapidly and eventually international vermiculturists were selling earthworms and vermicomposts to various outlets, irrespective of the purposes for which they were being used. Some introduced pyramid selling techniques, where profits depended on attracting others to grow earthworms. Inevitably, the claims made by many of these early entrepreneurs became wildly exaggerated and earned the disapproval of the scientific community (Bouché, 1987).

More conventional studies on the potential of earthworms for utilizing organic wastes began in Germany (Graff, 1974) and continued in the USA with Hartenstein and his associates (Hartenstein and Mitchell, 1977). Subsequently research in the United Kingdom (Edwards, 1982) began extensive studies of methods of using earthworms to break down animal and vegetable wastes into useful composts, and also to produce earthworm protein for animal feed.

## Soil improvement and reclamation

In the past, farming methods required that both crop and animal residues were incorporated back into the soil and, until the widespread introduction of synthetic fertilizers (in and around the 1950s), farmers depended on the natural recycling of organic residues to maintain soil fertility. However, with the demise of self-sufficient, mixed farming systems and the corresponding increase in intensive farming methods, the numbers of earthworms in arable soils have decreased and the importance of returning organic matter to the soil has been neglected (Swift *et al.*, 1979; Edwards and Lofty, 1982; St. Remy and Daynard, 1982).

Controlled introductions of several different earthworm species to sites that were otherwise earthworm-free have improved both the structure and fertility of some mineral soils (Nielson, 1951; Langmaid, 1964; Stockdill, 1982; Springett, 1984, 1987; Curry, 1988). The introduced species improved the rate at which organic residues were incorporated (Hoogerkamp, Rogaar and Eijsackers, 1983) and increased soil mixing and the formation of water-stable aggregates (van Rhee, 1977). Clearly, the introduction of suitable earthworm species and organic wastes would make an important contribution to soil improvement and reclamation.

## Vermicomposting and the stabilization of organic wastes

Many of the organic wastes produced by modern industrial technology are expensive to dispose of and cause odour problems or pollution of groundwater. Research findings suggest these problems could be alleviated by the use of earthworms, particularly *Eisenia foetida* (Savigny), to accelerate aerobic decomposition, thereby minimizing pollution and possibly producing a profit, both from the sale of surplus earthworms and the resulting vermicompost for use as a plant growth medium (Franz, 1978; Edwards *et al.*, 1985). Various human, animal and industrial organic wastes have been used for vermicomposting and the resulting composts marketed in the USA, Italy and the Philippines. Rather extravagant claims supported the effectiveness of some of these composts, but direct hormone-like effects on plant growth have been reported (Nielsen, 1965; Springett and Syers, 1979; Tomati, Grappelli and Galli, 1987).

Edwards (1983), using both animal and vegetable wastes, investigated the possibility of using earthworms to convert these wastes into useful materials that could be incorporated into agricultural soils or be used as plant growth media or as a component of commercial potting composts. If composts produced in this way were to be a commercial success, large quantities of homogeneous waste material must be available and strict controls of the nutrient status, pH and sterilization of the final product must be maintained.

Several species of earthworms have been used to stabilize organic solids from domestic sewage treatment plants. Research has shown that aerobic sewage sludge can supply the nutrients necessary for the growth and reproduction of earthworms of the genera *Eisenia, Eudrilus* and *Pheretima* (Hartenstein, Neuhauser and Kaplan, 1979; Kaplan *et al.*, 1980; Neuhauser *et al.*, 1980; Pincince, Donovan and Bates, 1980; Loehr, Martin and Neuhauser, 1988). Sewage treatment plants often combine sewage intake from both domestic and industrial sources, and the harmful effect of toxic contaminants on earthworm growth and reproduction were investigated (Hartenstein, Neuhauser and Collier, 1980; Malecki, Neuhauser and Loehr, 1982) and the limitations of using such contaminated materials documented. In the USA, attempts to process sewage sludge with vermicomposting techniques failed due to the onset of extreme anaerobic conditions. If vermiculture is to form part of the practical management of sewage sludge, much more research is required (Hartenstein and Mitchell, 1977; Birkner, 1986).

## Earthworms as animal feed protein

The fact that the body tissue of earthworms contains sufficient protein to be considered as a source of feed protein for domestic animals encouraged animal feeding trials (Sabine, 1978). Detailed analyses have shown earthworm dry-matter to consist of 60–70% protein, 7–10% fat, 8–20% carbohydrate, 2–3% minerals and a wide range of vitamins (Guerrero, 1983; Hilton, 1983; Tacon, Stafford and Edwards, 1983). Dietary feeding trials with laboratory animals, pigs, poultry and fish have shown that earthworm meal is an effective substitute for meatmeal or fishmeal (Sabine, 1988). The essential amino-acid content of earthworm tissues compares favourably with the FAO/WHO dietary recommendations for animal feeds. Additionally, earthworm tissues contain a wide variety of long-chain fatty acids essential to the diet of non-ruminant animals (Edwards, 1985). Many of these fatty acids, some of which are of medical importance, are at present extracted from fish oils.

The rate of conversion from organic waste material into earthworm protein varies, affected by the type of waste, the earthworm species, the environmental conditions and the methods and scale of production. Under optimal conditions in laboratory investigations using *E. foetida*, 10% of cattle and pig manure and 2% of poultry manure was converted into earthworm tissue (Edwards and Niederer, 1988). Other estimates of potential rates of production of earthworm protein range from 150 kg/cow/year (Fosgate and Babb, 1972) to 11 000 tonnes/year in the UK (Walsingham, 1981). These production rates are large enough to justify considerable investment in earthworm protein, but as yet no commerical schemes have been developed.

## Types of earthworms used in vermiculture

Most species of earthworm have clearly defined and highly specialized food preferences and consistent patterns of feeding behaviour (Wallwork, 1983; Lee, 1985). These preferences are related to the soil horizons in which they are most commonly found, e.g. litter species, or epigees, live only in the surface litter layers (Lee, 1959; Bouché, 1971); others are topsoil- or subsoil-dwelling earthworms that come to the surface to feed or ingest mineral soil enriched with organic matter. In general, the litter-dwelling species produce several generations each year, producing many cocoons and maturing rapidly, whereas the soil-dwelling species produce only one generation each year, with few cocoons and long maturation periods.

Almost invariably, the earthworm species used in the breakdown of organic wastes are of the type most commonly found in the surface litter layers (Bouché, 1987). Soil-dwelling species, although invaluable for the improvement and reclamation of soils, are not readily mass-cultured, although Springett (1987) describes practical methods for redistribution of the beneficial earthworms *Aporrectodea caliginosa* (Sav.) and *Lumbricus rubellus* Hoff. using turf cutting and spreading machinery developed in New Zealand by Stockdill and Cossens (1984).

The species most commonly used for accelerating the decomposition of organic wastes is *Eisenia foetida*, a peregrine species with two forms, variants or subspecies (Øien and Stenersen, 1984; Sheppard, 1988). It has a rapid reproduction rate and its tolerance over wide ranges of temperature and moisture content make it the most easily cultured species available. *Eisenia foetida* occurs naturally in stored organic wastes and its dispersal mechanisms are so effective that cultures of other earthworm species are often accidentally colonized by *E. foetida*. *Lumbricus rubellus* is a small red earthworm similar in appearance to *E. foetida* and often mistaken for it. Although reputedly used extensively by vermiculturists it shows a marked preference for mineral soils, where it feeds preferentially on surface animal faeces. *Dendrobaena veneta* (Rosa), a much larger earthworm, is slower growing and less prolific than either *E. foetida* or *L. rubellus*, particularly when reared exclusively in organic wastes.

Some earthworm species indigenous to much warmer climates have also been used successfully for organic matter breakdown. *Eudrilus eugeniae* (Kinberg), commonly known as the African night crawler but distributed world-wide, is particularly favoured by vermiculturists, as are *Pheretima* spp. and *Perionyx excavatus* (Perrier). All these species are fast growing and reproduce rapidly throughout the year but, if cultured in colder climates, care must be taken to maintain minimum temperatures of at least 10°C, with the optimum around 25°C.

## Methods of production in vermiculture

Vermiculturists have adopted a wide range of production methods, ranging from small-scale, labour-intensive, low-technology systems in simple containers, to large-scale, highly mechanized, continuous-flow reactors. The basic design of any system should ensure that the earthworms maximize both feeding and reproduction at all times. To achieve this, the environmental factors influencing the growth and survival of earthworm populations must be known (Edwards, 1988; Phillips, 1988). Both physical and chemical factors are important in maintaining the maximum productivity of earthworm beds or cultures and it is often difficult to balance all the interrelated factors in order to achieve optimum conditions.

Aerobic conditions are promoted by designing shallow beds with a high surface area to volume ratio, good drainage and ventilation. Under UK conditions the beds must be insulated, covered and thermostatically controlled if optimum temperatures are to be maintained. Moisture content of the waste material is another variable factor requiring control; too dry and the earthworms are unable to feed, too wet and the material becomes waterlogged and anaerobic. Wastes that contain even weak solutions of some organic salts or ammonia can be lethal if fed to earthworms, and these wastes must usually undergo some form of pretreatment to dilute or neutralize the toxicity. The most convenient forms of pretreatment are to prolong the period of storage and to encourage leaching. Earthworms are relatively tolerant of variations in pH (Edwards, 1988) but show a preference for neutral to slightly acid conditions. *Table 2* outlines some of the conditions necessary for breeding *E. foetida* in organic wastes; apart from temperature preferences, the conditions for other earthworm species are similar.

**Table 2.** Optimal conditions for breeding *Eisenia foetida* in organic wastes (after Phillips, 1988)

| |
|---|
| Oxygen: aerobic conditions essential |
| Temperature: 15–25°C preferred; 4–30°C tolerated |
| Moisture content: 80–90% preferred; 60–90% tolerated |
| Salt content: low (<0.5%) (measured as conductivity in mmhos) |
| Ammonia content: low (<0.5 mg/g) |
| pH: 5–6 preferred; 4–9 tolerated |

Earthworms have been grown in a wide range of organic wastes, but for successful large-scale operations a constant supply of waste of uniform quality is a prerequisite. Animal manures, particularly separated solids from farm slurry separators, have proved an ideal source material in this respect. Whole slurries cause anaerobic conditions in earthworm beds and cannot be used as such. Most straw-based animal manures would benefit from some form of mechanical shredding to promote comminution and homogenization before being used (Phillips, 1988). The application of these suitably prepared organic wastes is either by hand, using shovels and rakes, or by mechanical spreaders mounted on mobile gantries above the worm bed. The waste material is added in successive thin layers, 3–6 cm deep, both to maintain aerobicity and to allow the dilution or evaporation of any remaining substances toxic to the earthworms. Further layers are added only after the earthworms have ingested most of the waste present and this is usually detected by a change in surface texture of the previous layer of waste. *Figure 1* shows how earthworm density and biomass change with time and how introduction of additional waste material must be managed in order to maximize production of protein or compost.

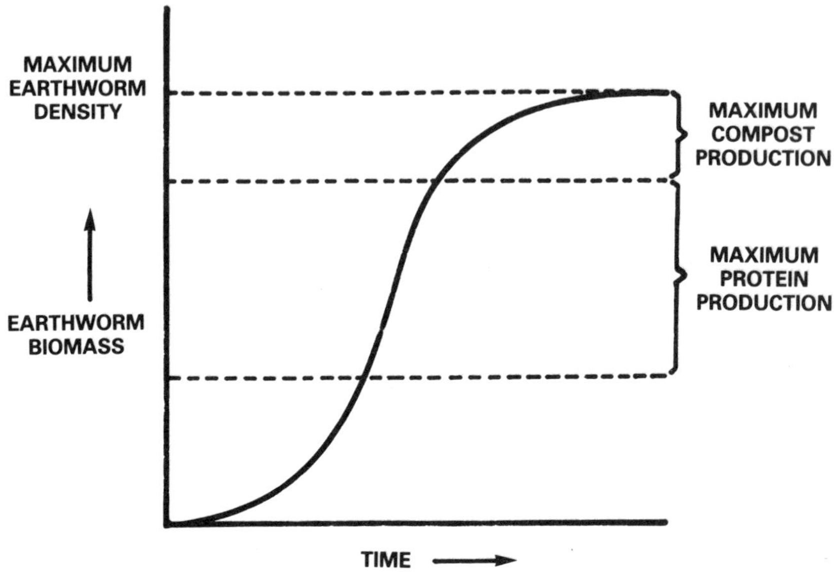


**Figure 1.** Earthworm population growth, showing maximum protein and compost production phases.

The earthworm species used are all surface dwellers and, given optimum conditions, restrict their activities to the top 10–15 cm of the earthworm bed. Layer feeding encourages a continuous vertical movement of earthworms upwards towards the newly applied layers of waste, leaving relatively worm-free compost underneath. This behaviour pattern encouraged the development of continuous-flow systems (Phillips, 1988) where the rates of addition of fresh wastes at the surface equalled the rate of removal of worm-worked material from the base of the reactor. Flow rates were adjusted to retain the maximum number of earthworms within the reactor, although the loss of cocoons in the worm-worked material may eventually necessitate re-stocking.

Methods to separate earthworms from the worm-worked material are also numerous and usually relate to the scale of operations. Hand sorting is too labour-intensive for all but the smallest systems. The simplest mechanical devices rely on reciprocating or revolving sieves that retain the earthworms but allow smaller particles to pass through, others employ comb-like structures which pass repeatedly through the worm-worked material selecting elongate particles, most of which are the earthworms. The efficiency of mechanical methods is decreased if wet material is used: sieves and combs become blocked if the moisture content exceeds 80%. On the other hand, methods that separate earthworms using external stimuli to attract or repel the mobile earthworms away from the worm-worked material are not adversely affected by moisture content. Light, heat, electric shock and repellent chemicals have all been used successfully to extract earthworms, but only from relatively small quantities of waste. Earthworm cocoons are extracted using washing and flotation methods.

To prepare composts, the worm-free waste is dried, sterilized and possibly blended with various bulking agents, such as sand, loam, peat or shredded bark, before being sieved and packaged. Earthworm protein requires that the earthworms are washed, cleansed internally, blanched and dried before being milled, ensiled or pelletized. The capital costs of some of the equipment required for these processes are quoted by Phillips (1988).

The composts produced must be of a standard quality, varying little in consistency or nutrient content, if they are to have commercial potential (Edwards and Burrows, 1988). A wide range of plants has been grown successfully in such composts; in some trials growth has exceeded that in recommended commercial composts (Scott, 1988) whereas in others seed germination improved (Edwards and Burrows, 1988).

## Conclusions

Ecological considerations suggest that not all the benefits arising from a viable earthworm industry may be measured in financial terms. A purely monetary analysis, undertaken by Fieldson (1988), on the economic viability of culturing earthworms on animal wastes, does, however, identify those areas within the industry which are critical to its survival. The analysis requires several assumptions on techniques, productivity and product value that can only be reconciled by commercial involvement and relevant research.

The potential of earthworms to process organic wastes and convert them into useful composts, and simultaneously provide a cheap form of animal protein, would appear to be a Utopian solution to the current problems of organic waste disposal. The problem of scale of these operations if they are to be effective, and the lack of expertise in manipulating mass cultures of earthworms are, however, major obstacles preventing exploitation of this innovative process. Vermiculture as a low-technology industry at individual or household level will thrive in the hands of the enthusiasts. The development of a commercial earthworm industry capable of contributing to large-scale waste-recycling programmes and environmental protection policies will require high levels of financial commitment and its realization remains speculative. The importance of earthworms in the production and maintenance of fertile soils and their prominent position in the food chains of numerous higher organisms indicate the significance of their environmental impact. Renewed interest in environmental problems, particularly the possible damage caused by some agricultural and industrial practices to agricultural soils, has heightened public awareness of the importance of soil organisms in general and may give impetus to further research in these areas.

## References


APPELHOF, M. (1982). *Worms eat my garbage*. Flower Press, 10332 Shaver Road, Kalamazoo, Michigan 49002, USA.

APPELHOF, M. AND WORDEN, D. D. (1981). *Workshop on the Role of Earthworms in the Stabilization of Organic Residues, Kalamazoo, Michigan, 9–12 April, 1980*, 2 volumes. Beech Leaf Press.

BARRETT, T. J. (1942). *Harnessing the Earthworm*. Faber and Faber, London.

BIRKNER, L. (1986). Earthworms, a new resource for production of protein and humus from sewage sludge. *Recycling in Chemical Water and Wastewater Treatment* **50**, 245–254. Schriftenreihe, ISWW Karlsruhe, Sweden.

BOUCHÉ, M. B. (1971). Relations entre les structures spatiales et fonctionelles des écosystèmes, illustrées par le rôle pedobiologique des vers de terre. In *La Vie dans les Sols* (C. Delamere Deboutville, Ed.), pp. 189–209. Gauthier Villars, Paris.

BOUCHÉ, M. B. (1987). Emergence and development of vermiculture and vermicomposting: from a hobby to an industry, from marketing to a biotechnology, from irrational to credible practices. In *On Earthworms* (A. M. B. Pagliai and P. Omodeo, Eds), pp. 519–531. Unione Zoologica Italiana (UZI), 2, Mucchi, Modena.

CURRY, J. P. (1988). The ecology of earthworms in reclaimed soils and their influence on soil fertility. In *Earthworms in Waste and Environmental Management* (C.A. Edwards and E. F. Neuhauser, Eds), pp. 251–261. SPB Academic Publishing, The Hague, The Netherlands.

DARWIN, C. R. (1881). *The formation of Vegetable Mould Through the Action of Worms, With Observations on Their Habits*. Murray, London.

EDWARDS, C. A. (1982). *The use of earthworms in waste disposal and protein production*. Rothamsted Experimental Station Report for 1981, Part 1, pp. 103–105.

EDWARDS, C. A. (1983). Earthworms, organic waste and food. *Span* **26**, 3.

EDWARDS, C. A. (1985). Production of food protein from animal wastes by earthworms. *Philosophical Transactions of the Royal Society, London* **310**, 299–307.

EDWARDS, C. A. (1988). Breakdown of animal, vegetable and industrial organic wastes by earthworms. In *Earthworms in Waste and Environmental Management* (C. A. Edwards and E. F. Neuhauser, Eds), pp. 21–31. SPB Academic Publishing, The Hague, The Netherlands.

EDWARDS, C. A. AND BURROWS, I. (1988). The potential of earthworm composts as plant growth media. In *Earthworms in Waste and Environmental Management* (C. A. Edwards and E. F. Neuhauser, Eds), pp. 211–219. SPB Academic Publishing, The Hague, The Netherlands.

EDWARDS, C. A. AND LOFTY, J. R. (1977). *Biology of Earthworms*, Chapman and Hall, London.

EDWARDS, C. A. AND LOFTY, J. R. (1982). The effect of direct drilling and minimal cultivation on earthworm populations. *Journal of Applied Ecology* **19**, 723–734.

EDWARDS, C. A. AND NEUHAUSER, E. F. (1988). *Earthworms in Waste and Environmental Management*. SPB Academic Publishing, The Hague, The Netherlands.

EDWARDS, C. A. AND NIEDERER, A. (1988). The production and processing of earthworm protein. In *Earthworms in Waste and Environmental Management* (C. A. Edwards and E. F. Neuhauser, Eds), pp. 169–179. SPB Academic Publishing, The Hague, The Netherlands.

EDWARDS, C. A., BURROWS, I., FLETCHER, K. E. AND JONES, B. A. (1985). The use of earthworms for composting farm wastes. In *Composting of Agricultural and Other Wastes* (J. K. R. Gasser, Ed.), pp. 229–242. Elsevier, Amsterdam.

FIELDSON, R. S. (1988). The economic viability of earthworm culture on animal wastes. In *Earthworms in Waste and Environmental Management* (C. A. Edwards and E. F. Neuhauser, Eds), pp. 145–156. SPB Academic Publishing, The Hague, The Netherlands.

FOSGATE, O. T. AND BABB, M. R. (1972). Biodegradation of animal wastes by *Lumbricus terrestris*. *Journal of Dairy Science* **55**, 870–872.

FRANZ, M. F. (1978). The economic future of the earthworm in recycling. *Compost Science – Land Utilisation* **19**, 23–26.

GRAFF, O. (1974). Gewinnung von Biomasse aus Abfallstoffen durch Kultur des Kompostregenwurms *Eisenia foetida* (Savigny 1826). *Landbauforschung Volkenrode* **24**, 137–142.

GUERRERO, R. D. (1983). The culture and use of *Perionyx excavatus* as a protein resource in the Philippines. In *Earthworm Ecology: from Darwin to Vermiculture* (J. E. Satchell, Ed.), pp. 309–313. Chapman and Hall, London.

HARTENSTEIN, R. AND MITCHELL, M. J. (1977). *Interim Report – Utilisation of soil invertebrates in stabilization, decontamination and detoxification of residual sludges for treatment of wastewater.* Report NSF-RANN. Grant Number ENV76–17225.

HARTENSTEIN, R., NEUHAUSER, E. F. AND COLLIER, J. (1980). Accumulation of heavy metals in the earthworm *Eisenia foetida*. *Journal of Environmental Quality* **9**, 23–26.

HARTENSTEIN, R., NEUHAUSER, E. F. AND KAPLAN, D. L. (1979). The reproductive potential of the earthworm *Eisenia foetida*. *Oecologia* **43**, 329–340.

HILTON, J. W. (1983). Potential of freeze-dried worm meal as a replacement for fish meal in trout diet formulations. *Aquaculture* **32**, 277–283.

HOOGERKAMP, M., ROGAAR, H. AND EIJSACKERS, K. J. P. (1983). Effect of earthworms on grassland on recently reclaimed polder soils in the Netherlands. In *Earthworm Ecology* (J. E. Satchell, Ed.), pp. 85–105. Chapman and Hall, London.

KAPLAN, D. L., NEUHAUSER, E. F., HARTENSTEIN, R. AND MALECKI, M. R. (1980). Physicochemical requirements in the environment of the earthworm *Eisenia foetida*. *Soil Biology and Biochemistry* **12**, 347–352.

LANGMAID, K. K. (1964). Some effects of earthworm invasion in virgin podzols. *Canadian Journal of Soil Science* **44**, 34–37.

LARKIN, S. B. C., MORRIS, R. M., NOBLE, D. M. AND RADLEY, R. W. (1981). *Resource Mapping of Agricultural Wastes and Residues*. NCAE, Silsoe, Bedfordshire, UK.

LEE, K. E. (1959). The earthworm fauna of New Zealand. *New Zealand Department of Science and Industrial Research Bulletin* **130**, 1–486.

LEE, K. E. (1985). *Earthworms: Their Ecology and Relationships with Soils and Land Use*. Academic Press, Sydney.

LOEHR, R. C., MARTIN, J. H. AND NEUHAUSER, E. F. (1988). Stabilization of liquid municipal sludge using earthworms. In *Earthworms in Waste and Environmental Management* (C. A. Edwards and E. F. Neuhauser, Eds), pp. 95–110. SPB Academic Publishing, The Hague, The Netherlands.

LOEHR, R. C., MARTIN, J. H., NEUHAUSER, E. F. AND MALECKI, M. R. (1984). *Waste management using earthworms. Engineering and scientific relationships.* PB84–193218. NTIS, Springfield, Virginia.

LOFS-HOLMIN, A. (1985). *Vermiculture. Present knowledge of the art of earthworm farming. A summary of recent literature*. Swedish University of Agricultural Sciences, Department of Ecology and Environmental Research, Report 20, Uppsala.

LOFTY, J. R. (1974). Oligochaetes. In *Biology of Plant Litter Decomposition*, volume 2 (G. J. F. Pugh and C. H. Dickinson, Eds), pp. 467–488. Academic Press, London.

MACDONALD, D. W. (1983). Predation on earthworms by terrestrial vertebrates. In *Earthworm Ecology: from Darwin to Vermiculture* (J. E. Satchell, Ed.), pp. 393–414. Chapman and Hall, London.

MAFF (1980). *The storage of farm manures and slurries: Farm waste management.* ADAS Booklet No. 2237, HMSO.

MAFF (1982). *Farm waste management: Index of current research and development projects in the British Isles.* ADAS Farm Waste Unit.

MAFF (1983). *Profitable utilisation of livestock manures: Farm waste management.* ADAS Booklet No. 2081, HMSO.

MAFF (1986). *Organic manures.* Reference Book 210, HMSO.

MAFF (1987). *Storage of farm manures and slurries: Choosing a storage system.* ADAS Leaflet No. P3042, HMSO.

MALECKI, M. R., NEUHAUSER, E. F. AND LOEHR, R. C. (1982). The effect of metals on the growth and reproduction of *Eisenia foetida. Pedobiologia* **24**, 129–139.

MICHAELSON, W. (1903). *Die geographische Verbreitung der Oligochaeten*, pp. 1–186. Friedlander und Sohn, Berlin.

MILOCCO, A. (1782). Della maniera di nuticare e di far propagare le galline. In *Il gentiluomo coltivatore etc. Venezia*, volume X, pp. 3–16.

NEUHAUSER, E. F., HARTENSTEIN, R. AND KAPLAN, D. L. (1979). A second progress report on the potential use of earthworms in sludge management. In *Proceedings of the Conference on Sludge Composting, 14–17 November.* Sponsored by Information Transfer Inc., Silver Spring, Maryland, USA.

NEUHAUSER, E. F., KAPLAN, D. L., MALECKI, M. R. AND HARTENSTEIN, R. (1980). Materials supporting weight gain by the earthworm *E. foetida* in waste conversion systems. *Agricultural Wastes* **2**, 43–60.

NIELSEN, R. L. (1951). Earthworms and soil fertility. *Proceedings of the New Zealand Grassland Association* **13**, 158–167.

NIELSEN, R. L. (1965). Presence of plant growth substances in earthworms demonstrated by paper chromatography and the Went Pea test. *Nature* **208**, 1113.

ØIEN, N. AND STENERSEN, J. (1984). Esterases of earthworms III. Electrophoresis reveals that *Eisenia foetida* (Savigny) is two species. *Comparative Biochemistry and Physiology* **78C**, 277–282.

OLIVER, G. S. (1937). *Our Friend the Earthworm.* Gardener's Book Club, volume 8, California.

ORSKOV, E. R. (1980). Nutritional principles and utilization of waste and by-products. In *By-products and wastes in animal feeding. Occasional Symposium No. 3* (E. R. Orskov, Ed.). British Society of Animal Production.

PAGLIAI, A. M. B. AND OMODEO, P. (1987). *On Earthworms. Selected Symposia and Monographs*, pp. 562. Unione Zoologica Italiana (UZI), 2, Mucchi, Modena.

PHILLIPS, V. R. (1988). Engineering problems in the breakdown of animal wastes by earthworms. In *Earthworms in Waste and Environmental Management* (C. A. Edwards and E. F. Neuhauser, Eds), pp. 111–118. SPB Academic Publishing, The Hague, The Netherlands.

PINCINCE, A. B., DONOVAN, J. F. AND BATES, J. E. (1980). Vermicomposting municipal sludge: an economical stabilization alternative. *Sludge* **3**, 26–30.

RUSSEL, E. J. (1910). The effect of earthworms on soil productiveness. *Journal of Agricultural Science* **2**, 245–257.

RUSSEL, S. H. (1982). *Resources Recovery Economics.* Marcel Dekker, New York.

SABINE, J. (1978). The nutritive value of earthworm meal. In *Proceedings of the Conference on the Utilization of Soil Organisms in Sludge Management, Syracuse, New York* (R. Hartenstein, Ed.), pp. 122–130.

SABINE, J. R. (1988). Vermiculture: bring on the future. In *Earthworms in Waste and Environmental management* (C. A. Edwards and E. F. Neuhauser, Eds), pp. 3–6. SPB Academic Publishing, The Hague, The Netherlands.

ST. REMY, E. A. AND DAYNARD, T. B. (1982). Effects of tillage methods on earthworm populations in monoculture corn. *Canadian Journal of Soil Science* **62**, 699–703.

SALISBURY, E. J. (1925). The influence of earthworms on soil reaction and the stratification of undisturbed soil. *Journal of the Linnean Society (Botany)* **46**, 415–425.

SATCHELL, J. E. (1983). *Earthworm Ecology: From Darwin to Vermiculture*. Chapman and Hall, London.

SATCHELL, J. E. AND MARTIN, K. (1981). *A Bibliography of Earthworm Research*. Institute of Terrestrial Ecology, Merlwood Research Station, Cumbria.

SATCHELL, J. E. AND MARTIN, K. (1984). *A Bibliography of Earthworm Research, Supplement*. Institute of Terrestrial Ecology, Merlwood Research Station, Cumbria.

SCOTT, M. A. (1988). The use of worm-digested animal waste as a supplement to peat in loamless composts for hardy nursery stock. In *Earthworms in Waste and Environmental Management* (C. A. Edwards and E. F. Neuhauser, Eds), pp. 221–229. SPB Academic Publishing, The Hague, The Netherlands.

SHEPPARD, P. S. (1988). Specific differences in cocoon and hatchling production in *Eisenia foetida* and *E. andrei*. In *Earthworms in Waste and Environmental Management* (C. A. Edwards and E. F. Neuhauser, Eds), pp. 83–92. SPB Academic Publishing, The Hague, The Netherlands.

SIMS, R. W. AND GERARD, B. M. (1985). *Synopsis of the British Fauna (31) – Earthworms*. Linnean Society, London.

SPRINGETT, J. A. (1984). The effect of introducing *Allolobophora longa* Ude on root distribution and on some soil properties in New Zealand pastures. In *Ecological Interactions in the Soil Environment* (A. H. Fitter, Ed.), pp. 399–405. Blackwell Scientific Press, Oxford.

SPRINGETT, J. A. (1987). National assessment of soil forming Lumbricidae in New Zealand. In *On Earthworms. Selected Sympsia and Monographs* (A. M. B. Pagliai and P. Omodeo, Eds), pp. 497–504. Unione Zoologica Italiana (UZI), 2, Mucchi, Modena.

SPRINGETT, J. A. AND SYERS, J. K. (1979). The effect of earthworm casts on ryegrass seedlings. In *Proceedings of the Second Australian Conference on Grassland Invertebrate Ecology* (T. K. Crosby and R. P. Pottinger, Eds), pp. 44–47. Government Printers, Wellington.

STEPHENSON, J. (1930). *The Oligochaeta*. Oxford University Press.

STOCKDILL, S. M. J. (1982). Effects of introduced earthworms on the productivity of New Zealand pastures. *Pedobiologia* **24**, 29–35.

STOCKDILL, S. M. J. AND COSSENS, G. G. (1984). *Soils: earthworm introductions, methods and effects on production*. MAF Aglink FP211, New Zealand.

SWIFT, M. J., HEAL, D. W. AND ANDERSON, J. M. (1979). *Decomposition in Terrestrial Ecosystems*. Blackwell Scientific, Oxford.

TACON, A. G. J., STAFFORD, E. A. AND EDWARDS, C. A. (1983). A preliminary investigation of the nutritive value of three terrestrial lumbricid worms for rainbow trout. *Aquaculture* **35**, 187–199.

TOMATI, U., GRAPPELLI, A. AND GALLI, E. (1987). The presence of growth regulators in earthworm worked wastes. In *On Earthworms. Selected Syposia and Monographs* (A. M. B. Pagliai and P. Omodeo, Eds), pp. 423–435. Unione Zoologica Italiana (UZI), 2, Mucchi, Modena.

VAN RHEE, J. A. (1977). A study of the effect of earthworms on orchard productivity. *Pedobiologia* **17**, 107–114.

WALLWORK, J. A. (1983). *Earthworm Biology*, pp. 10–19. Institute of Biology's Studies in Biology No. 161. Edward Arnold, London.

WALSINGHAM, J. M. (1981). The use of invertebrates in recycling. In *Vegetable Productivity* (C. R. W. Spedding, Ed.), pp. 239–246. Institute of Biology Symposium 25. Macmillan, London.

WALTON, I. (1653). *The Compleat Angler*. (Facsimile of first edition reprinted in 1928. A. and C. Black, London.)

# Index


*Abelmoschus esculentus*, 182
*Acalypha* sp., 99
Acaricide, 37, 79
Acetone, 152
*Acremonium coenophialum*, 219
*Acridotheres tristis*, 24
*Acrosternum hilare*, 14, 16
*Aenasius* sp., 118
  *A. advena*, 118
Africa
  animal feed, 23
  ants, 39
  blister beetles, 47
  cassava, 97–127
    green mite, 122, 126
    mealybug, 98, 126
      parasitoids of, 102
      hyperparasitoids, 105
  *Creontiades pallidus*, 46
  crop damage by nematodes, 214
  cyst nematodes, 217
  earhead caterpillar, 45
  *Eurystylus marginatus*, 46
  grasshoppers, 30
  leaf caterpillars, 30
  maize stem borer, 40
  nematodes, 177–180
  shoot bug, 27, 36
  sorghum in, 2, 23–25, 27, 30, 34, 36–37
  sorghum stem borers, 37–38
  sugarcane aphid, 34
  tsetse fly, 137–154
    area infested, 138
  variegated grasshopper, 118
Agricultural soils, 241
*Agrobacterium rhizogenes*, 231
  *A. tumefaciens*, 190
*Agrotis infusca*, 24
  *A. ypsilon*, 24
*Agrypmus variabilis*, 25
Aldicarb, 35, 191–192, 199
Aldrin, 24, 118
Alfafa, 219, 226
Algae, 161
*Allium cepa*, 182
*Allotropa* sp., 118
*Amblyseius fustis*, 124
America
  Central
    Chinch bug, 35
    earthworms, 242
    soybean production, 2
    stem borers, 37
    sugarcane aphid, 34
  North
    European corn borer, 41
    sorghum, 23
    soybean production, 1–2
    stemborers, 37
    white grubs, 24
  South
    earthworms, 242
    nematodes, 177, 214
    stemborers, 37
American bollworm, 44
American Soybean Association, 1
*Amitermes evuncifer*, 118
*Anagasta kuehniella*, 73
Anagyrus spp., 102–103, 105, 114, 118
  *A. bugandaensis*, 102
  *A. pseudococci*, 102
  *A. pullus*, 102
*Ananas comosus*, 177
*Ancistrotermes* sp., 118
Angola
  cassava mealybugs, 98
    parasitoids of, 103
*Anguina tritici*, 163–164, 166, 170
Anhydrobiotic nematodes (*see* Nematodes)
Animal feed protein, 241
*Anisochrysa boninensis*, 108
Annelids, 161
*Anomala* spp., 24
*Anticarsia gemmatalis*, 2
Anthocorids, 41, 43, 108
  *Orius maxidentex*, 43
Anthroinose disease, 117
Ants
  *Atherigona soccata*, 25–26
    *A. yorki*, 25
  *Monomorium salomonis*, 25
  *Pheidole megacephala*, 39
    *P. sulcatipes*, 25
  *Solenopsis invicta*, 41
*Aonidomytilus albus*, 117
*Aphanogmus* sp., 112
*Aphelenchoides besseyi*, 170
*Aphelenchus avenae*, 166, 169–170
  membranes, 170
Aphelinids, 103
  parasitoids of, 117
Aphids
  *Aphis (Melanaphis) sacchari*, 34
  control strategies, 31
  parasitoids
    *Aphelinus nigritus*, 33
    *Diaeretiella rapae*, 33
    *Lysiphlebus testaceipes*, 33
  predators, 33
    *Chrysopa basilis*, 33
    *Coleomegilla masculata*, 33
    *Hippodamia convergens*, 33

*H. sinuata*, 33
*Scymnus* spp., 33
*Syrphis batteatus*, 33
resistance to insecticides, 31
*Rhopalosiphum maidis*, 33
*Schizaphis graminum*, 31–32, 34
*Sipha flava*, 34
*Arachis hypogaea*, 180
*Araecerus fasciculatus*, 116
Argentina, 1
soybean production in, 1
sugarcane aphid, 34
Arkansas
greenbug, 31
Armyworms, 28–29
effect of braconid wasps, 30
fall army worm (*Spodoptera frugiperda*), 28
in sorghum, 28–30
*Mythimna loreyi*, 30
parasites on, 29
*Pseudoletia convecta*, 30
rice armyworm (*Mythimna separata*), 30
*Spodoptera exigua*, 30
*S. exempta*, 30
susceptibility to entomogenous pathogens, 29
*Artemia*, 165
Artificial infestation, 41
Asia
East
stemborers, 37
sorghum, 25
South East
earthworms, 242
nematodes, 177
sorghum shoot fly, 25
stored products, 75
*Atherigona soccata*, 25–26
*A. yorki*, 25
Australia
earthworms, 242
nematodes, 177
sorghum, 25, 45
sorghum midge, 42
*Avena sativa*, 216
*Azadirachta indica*, 193
*Azospirillum*, 26

*Bacillus thuringiensis*, 82, 231
Bacterial spores, 161
Banana, 98, 192
Bangladesh
stored produces, 75, 76
Banks green mite, 36
Barley
greenbug on, 31
resistance to nematodes, 215, 217, 219, 232
Bean, 98, 219, 223, 227
Beetles
*Chaetocnema india*, 28
*C. tibialis*, 28
leaf, 28
*Lema planifrons*, 28
*Longitarsus* spp., 28
*Menolepta* spp., 28
*Phyllotreta chotanica*, 28
root, 25
*Bemisia tabaci*, 98, 117
Bendiocarb, 78
Benfuracarb, 27
Benin
cassava mealybug, 98
parasitoid, 103
grain borers, 116
Benomyl, 27
Betelvine, 185
Bioregulators, 33
Bioresmethrin, 78
Black field earwig, 25
Blattidae, 73
*Blepyrus insularis*, 118
Blister beetles, 47
*Bolothrips marshalli*, 119
*Brachiaria* spp., 26
Braconid wasps
*Apanteles* sp., 30
*A. flavipes*, 39, 41
*A. sesamiae*, 39
*Bracon* sp., 39
*Chelonus* sp., 39
*C. curvimaculatus*, 39
*Iphiaulax* sp., 39
*Myosoma nyanzaensis* sp., 39
*Brassica campestris*, 189, 198
*B. juncea*, 198
*B. oleracea*
var. *acephala*, 11
var. *botrytis*, 183
Brazil
greenbug in, 32
sorghum in, 32, 41
sorghum midge, 42, 44
soybean production, 1
*Thyanta preditor*, 46
Brine shrimp, 165
Broad beans, 185, 194
Burundi
cassava green mite, 119
Bushbean, 197

Cadelles, 74
*Caenorhabditis elegans*, 232
*Cajanus cajan*, 185, 227
*Callosobruchus analis*, 80
*C. chinensis*, 80
*C. maculatus*, 80
*Calotropis*, 193
Cameroon
cassava, 97
mealybug parasitoid, 103
Cantaloupe, 185, 190
*Capsicum annuum*, 182, 219, 222, 227
Captan, 27
Carabid larvae, 41

Carbamates, 78, 140, 191
Carbaryl, 28–29, 40, 46, 48, 114
Carbendazim, 27
Carbofuran, 26–28, 35, 40, 185, 191–192, 198–199
*Cardiastethus exiguus*, 108
*Carica papaya*, 189
Cassava mosaic virus, 117
Cassava pest management, 97–127
  bionomics, 99, 120
  cassava cultivars, grasshopper attacks on, 119
  chemicals, 116
  damage to cassava plants, 100, 120
  factors affecting population abundance and distribution, 101, 102
  green mite, 98, 119, 121–126
  hyperparasitoids, 105
  insect diseases, 112
  insect pests
    Coleoptera, 116
    Heteroptera, 117
    Homoptera, 117
    Isoptera, 118
    Lepidoptera, 118
    Orthoptera, 118
    Thysanoptera, 119
  parasitoids, 102
  pest biology and ecology, 100, 120
  predators of pests, 105
  red mites, 124
  resistant cultivars, 115, 124
Castor, 180, 182
Catechol, 193
Cauliflower, 183
Ceava rubber tree, 99
Cereals, 229
  resistance to cyst nematodes, 219, 223
*Cestrinus trivialis*, 25
Charcoal rot, 31
*Chartocerus* spp., 105
Chick peas, 80, 185, 192, 194
Chilli, 182, 199
China
  nematodes, 177, 186
  *Nysius plebejus*, 46
  sorghums, 25, 42
    midge, 42
    shoot fly, 25
  sugarcane aphid, 34
Chinch bug, 35
Chlordane, 24, 118
Chloropids
  *Anatrichus erinaceum*, 28
  *Elachipterus abessynicus*, 28
  *Scolipthalmus* sp., 28
    *S. mecantipennis*, 28
Chlorpyrifos, 27, 29, 44, 46, 78
Chlorpyrifos-methyl, 78
*Chromolaena odorata*, 118
Chrysopids, 33, 41, 108, 112
*Cicer arietinum*, 185
*Clitostethus neuenschwanderi*, 117
Clover, 185, 194
Cluster bean, 184
Coccinellids, 33–34, 41, 100, 105, 108, 112, 115, 117–118, 122
  *Coccinella undemcipunctata*, 39
  *Diomus*, 112
  *Menochilus sexmaculatus*, 39
  *Scymnus babai* Sassji, 34
    *S. gracilis*, 37
*Coccodiplosis citri*, 108, 112
*Coccophagus* sp., 103
Cocoyam, 98
Coleoptera, 47, 67
Collards, 23
Collembola, 68
*Colletotrichum gloeosporioides*, 117
Columbia
  sorghum midge, 42
    predated by spider, 43
Common crow (*Corvus splendens*), 24
Composts, 241
Congo
  cassava mealybug, 98
    parasitoids of, 103
  coreid bug, 117
  *Hyperaspis raynevali*, 108
  mite predators, 123
  thrips, 119
Control strategies
  greenbug, 31
  soybean pests, 2, 4–5
*Corcyra cephalonica*, 79
Coreid bug, 117
Corn, 98
Corn earworm, 8, 44
Corn leaf aphids, 33
Cotton, 44, 180, 190–191, 219, 222
Cottonseed, 1, 192
Cowpea, 39, 98, 180, 182, 190, 195, 197, 231
*p*-Cresol, 193
Crop injury, 2–10
*Croton hirtus*, 99
Cryptophagidae, 73
Cucumber, 197
*Cucumis* sp., 186
  *C. melo*, 186, 197
  *C. sativus*, 197
Curcurbits, 185, 218
Cutworms, 24
  *Agrotis infusca*, 24
    *A. ypsilon*, 24
*Cyamopsis tetragonaloba*, 184
*Cymbopogon flexuosus*, 198
*Cynodon dactylon*, 26
Cypermethrin, 44, 78

Damage
  boundary, 8–9
  by nematodes, 214
  curve, 7–9, 11–12, 15
  on sorghum, 25

on soybean, 5, 7, 9, 11–13
to cassava, 100
DDT, 28, 40
Decision-making algorithms, 2–4
*Delia platura*, 8
Deltamethrin, 78
Demeton-S-methyl, 46
*Dendrobaena veneta*, 246
Dermestid larvae, 74
Desiccation, 169
*Diabrotica* spp., 25
Diaspid scale, 117
Diazinon, 24, 44, 116
Dichlorvos, 78
*Dicrodiplosis manihoti*, 108, 114
Dictyoptera, 67
Dimethoate, 44, 116
*N-N*-Dimethylpiperidinium chloride, 33
Dimlin, 78
*Diomus* sp., 116
*Diopsis* sp., 28
Diptera, 67
Disulfoton, 27, 199
*Ditylenchus angustus*, 170
*D. dipsaci*, 162–166, 168–171, 187
*D. myceliophagus*, 166
*Dolichos lablab*, 80
Dominican Republic
nematodes, 180

Earhead beetles
*Anomala* sp., 47
*Chiloloba acuta*, 47
*Cyanolylta* sp., 47
*Cylindrothorax* sp., 47
*Mylabris* sp., 47
*Psalydolytta fusca*, 47
*Rhinyptia* sp., 47
*Schizonycha* sp., 47
Earhead bugs
*Calocoris angustatus*, 45
Earhead caterpillars
*Celama sorghiella*, 45
*C. analis*, 45
*Cryptoblabes gnidiella*, 45
*Eublemma gayneri*, 45
*E. silicula*, 45
*Euproctis subnotata*, 45
*Heliothis armigera*, 44
*H. zea*, 44
*Pyroderces simplex*, 45
*Spodoptera frugiperda*, 45
Earhead pests
*Contarinia sorghicola*, 41
Earthworms
*Aporrectodea caliginosa*, 246
as animal feed, 245
*Dendrobaena veneta*, 246
*Eisenia foetida*, 244–247
*Eudrilus eugeniae*, 245–246
farming of, 242
fauna, 242
*Lumbricus rubellus*, 246
methods of production of, 246
origins of vermiculture, 243
*Perionyx excavatus*, 246
*Pheretima*, 245, 246
problems of organic waste disposal, 242
soil improvement and reclamation, 244
types of earthworms used, 246
utilization of organic wastes, 241–249
vermicomposting and stabilization of wastes, 244
Earwigs, 25
*Diaperasticus erythrocephala*, 43
*Nala lividipes*, 25
Economic Injury Level (EIL), 2–5, 7, 9, 11–14, 18, 40
egg-plant, 182, 184–185, 191, 193, 198
Egypt
nematodes, 185
*Eichhornia crassipes*, 193
*Empoasca fabae*, 8
Encyrtids, 102, 118
Endosulphan, 30, 40, 44, 46, 140
Endrin, 40
*Entomophthora fumosa*, 112
*Epidinocarsis (Apoanagyrus) lopezi*, 103–105, 113–116, 126
*Eragrostis curvula*, 218
*Eretmocerus* sp., 117
*Eruca sativa*, 198
Ethephon, 192
Ethiopia
armyworm in sorghum, 29
maize stem borer, 40
*Euchistus servus*, 14
*Euphorbia* sp., 193
*Euproctis fasciata*, 118
Europe, Northwest
cyst nematodes, 216–217, 232
European corn borer, 41
European Economic Community
soybean production in, 1
*Euseius* spp., 122
*E. concordis*, 123
*Exochomus* spp., 108, 114–115
*E. concavus*, 105
*E. flavipes*, 105
*E. flaviventris*, 105, 108

Fenamiphos, 191–193
Fenitrothion, 48, 78, 192
Fenpropathion, 37
Fensulphotron, 27
Fenthion, 30
Fenvalerate, 40
*Ferrisia virgata*, 117–118
Fertilizers, 36, 39, 98
ammoniacal, 194
effect on sorghum mites, 36
for sorghum shoot fly, 36
phosphatic, 194
NPK, 199

Fescue, 219
Florida
  armyworms, 29
Forest areas, 114–115
Freeze fracture, 170
Fumigants, 78, 191
Fungicide, 27, 199
Fungus
  *Acremonium coenophialum*, 219
  as treatment for white grubs, 24
  *Aspergillus niger*, 198
  *Beauveria bassiana*, 35
  *Fusarium* sp., 198
    *F. oxysporum*, 190
  *Glomus fasciculatum*, 197
    *G. mosseae*, 197
  *Helminthosporium* sp., 198

Gabon
  cassava mealybug, 98
    parasitoid of, 103
Gain threshold (GT), 4–5, 13
*Galendromus annectens*, 122–123
Gallic acid, 195
Gambia
  cassava mealybug, 98
    parasitoid of, 103
Gene transfer, 231
Genetic engineering, 231
Genetics of resistance
  sorghum, 27
Genetics of variability
  sorghum, 27
Geometrid
  *Pleuroprucha insulsaria*, 45
Germany
  earthworms, 244
Ghana
  cassava, 97
    mealybug, 99, 114
      parasitoid of, 103
  grain borer, 116
  Phytoseiid mites, 122
  termites, 118
*Globodera* spp., 213, 219, 225
  *G. rostochiensis*, 223, 225, 227, 231
*Glossina* spp., 137
  *G. medicorum*, 149
  *G. morsitans*, 142, 147, 149
    *G. m. morsitans*, 147–149, 152
  *G. pallidipes*, 145, 147–150, 152
  *G. palpalis palpalis*, 139, 141
  *G. swynnertoni*, 149
see also Tsetse fly
Glossoscolecidae, 241–242
β-Glucosidase, 229
*Glycine max*, 1, 184, 219
β-Glycosidase, 229
*Gnathocerus cornutus*, 73
*Gonocephalum macleayi*, 25
  *G. carpentariae*, 25
*Gossypium hirsutum*, 219
Grain borer, 74, 98, 116
Grapevines, 183, 185, 190–191
Grasshoppers
  *Chrotogonus* spp., 30
  *Hieroglyphus* spp., 30
  *Kraussaria angulifera*, 30
  *Oedaleus senegalensis*, 30
  on sorghum, 30
Greenbug, 31, 32, 34
Green clover worm, 4, 10
  potential consumption of, 10, 13
Greengram, 185
Grey weevil
  *Myllocerus undecimpunctata maculosus*, 28
Groundnut, 98, 180
Guilds
  injury, 1–18
    animal, 11
    insect, 11
Guinea Bissau
  cassava mealybug parasitoid, 103

Hawaii
  nematodes, 177, 199
HCH, 44
*Heliothis* spp., 2
  host response to injury, 5, 7–9, 11–13
    compensation, 7, 8
    desensitization, 7, 8
    inherent impunity, 7, 9
    linearity, 7, 8
    overcompensation, 8
    tolerance, 7
  *H. virescens*, 14, 16
  *H. zea*, 8, 14, 16
Heptachlor, 24, 118
Herbicides, 195
*Heterodera glycines*, 195, 213, 218–219, 225
  *H. avenae*, 215–217, 219, 224–225, 229, 232
  *H. schachtii*, 220–221, 225, 230–231
*Hibiscus* sp., 99
Histerid beetle, 116
*Holobus* spp., 122
*Holotrichia serrata*, 24
*Homalotylus* spp., 112
Honduras
  armyworm, 27
Hoplolaimidae, 179
*Hoplolaimus indicus*, 190
*Hordeum vulgare*, 215
Hydrocarbons, 191
Hydroquinone, 193
Hymenopterous parasitoids, 67, 110–112
  *Eupelmus popa*, 44
  *Tetrastichus* spp., 44
    *T. diplosidis*, 44
*Hyperaspis* spp., 108, 114–115
  *H. jucunda*, 108
  *H. marmottani*, 105
  *H. pumila*, 105, 108
  *H. raynevali*, 108
  *H. senegalensis*, 105, 108

Impact index, 2
India
  armyworms, 30
  brachonid wasp, 30
  corn leaf aphid, 33
  cyst nematodes, 217
  earhead beetles, 45
  earhead bug, 45
  earhead caterpillar, 45
  grasshoppers, 30
  grey weevils, 28
  leaf beetles, 28
  mirids, 46
  nematodes, 180, 184, 189
  rice armyworms, 30
  shoot fly, 27
    control by insecticide, 27
  sorghum mite, 36
  sorghum shoot bug, 35
  sorghum stem borer, 37–39
  sorghums, 26, 45
    white grubs in, 24
  stemborers, 37
  stored products, 75
  sugarcane pyrilla, 35
Injury, 2, 6–7
  equivalency, 9–10
  guilds, 2, 9, 11, 13–15, 17
    impact on soybean, 2–24
  insect-induced, 2, 5, 7–9
  plant response to, 5, 7
  seasonal development, 6
  simulated techniques, 17
  to fruit, 14, 16–17
  to leaf-tissue, 2, 11–13
  to leaf-mass, 17
Insect pests (soybean), 1–22
  architecture modifiers, 7, 17
  assimilate sappers, 7
  consumption potential, 6–7, 9, 11
  feeding, 17
    on sorghum, 23
  fruit feeders, 7, 14, 16
  herbivorous, 11
    pit feeders, 11
    sap feeders, 11
    strip feeders, 11
  larvae, 9–10
  leaf-mass consumers, 7, 10, 12
  seasonal development, 6
  size-class differences, 9, 10–12, 17
  species differences, 11, 17
  stand reducers, 7
  turgor reducers, 7
insecticide residues, 48
insecticides, 25–28, 30–31, 35, 44–45, 79, 140–141, 145–146, 152–153
Integrated Pest Management (IPM), 4, 24, 85
Iowa
  soybean production, 17
*Ipomoea batatas*, 183, 218
Isofenphos, 26–27, 48, 191
Isozophos, 192
Italy
  earthworms, 244
  lygaeids, 46
Ivory coast
  cassava mealybug, 99
    parasitoid of, 103

Jamaica
  armyworms, 29
Japan
  nematodes, 177, 180, 184, 186
Johnson grass, 42, 44

*Kalodiplosis* sp., 118
Kansas
  greenbug, 31
Karanj, 193
Kentucky blue grass
  greenbug on, 31
Kenya
  braconid wasp, 39
  cassava green mite, 119
  crop pest resistance, 117
  grain borer, 116
  mite predators, 123
  red spider mite, 124
  sorghum midge, 42
    predated by earwig, 43
  tsetse fly, 153
    targets, 146–147
Khapra beetle, 83

Lablab, 39
Lacewing, 43
*Lachnosterna consanguinea*, 24
Laos weed, 118–119
*Lasioderma serricorne*, 116
Leaf
  necrosis, 17
  production, 2
  tissue, 2
Leaf caterpillars, 30
  *Amsacta albistriga*, 30
    *A. moloneyi*, 30
    *A. moorei*, 30
  *Euproctis subnotata*, 30
    *E. virguncula*, 30
  *Marasmia trapezalis*, 30
  *Spilosoma obliqua*, 30
Legumes, 44, 191, 219
Lemon grass, 30
Lepidoptera, 47, 67
*Leptomastix* sp., 118
  *L. dactylopii*, 118
Lettuce, 100
Lichens, 161
Lima bean, 214
Lindane, 40
Lumbricidae, 241–242
*Lumbricus rubellus*, 246
Lycaenid caterpillar, 108
*Lycopersicon esculentum*, 180, 214

*L. peruvianum*, 219
Lygaeids
  *Nysius graminicolus*, 46
    *N. plebejus*, 46

*Macrophomina phaseolina*, 190
*Macrotermes subhyalinus*, 118
  *M. bellicosus*, 118
Madagascar
  cassava, 97
  grasshopper, 119
  red spider mite, 124
Maize, 30, 37, 40, 44, 75, 80, 192, 227
  dwarf mosaic virus, 31, 34
  greenbug on, 31
  stem borer, 40
Malathion, 30, 44, 46, 78, 116
Malawi
  cassava mealybug, 98, 114
    parasitoid of, 103
  green mite, 119
*Mangifera indica*, 192
Mango, 192
*Manihot esculenta*, 97, 99
  *M. glaziovii*, 99
  *M. utillissima*, 97
Manure
  animal, 193
  cattle, 26
  chicken, 193, 199
  green, 192
Marigold, 192, 198
  mottle virus, 190
Mealybug, 98–102, 105, 125–126
  parasitoids of mealybug predators, 112, 118
*Medicago sativa*, 218
*Melia azedarach*, 193
*Meloidogyne* spp. (*see* Nematodes)
Methacrifos, 78
Methadathion, 116
Methomyl, 29
Methoprene, 78
Methyl bromide, 47, 78, 116
Methyl iso-thiocyanates, 191
*Microtermes* spp., 25, 118
  *M. nadjensis*, 25
Middle East
  nematodes, 177
  sorghums, 25
Midges, 105, 108, 112, 118
Millet red leaf virus, 34
Mirids
  *Creontiades pallidus*, 46
  *Eurystylus bellevoyi*, 46
    *E. marginatus*, 46
*Mirperus torridus*, 117
Mississippi
  armyworms, 29
Mites, 41, 43, 79–80, 120–122
  cassava green mites, 119, 121–122, 124–126
  *Oligonychus indicus*, 37
    predators, 37
    *O. pratensis*, 36
  phytoseiid mites, 122–124
  red mites, 124
Monocrotophos, 28, 30, 37, 116
*Monomorium salomonis*, 25
*Mononychellus progresivus*, 119
  *M. tanajoa*, 98, 119, 124
    predators of, 123
Mosses, 161, 169
Mothbean, 185
Mozambique
  cassava, 97
Mungbean, 80, 191
*Musa* spp., 192
Musk melon, 197
Mustard, 189, 192, 198
Mynah, 24

*Nala lividipes*, 25
Napier grass, 30
Natural plant products, 79
Nebraska
  greenbug, 31
Neem, 193, 198
Nemacur, 191
Nematicides, 163, 186, 191–192, 199
Nematodes
  anhydrobiotic, 161–172
    characteristics of, 162
    commercial aspects, 170
    membrane function, 169
    metabolic changes during desiccation, 165
      during rehydration, 169
    structural adaptations, 164
  *Caenorhabditis elegans*, 232
  cereal cyst, 216
  cyst and root-knot forming, 213–232
    biochemical changes in plant, 221
    damage to crops, 214
    elicitors and defence, 228
    enzyme systems, 229
    genetic engineering 231
    genetic interactions, 227
    multiple resistance, 226
    non-host resistance, 226
    occurrence in plants, 213
    plant growth substances, 224
    plant-nematode recognition, 227
    progress after invasion, 220
    resistance of plants, 213–232
    resistance phenomena, 217, 230, 232
    root invasion, 218
    temperature effects, 223
    ultrastructural responses to invasion, 221
  Globodera (cyst forming), 181, 213, 218
  Hetorodera (cyst forming), 181, 213, 218
  *Meloidogyne* spp. (root-knot forming), 178, 183, 188, 190, 193, 213, 216, 218–219, 222, 226
    *M. arenaria*, 222, 226–227
    *M. graminicola*, 219

*M. hapla*, 218–219, 226–227
*M. incognita*, 188–190, 198–199, 218, 220, 222–223, 226–227
*M. javanica*, 189, 199, 221, 223, 226–227, 231
*M. marylandi*, 219
*Pratylenchus penetrans*, 230
*P. scribneri*, 231
*P. sefaensis*, 189–190, 199
*Radopholus similis*, 189
*Rotylenchulus borealis*, 179
*R. macrodoratus*, 179, 183
*R. macrosoma*, 179
*R. parvus*, 179
*R. reniformis* (reniform nematode), 162, 170, 177–200
biological races, 180
control, 190
biological, 197
chemical methods, 191
cultural methods, 192–195
host resistance, 195–197
integrated, 198
non-conventional, 198
crop resistance to, 196–197
ecology, 186–189
gross morphology and systematics of, 178
infectivity and histopathology, 183
interaction with other soil micro-organisms, 189
life cycle, 184
life history, 181
pathogenicity and crop losses, 184–185
*R. sacchari*, 179
soybean cyst, 218
tylenchid nematodes, 178
*Tylenchorhynchus* spp., 190
*T. vulgaris*, 188
Neoaplectana, 171
*Neoseiulus anonymus*, 122–123
*N. idaeus*, 122–125
*Neozygites fumosa*, 112
*Nephus flavimaculatus*, 118
*N. phenacoccophagus*, 105, 118
*N. vetustus*, 105
*Nesothrips marshalli*, 119
New Zealand
earthworms, 246
*Nezara viridula*, 2, 14, 16
*Nicotiana repanda*, 222
*N. tabacum*, 184, 222
Nigeria
cassava, 97
cassava green mite, 120
coccinellids, 112
*Dysdercus voelkeri*, 46
*Hyperaspis raynevali*, 108
mealybug, 98, 100, 114
parasitoids of, 103
phytoseiid mites, 122
red spider mite, 124
termites, 118
Nimbidine, 193
Nitidulidae, 73

Oats
greenbug on, 31
invasion by *H. avenae*, 219, 222
Octenol, 152
*Odontotermes* spp., 25, 118
Oklahoma
greenbug, 31
Okra, 100, 182, 184, 190, 193
Oligochaeta, 241
*Oligonychus gossypii*, 124
Onion, 182, 185, 194
Organic wastes, 241
Organochlorines, 140
Organophosphates, 78, 191
*Oryza sativa*, 219
Oxamyl, 191–192

*Paecilomyces lilacinus*, 197
Palaeoarctic zone, 241
*Pangamia glabra*, 193
Pangolan grass, 191
*Panicum* spp., 26
Papaya, 189
Parasitoids, 45
of greenbug, 33
of maize stem borer, 40
Parathion, 118
Peanuts, 191
Pearl millet, 30, 37
Peas, 80
*Pennisetum purpureum*, 40
Pentatomids
*Thyanta preditor*, 46
Pepper, 100
*Perionyx excavatus*, 246
Permethrin, 29, 78
Persian lilac, 193, 198
Pest management, 1–4, 6–7, 13–14, 17, 34, 40, 45
by soil compaction, 25
control in sorghum, 24
decision algorithms, 3, 9
minimizing economic losses, 3
Pesticides, 3–4, 17, 28, 30, 98
use in stored products, 79
Pests (in soybean), 1–4, 6, 8–18
agricultural, 11
architecture modifiers, 7, 17
arthropod, 1, 11, 14
artificial control of, 4
assilimate sappers, 7
bionomics, 24
corn earworm, 8
fruit feeding, 2, 7, 11, 14, 16
green clover worm, 10, 11
herbivorous, 11
pit feeders, 11
sap feeders, 11

strip feeders, 11
in soil, 24
control by compaction, 25
leaf-consuming, 2, 6–7, 10–12, 28
lepidopterous, 9, 10
mortality factors, 26
natural control of, 4
on sorghum, 23
parasites, 29
on armyworm, 29
*Archytos marmoratus*, 29
*Chelonus insularis*, 29
*Diapetimorpha introita*, 29
on shoot fly, 27
phytophagous, 9
podworm, 2, 14–15, 17
seedcorn maggot, 8
size-class differences, 10–12, 17
soybean looper, 2, 10–11, 14
species differences, 11, 17
stand-reducers, 7
stinkbug, 2, 14–17
turgor reducers, 7
velvetbean caterpillar, 2, 10–11
*Phaseolus aconitifolius*, 185
*P. aureus*, 185, 191
*P. lunatus*, 214
*P. vulgaris*, 197, 219, 223, 227
*Pheidole sulcatipes*, 25
*Phenacoccus citri*, 112, 117–118
*P. madeirensis*, 117–118
*P. manihoti*, 98–99, 103, 105–107, 109–113, 116–117
Phenthoate, 40
Phenyl ammonium lyase (PAL), 229
Pheretima, 245
Pheromones
use in stored products, 80–81, 85
Philippines
earthworms, 244
Phorate, 27, 35, 191–192, 199
Phosalone, 44, 46
Phosphamidon, 116
Phosphine, 48, 78, 116
*Phyllophaga crinita*, 24
*Phymateus puniceus*, 119
*Phytophthora megasperma* f. sp. *glycinea*, 195
Pigeonpea, 185, 194–195, 227
Pineapple, 177, 191, 199
damage by nematodes, 177
Piper beetle, 185
Pirimiphos-methyl, 48, 78
Plantain, 98
Plant bugs
*Blissus leucopterus leucopterus*, 35
*Peregrinus maidis*, 36
*Poophilus costalis*, 36
*Pyrilla perpusilla*, 35
Plant parasites, 213
Plant parasitic nematodes, 162
*Plathypena scabra*, 4, 17
Platigasterid parasitoids, 118
*Plodia interpunctella*, 73, 79
Podworm, 2, 14–15, 17
Potato, 219, 222–223, 225, 227, 229
Potato leaf hopper, 8
*Pratylenchus penetrans*, 230
*P. scribneri*, 231
Predators
of greenbug, 33
of maize stem borer, 40
Principe Island
tsetse fly, 139
*Prochiloneurus* spp., 105
*P. insolitus*, 105
Propoxur, 78
*Prostephanus truncatus*, 98, 116
Protozoans
in controlling storage pests, 82–83
*Pseudoplusia includens*, 2
*Pseudotheraptus devastans*, 117
Psocids, 67
Psocoptera, 68
*Pterohelaeus alternatus*, 25
*P. darlingensis*, 25
Ptinidae, 73
Puerto Rico
nematodes, 180
sorghum midge, 42
Pyrethroids, 140
Pyriproxyfen, 152
Pyrogallol, 193
Pyrrhocorids
*Dysdercus voelkeri*, 46

Quinalphos, 27, 30, 44

Rainforest areas, 99
Ratoon crops
effect of sugarcane pyrilla on sorghum, 35
*Retithrips syriacus*, 119
*Rhizoctonia solani*, 190
*Rhizopertha dominica*, 73, 79, 116
Risk averse strategy, 3
Root beetles, 25
*Cestrinus trivialis*, 25
*Diabrotica* spp., 25
*Gonocephalum carpentariae*, 25
*G. macleayi*, 25
*Pterohelaeus alternatus*, 25
*P. darlingensis*, 25
Root fungi, 190
Root-knot nematode (*see* Nematodes)
Root rot, 190
Rotifers, 161
*Rotylenchulus reniformis* (*see* Nematodes)
*Rotylenchus robustus*, 162
Rwanda
cassava mealybug, 114
parasitoids of, 103
phytoseiid mites, 122

*Saccharum* spp., 180
*Salmonella*, 75
Savanna zones, 99, 114–115

*Schizonycha* spp., 24
*Scirtothrips manihoti*, 119
*Scutellonema cavenessi*, 163, 170
Seed corn maggot, 8
Senegal
  cassava mealybug, 98
    parasitoids of, 103
  nematodes, 180
  red spider mite, 124
  sorghum, 26
*Serratia marcescens*, 40
Sierra Leone
  cassava mealybug, 98–99
  thrips, 119
Silverfish, 68
*Sitophilus* spp., 79, 80, 116
Sleeping sickness, 137, 153
Soil fertility, 241
Soil insecticides, 24–25
  compaction, 25
*Solanum* spp., 219
  *S. melongena*, 182
  *S. tuberosum*, 219
Sooty mould, 31, 34, 44, 100
Sorghums, 23–39, 80
  animal consumption, 33
  control agents, 29
  cultural practices, 29
  earhead pests, 41–47
    earhead beetles, 47
    earhead bugs, 45–46
    earhead caterpillars, 41
    sorghum midge, 41
    predators of, 43
  entomology research, 23–39
  foliage pests, 28–37
    aphids, 31–34
      corn leaf aphid, 33
      green bug, 31–32
      sugarcane aphid, 34
      yellow sugarcane aphid, 34
    armyworms, 28–30
      fall armyworm, 30
      rice armyworm, 31
    grasshoppers, 30
    leaf caterpillars, 30
    mites, 36–37
      Banks green mite, 36
      sorghum mite, 37
    plant bugs, 35–36
      chinch bug, 35
      shoot bugs, 36
      sugarcane pyrilla, 35
        cultivar resistance, 35
  human consumption of, 23
  insecticide residues, 48
  seedling pests, 25–28
    chloropids, 28
    diopsid, 28
    grey weevil, 28
    leaf beetles, 28
    sorghum shoot fly, 25–28
      parasites of, 27
      resistance to, 27
  soil pests, 24–25
    cutworms, 24
    white grubs (beetle larvae), 24
  sooty mould, 31, 34
  *Sorghum bicolor*, 23–39
    *S. halepense*, 26–27, 32
    *S. sudanense*, 27
    *S. verticilliflorum*, 40
    *S. virgatum*, 39
  Stem borers, 37–41, 49
    European corn borer, 41
    maize stem borer, 40
      parasitoids of, 40
      predators of, 40
    sorghum stem borer (spotted stalk
      borer), 37
      predators of, 39
    sugarcane borer, 40
      predators of, 41
  stored grain pests, 47, 48
Sorghum earhead bug, 45
  (*see* Earhead bugs, *Calocoris angustatus*)
Sorghum midge, 41
  (*see* Earhead pests, *Contarinia sorghicola*)
Sorghum mite, 37
  (*see* Mites, *Oligonychus indicus*)
Sorghum shoot bug, 36
  (*see* Plant bugs, *Peregrinus maidis*)
Sorghum stem borer (spotted stem borer), 37
  (*see* Stem borers, *Chilo partellus*)
Soybean, 1–21, 184, 191
  fruit, injury to, 2, 17
  genetics of resistance against
    *R. reniformis*, 195
  geographical distribution, 1
  leaf, injury to, 2, 8–11, 13–14
  managing pest losses to, 1–4, 7, 13, 17
  pests, 1–3, 14, 17
    relationship to, 3
  primary pest feeding sites, 6–7
  resistance to cyst nematodes, 219, 223, 225
  yield, loss of, 7
Soybean looper, 2, 10, 14
  consumption potential of, 10–11
*Spalgis lemolea*, 108
Spiders
  *Argyopes* sp., 37
  *Cheiracanthium*, 37
  *Gregalis* sp., 43
  *Metazygia*, 43
  *Theridion* sp., 37
  *Theridula* sp., 37
  *Uloborus* sp., 37
Spodoptera, 118
Springtails, 68
Standard equivalency species, 12
Staphylinid beetles, 122
*Stegobium paniceum*, 73
*Steinernema*, 171
Stemborers, 49

*Acigona ignefusalis*, 37
*Busseola fusca*, 37
*Chilo partellus*, 37
*Diatraea grandissella*, 37
*D. lineolata*, 37
*D. saccharalis*, 37
*Elasmopalpus lignosellus*, 37
*Eldana saccharina*, 37
*Eoreuma loftini*, 37
*Ostrina furnacallis*, 37
*O. nubilalis*, 37
*Sesamia calamistis*, 37, 41
*S. cretica*, 41
*S. inferens*, 37, 41
Sterile control of pests, 83
*Stethorus* spp., 122
Stink bugs, 2, 14–17
Storage moths, 82
*Cadra cautella*, 82
*Ephestia kuehniella*, 82
*Plodia interpunctella*, 82
Storage pests, 68–73
constraints in reducing loss, 76
control of, 76
biological, 80
chemical, 78
legislative, 83
mechanical, 77
physical, 77
damage, 74
contamination, 74
dry heating and moisture migration, 74
fungi and other micro-organisms, 73
grain dust, 74
insect parts in food products, 75
lowered seed germination, 75
to wooden structures and containers, 74
problems in post-harvest phase, 67
taxonomy of, 73–74
Stored grain pests
*Corcyra cephalonica*, 47–48, 82
*Ephestia cautella*, 47
*Oryzaephilus surinamensis*, 47
*Prostephanus truncatus*, 47
*Rhizopertha dominica*, 47
*Sitophilus oryzae*, 47–48, 76, 80
*S. zeamais*, 47–48
*Sitotroga cerealella*, 47–48, 75–76, 80, 82
*Tribolium castaneum*, 47, 77
*T. confusum*, 47–48
*Trogoderma granarium*, 77
Stored products, 67–85
ants associated with, 72
cockroaches associated with, 73
insects associated with, 69–71
mites associated with, 72
Sudan
sorghum, 41
sugarcane, 41
Sudan grass
greenbug on, 31
Sugarcane, 180, 192
aphid (*see* Aphids, *Aphis (Melanaphis) sacchari)*, 34
pyrilla (*see* Plant bugs, *Pyrilla perpusilla*), 35
Sweet potato, 98, 183, 185, 218, 226
Synthetic pyrethroids, 78
Syrphids, 33

*Tagetes erecta*, 192
*T. minuta*, 198
*T. patula*, 198
*T. tenuifolia*, 198
Tanzania
cassava, 97
grain borer, 116
Tardigrades, 161
*Tenebrio molitor*, 73
Terbufos, 35, 191
*Teretriosoma nigrescens*, 116
Termites, 25, 118
*Microtermes* spp., 25
*M. nadjensis*, 25
*Odontotermes* spp., 25, 118
*Tetranychus neocaledonicus*, 124
*T. lombardinii*, 124
*Tetrastichus* spp., 105, 112
Texas
ant, 41
greenbug, 31–32
sorghum, 31–32, 41
Thiodicarb, 28
Thiometon, 37
Thionemone, 193
Thiram, 27–28
Thrips, 119
Thysanura, 68
Tobacco, 184, 191, 222
Tobacco white fly, 98, 117
Togo
cassava mealybug parasitoid, 103
grain borer, 116
Tomato, 180, 182–185, 193–194, 197–199, 214, 220, 222–223, 226–227
Tortula, 169
Traps, 81, 140–144, 147–149, 151–152
'biconical', 141, 148
electric, 77
electrified nets, 147
'epsilon', 144, 148
'F3', 143, 148
mechanical, 77
'monoconical', 141, 153
Morris, 140
'Ngu', 146, 148
pan, 26
paper, 38
pheromone, 45
suction, 42
yellow sticky, 26
*Tribolium* spp., 116
*T. castaneum*, 73
Trichlorfon, 40

*Trichostrongylus colubriformis*, 162, 164
*Triticum aestivum*, 189
*Trogoderma granarium*, 80, 83–84
Trypanocidal drugs, 153–154
Trypanosomes, 137, 153–154
  chemotherapeutic treatment, 138
Tsetse fly, 137–154
  autosterilizing techniques, 151
  control with traps and targets, 152–153
  early control, 139
  food aquisition, 150
  insecticidal control, 140
  life cycle, 138–139
  mating behaviour, 150
  olefactory response, 147–148
  on Principe Island, 139
  population management, 137–154
  response to traps and targets, 150–151
  role of vision, 148–150
  savannah species, 142, 144
  sterile male control, 141
  targets for, 146, 151–153
  trap development, 141–146
  traps for, 140–152
*Tylenchorhynchus vulgaris*, 188
*Typhlodromalus (Amblyseius) limonicus*, 123

Uganda
  cassava, 97
    green mite, 119
  phytoseiid mites, 122
  thrips, 119
United Kingdom
  earthworms, 242, 244, 247
United States of America
  Banks grass mite, 36
  chinch bug, 35
  corn earworm, 44
  corn leaf aphid, 33
  earthworms, 243–244
  earhead caterpillar, 25
  greenbug, 31–32
  nematodes, 177, 186, 188
  sorghum, 24, 29, 33
  Yellow sugarcane aphid, 34
  *see also* America, North
Uruguay
  sugarcane aphid, 34

Variegated grasshopper, 98, 118
Venezuela
  nematodes, 180
Velvetbean caterpillar, 2, 10
Vermicompost, 242–244
Vermiculture, 242–243, 246
*Verticillium* sp., 190
  *V. dahliae*, 190
  *V. lecanii*, 24
*Vicia faba*, 185
*Vigna unguiculata*, 223
*Vitis vinifera*, 183

Wasps, 103–105
Water hyacinth, 193
Wheat, 75, 80, 189
  aphid, 34
  greenbug on, 31
White grubs
  *Anomala* spp., 24
  *Holotrichia serrata*, 24
  *Lachnosterna consanguinea*, 24
  *Phyllophaga crinita*, 24
  *Schizonycha* spp., 24
Wild spinach, 100
Wireworms, 25
  *Agrypmus variabilis*, 25

*Xanthomonas phaseoli* var. *sojensi*, 195

Yam, 98
Yellow mosaic virus, 190
Yellow sugarcane aphid, 34
Yemen Arab Republic
  armyworm in, 29
  sorghum, termites in, 25

Zaire
  cassava, 97
    mealybug, 98, 114
    parasitoid of, 103
  coccinellids, 112
  coreid bug, 117
  mite predators, 123
  phytoseiid mites, 122
Zambia
  cassava mealybug, 114
    parasitoid of, 103
  phytoseiid mites, 122
*Zea mays*, 189, 227
Zimbabwe
  *Eretmocerus* sp., 117
  tsetse fly, 140, 147, 152–153
    targets for, 145–146
    traps, 148
*Zinnia elegans*, 192
*Zonocerus variegatus*, 98, 118

*Agricultural Zoology Reviews* VOLUME 1 (November 1986);
ISBN 0 946707 09 X; ISSN 0269–0543; xii + 401 pp; 234 × 156 mm;
illustrated; comprehensive index; hardback.

## CONTENTS

**THE BIOLOGY, PEST STATUS AND CONTROL OF LEAF-CUTTING ANTS**
J.M. Cherrett (UK)

**GENETICS OF THE FLOUR MOTH, *EPHESTIA KÜHNIELLA***
F. Leibenguth (FRG)

**GENETICS OF THE MEDITERRANEAN FRUIT FLY (*CERATITIS CAPITATA*) (WEIDEMANN)**
Stephen H. Saul (USA)

**THE BOLL WEEVIL: RECENT RESEARCH DEVELOPMENTS AND PROGRESS TOWARDS ERADICATION IN THE USA**
E.P. Lloyd (USA)

**MANAGEMENT OF TEMPERATE-ZONE DECIDUOUS FRUIT PESTS: APPLIED BEHAVIOURAL ECOLOGY**
Bernard D. Roitberg and Nello P.D. Angerilli (Canada)

**BIOLOGY, SAMPLING AND POPULATION DYNAMICS OF *BEMISIA TABACI***
G.D. Butler, Jr, T.J. Henneberry and W.D. Hutchison (USA)

**EFFECTS OF TOXICANTS ON INSECT CELLS**
Harald C. von Keyserlingk (FRG)

**THE BIOLOGY AND BEHAVIOUR OF SLUGS IN RELATION TO CROP DAMAGE AND CONTROL**
C.M. Port and G.R. Port (UK)

**RECENT DEVELOPMENTS IN THE BIOCHEMICAL TAXONOMY OF PLANT-PARASITIC NEMATODES**
P.C. Fox and H.J. Atkinson (UK)

**ANTHELMINTIC RESISTANCE IN NEMATODE PARASITES OF SHEEP**
P.J. Waller (Australia)

*Agricultural Zoology Reviews* VOLUME 2 (November 1987);
ISBN 0 946707 11 1; ISSN 0269–0543; xii + 412 pp; 234 × 156 mm;
illustrated; comprehensive index; hardback.

**CONTENTS**

**CEREAL APHIDS AS AN APPLIED PROBLEM**
A.F.G. Dixon (UK)

**DEVELOPMENTAL AND SEASONAL BIOLOGY OF *OSTRINIA NUBILALIS***
Stanley D. Beck (USA)

**BIOLOGY AND CONTROL OF THE CARROT FLY, *PSILA ROSAE* (F.)**
C.P. Dufault and T.H. Coaker (UK)

**BEET CYST NEMATODE (*HETERODERA SCHACHTII* SCHMIDT) AND ITS CONTROL ON SUGAR BEET**
D.A. Cooke (UK)

**ECOLOGICAL AND AGRICULTURAL CONSIDERATIONS IN THE MANAGEMENT OF TWOSPOTTED SPIDER MITE (*TETRANYCHUS URTICAE* KOCH)**
R.L. Brandenburg and G.G. Kennedy (USA)

**BIOLOGY OF POLYPHAGOUS GROUND BEETLES IN AGRICULTURE**
M.L. Luff (UK)

**NEW POLLINATORS FOR OUR CROPS**
F.D. Parker, S.W.T. Batra and V.J. Tepedino (USA)

**BIOCHEMICAL CHARACTERS IN AGRICULTURAL ENTOMOLOGY**
Steph B.J. Menken and Sandrine A. Ulenberg (The Netherlands)

**BIOSYSTEMATIC AND BIOLOGICAL CONTROL INFORMATION SYSTEMS IN ENTOMOLOGY**
L. Knutson, F.C. Thompson and R.W. Carlson (USA)

*Agricultural Zoology Reviews* VOLUME 3 (May 1989);
ISBN 0 946707 20 0; ISSN 0269–0543; x + 355 pp; 234 × 156 mm;
illustrated; comprehensive index; hardback.

**CONTENTS**

**ECOLOGICAL EFFECTS OF MANURE, SEWAGE SLUDGE AND OTHER ORGANIC WASTES ON ARTHROPOD POPULATIONS**
David Pimentel and Alan Warneke (USA)

**GENETICS OF THE PINK BOLLWORM, *PECTINOPHORA GOSSYPIELLA***
Alan C. Bartlett (USA)

**SEVENTY YEARS OF EUROPEAN CORN BORER (*OSTRINIA NUBILALIS*) RESEARCH IN NORTH AMERICA**
Marcel Hudon, E.J. LeRoux and D.J. Harcourt (Canada)

**FACTORS AFFECTING RESISTANCE OF RICE VARIETIES TO PLANTHOPPER AND LEAFHOPPER PESTS**
R.C. Saxena (Philippines) and Z.R. Khan (Kenya)

**BIOLOGY AND DYNAMICS OF NORTH AMERICAN CONIFEROPHAGOUS *CHORISTONEURA* POPULATIONS**
W. Jan A. Volney (Canada)

**COLD HARDINESS AND OVERWINTERING OF INSECTS**
J.S. Bale (UK)

***NOSEMA* SPP. (MICROSPORA: MICROSPORIDA: NOSEMATIDAE) OF STORED-PRODUCT COLEOPTERA AND THEIR POTENTIAL AS MICROBIAL CONTROL AGENTS**
A.R. Khan and B.J. Selman (UK)

**A REVIEW OF VARIABILITY IN FOUR *MELOIDOGYNE* SPP. MEASURED BY REPRODUCTION ON SEVERAL HOSTS INCLUDING *LYCOPERSICON***
Philip A. Roberts and Ivan J. Thomason (USA)

**POPULATION DYNAMICS AND MANAGEMENT OF *HETERODERA GLYCINES***
D.P. Schmitt and R.D. Riggs (USA)

**FLOCKING AND SEASONAL MOVEMENTS OF *QUELEA QUELEA* AND *AGELAIUS PHOENICEUS* IN RELATION TO CROP DAMAGE**
Michael M. Jaeger and Richard L. Bruggers (USA)

**THE CHANGING FORTUNES OF AN AGRICULTURAL BIRD PEST: THE EUROPEAN STARLING**
C.J. Feare (UK)